Materials under High Pressure

Materials under High Pressure

Editors

Chuanting Wang
Yong He
Yuanfeng Zheng
Xiaoguang Qiao
Wenhui Tang
Shuhai Zhang

Basel • Beijing • Wuhan • Barcelona • Belgrade • Novi Sad • Cluj • Manchester

Editors

Chuanting Wang
School of Mechanical
Engineering
Nanjing University of Science
Technology
Nanjing
China

Yong He
School of Mechanical
Engineering
Nanjing University of Science
Technology
Nanjing
China

Yuanfeng Zheng
State Key Laboratory of
Explosion Science and
Technology
Beijing Institute of
Technology
Beijing
China

Xiaoguang Qiao
School of Materials Science
and Engineering
Harbin Institute of
Technology
Harbin
China

Wenhui Tang
College of Science
National University of
Defense Technology
Changsha
China

Shuhai Zhang
School of Environment and
Safety Engineering
North University of China
Taiyuan
China

Editorial Office
MDPI
St. Alban-Anlage 66
4052 Basel, Switzerland

This is a reprint of articles from the Special Issue published online in the open access journal *Materials* (ISSN 1996-1944) (available at: www.mdpi.com/journal/materials/special_issues/Mater_High_Press).

For citation purposes, cite each article independently as indicated on the article page online and as indicated below:

Lastname, A.A.; Lastname, B.B. Article Title. *Journal Name* **Year**, *Volume Number*, Page Range.

ISBN 978-3-7258-0394-1 (Hbk)
ISBN 978-3-7258-0393-4 (PDF)
doi.org/10.3390/books978-3-7258-0393-4

Contents

Chuanting Wang, Yuanfeng Zheng, Xiaoguang Qiao, Wenhui Tang, Shuhai Zhang and Yong He
Editorial for the Special Issue "Materials under High Pressure"
Reprinted from: *Materials* **2023**, *17*, 17, doi:10.3390/ma17010017 . . . . **1**

Shuaizhuo Wang, Haotian Yan, Dongmei Zhang, Jiajun Hu and Yusheng Li
The Microstructures and Deformation Mechanism of Hetero-Structured Pure Ti under High
Strain Rates
Reprinted from: *Materials* **2023**, *16*, 7059, doi:10.3390/ma16217059 . . . . **5**

Zhenwei Zhang, Yong He, Yuan He, Lei Guo, Chao Ge and Haifu Wang et al.
Compressive Mechanical Properties and Shock-Induced Reaction Behavior of Zr/PTFE and
Ti/PTFE Reactive Materials
Reprinted from: *Materials* **2022**, *15*, 6524, doi:10.3390/ma15196524 . . . . **18**

Binfeng Sun, Chunhua Bai, Caihui Zhao, Jianping Li and Xiaoliang Jia
Dispersal Characteristics Dependence on Mass Ratio for Explosively Driven Dry Powder
Particle
Reprinted from: *Materials* **2023**, *16*, 4537, doi:10.3390/ma16134537 . . . . **31**

Yuan He, Lei Guo, Chuanting Wang, Jinyi Du, Heng Wang and Yong He
Study on Axial Dispersion Characteristics of Double-Layer Prefabricated Fragments
Reprinted from: *Materials* **2023**, *16*, 3966, doi:10.3390/ma16113966 . . . . **44**

Yuanpei Meng, Yuan He, Chuanting Wang, Yue Ma, Lei Guo and Junjie Jiao et al.
The Effect of Surface Electroplating on Fragment Deformation Behavior When Subjected to
Contact Blasts
Reprinted from: *Materials* **2023**, *16*, 5464, doi:10.3390/ma16155464 . . . . **63**

Bin Ma, Zhengxiang Huang, Yongzhong Wu, Yuting Wang, Xin Jia and Guangyue Gao
Research on the Formation Characteristics of the Shaped Charge Jet from the Shaped Charge
with a Trapezoid Cross-Section
Reprinted from: *Materials* **2022**, *15*, 8663, doi:10.3390/ma15238663 . . . . **82**

Yizhen Wang, Jianping Yin, Xuepeng Zhang and Jianya Yi
Study on Penetration Mechanism of Shaped-Charge Jet under Dynamic Conditions
Reprinted from: *Materials* **2022**, *15*, 7329, doi:10.3390/ma15207329 . . . . **97**

Zhifan Zhang, Hailong Li, Longkan Wang, Guiyong Zhang and Zhi Zong
Formation of Shaped Charge Projectile in Air and Water
Reprinted from: *Materials* **2022**, *15*, 7848, doi:10.3390/ma15217848 . . . . **112**

Mengmeng Guo, Yanxin Wang, Haifu Wang and Jianguang Xiao
The Mechanical and Energy Release Performance of THV-Based Reactive Materials
Reprinted from: *Materials* **2022**, *15*, 5975, doi:10.3390/ma15175975 . . . . **133**

Ying Yuan, Yiqiang Cai, Dongfang Shi, Pengwan Chen, Rui Liu and Haifu Wang
Controlling Shock-Induced Energy Release Characteristics of PTFE/Al by Adding Oxides
Reprinted from: *Materials* **2022**, *15*, 5502, doi:10.3390/ma15165502 . . . . **153**

Tao Sun, Aoxin Liu, Chao Ge, Ying Yuan and Haifu Wang
Mechanical Properties, Constitutive Behaviors and Failure Criteria of Al-PTFE-W Reactive
Materials with Broad Density
Reprinted from: *Materials* **2022**, *15*, 5167, doi:10.3390/ma15155167 . . . . **166**

Guancheng Lu, Peiyu Li, Zhenyang Liu, Jianwen Xie, Chao Ge and Haifu Wang
Theoretical Model for the Impact-Initiated Chemical Reaction of Al/PTFE Reactive Material
Reprinted from: *Materials* **2022**, *15*, 5356, doi:10.3390/ma15155356 . . . . **181**

Xiangrong Li, Guohui Wang, Yongkang Chen, Bo Zhao and Jianguang Xiao
Failure Mechanism of the Fire Control Computer CPU Board inside the Tank under Transient
Shock: Finite Element Simulations and Experimental Studies
Reprinted from: *Materials* **2022**, *15*, 5070, doi:10.3390/ma15145070 . . . . **199**

Wei Zhao, Zhaojun Pang, Zhen Zhao, Zhonghua Du and Weiliang Zhu
A Simulation and an Experimental Study of Space Harpoon Low-Velocity Impact, Anchored
Debris
Reprinted from: *Materials* **2022**, *15*, 5041, doi:10.3390/ma15145041 . . . . **214**

 *materials*

Editorial

Editorial for the Special Issue "Materials under High Pressure"

Chuanting Wang [1], Yuanfeng Zheng [2], Xiaoguang Qiao [3], Wenhui Tang [4], Shuhai Zhang [5] and Yong He [1,*]

1   School of Mechanical Engineering, Nanjing University of Science & Technology, Nanjing 210094, China; ctwang@njust.edu.cn
2   State Key Laboratory of Explosion Science and Technology, Beijing Institute of Technology, Beijing 100081, China; zhengyf@bit.edu.cn
3   School of Materials Science and Engineering, Harbin Institute of Technology, Harbin 150001, China; xgqiao@hit.edu.cn
4   College of Science, National University of Defense Technology, Changsha 410073, China; wenhuitang@163.com
5   School of Environment and Safety Engineering, North University of China, Taiyuan 030051, China; zsh93y@nuc.edu.cn
*   Correspondence: yonghe1964@163.com

Citation: Wang, C.; Zheng, Y.; Qiao, X.; Tang, W.; Zhang, S.; He, Y. Editorial for the Special Issue "Materials under High Pressure". *Materials* **2024**, *17*, 17. https://doi.org/10.3390/ma17010017

Received: 4 December 2023
Accepted: 15 December 2023
Published: 20 December 2023

1. Introduction

The high-pressure-related problems of materials constitute a field at the confluence of several scientific disciplines. High pressure can be generated by die compression, high-velocity impact, or explosions. The understanding of materials under extreme high pressure, including flow, plastic deformation, phase transformation, fracture, temperature rise, and chemical reactions.

This Special Issue on the "Materials under High Pressure" collected recent research findings on the high-pressure-related problems of various materials. A collection of fourteen peer-reviewed research articles was included in this Special Issue. The main topics covered include processing technology, characterization, testing, theoretic modeling, and simulation.

2. Dynamic Mechanical Properties and Constitutive Model of Materials

Wang et al. [1] investigated the microstructures and deformation mechanism of hetero-structured pure Ti under various high strain rates. The results indicated that as the strain rate increased, the dominance of the dislocation slip decreased, while deformation twinning became more prominent. In addition, nanoscale twin lamellae were activated within the grain with a size of 500 nm at a strain rate of 2000 s^{-1}. The modified Hall–Petch model was obtained, with the obtained value of K_{twin} serving as an effective metric for this relationship. Zhang et al. [2] prepared Zr/PTFE and Ti/PTFE composites via cold isostatic pressing and vacuum sintering. They investigated the static and dynamic compressive mechanical properties of these composites at various strain rates. The results showed that the introduction of zirconium powder and titanium powder could increase the strength of the material under dynamic loading. Meanwhile, a modified Johnson–Cook (J–C) model considering strain and strain rate coupling was proposed.

3. Dispersion Characteristics of Materials Driven by Explosion

Materials would exhibit various dispersion characteristics when driven by explosion. Sun et al. [3] presented an investigation on the dispersal characteristics of the cylindrically packed material of dry powder particles driven by explosive load. By establishing a controllable experimental system under laboratory conditions and combining with near-field simulation, the particle dispersal process was described. The characteristic parameters of radially propagated particles were explored under different mass ratios of particle to charge (M/C). Results indicate that when the charge mass remains constant, an increase in M/C leads to a decrease in dispersed jet number, void radius, and maximum velocity, wherein the maximum velocity correlates with calculations by the porous Gurney model. The case of the smaller M/C always has a higher outer-boundary radius and area expansion factor.

He et al. [4] investigated the axial distribution of the initial velocity and direction angle of double-layer prefabricated fragments after an explosion. A three-stage detonation driving model of double-layer prefabricated fragments was proposed. It was shown that the energy utilization rate of detonation products acting on the inner-layer and outer-layer fragments were 69% and 56%, respectively. The deceleration effect of sparse waves on the outer layer of fragments was weaker than that on the inner layer. The maximum initial velocity of fragments was located near the center of the warhead where the sparse waves intersected, located at around 0.66 times the full length of the warhead.

4. Detonation-Driven Deformation and Jet Formation

High pressure on materials could be generated by an explosion. The interaction between explosion products and materials is very quick, and the strain rate could be extremely high. Preformed fragments can deform or even fracture when subjected to contact blasts, which might lead to a reduction of the terminal effect [3–5]. To solve this problem, Meng et al. [5] analyzed the effect of surface electroplating on the fragment deformation behavior under contact blasts. The results showed that the pressure amplitude of the uncoated samples instantly dropped to zero after the shock wave passed through the far-exploding surface, which resulted in the formation of a tensile zone. But the pressure amplitude of the coated samples increased, transforming the tensile zone into the compression zone, thereby preventing the fracture of the fragment near the far-exploding surface, which was consistent with the test and simulated results.

In addition to deformation and fracture, materials under extreme high pressure also undergo flow and phase changes. Ma et al. [6] analyzed the formation characteristics of the shaped charge jet (SCJ) from the shaped charge with a trapezoid cross-section. A theoretical model was developed to analyze the collapsing mechanism of the liner driven by the charge with a trapezoid cross-section. The results showed that the influence of the angle of the trapezoidal charge on the axial velocity of the SCJ was not distinct, whereas the variation of the radial velocity of the shaped charge jet was obvious as the change in the angle of the trapezoidal charge. Aiming at the dynamic penetration process of a shaped-charge jet, Wang et al. [7] proposed a mathematical model for the penetration of a jet under dynamic conditions, based on the theory of the virtual origin and the Bernoulli equation taking into account the jet and target intensities. The dynamic penetration process of the jet was divided according to the penetration channel. The dynamic penetration model of the jet based on the unperturbed section and perturbed section was established. Zhang et al. [8] studied the forming process of explosively formed projectiles (EFPs) in air and water. They used Euler governing equations to establish numerical models of EFPs subjected to air and underwater explosions. The fitting formulae of velocity attenuation of EFPs, which form and move in different media, were gained.

5. Shock-Induced Chemical Reaction

Metal/fluoropolymer-based reactive materials (RMs) would undergo violent energy release reactions under high-velocity impact and high pressure, and their impact-induced chemical reactions have attracted extensive research. Guo et al. [9] systematically researched the mechanical performances, fracture mechanisms, thermal behavior, energy release behavior, and reaction energy of four types of RMs (26.5% Al/73.5% PTFE; 5.29% Al/80% W/14.71% PTFE; 62% Hf/38% THV; 88% Hf/12% THV) by conducting compressive tests, scanning electron microscope (SEM), differential scanning calorimeter, thermogravimetric (DSC/TG) tests and ballistic experiments. The results show that the THV-based RMs have a unique strain-softening effect, whereas the PTFE-based RMs have a remarkable strain-strengthening effect, which is mainly caused by the different glass transition temperatures. Yuan et al. [10] studied the control of the shock-induced energy release characteristics of PTFE/Al-based energetic material by adding oxides (bismuth trioxide, copper oxide, molybdenum trioxide, and iron trioxide) via experimentation and theoretical analysis. Based on these experimental results, an analytical model was developed, indicating that the apparent activation energy and impact shock pressure dominated the energy release characteristic of PTFE/Al/oxide. This controlling mechanism indicated that oxides enhanced the reaction after shock wave unloading, and the chemical and physical properties of the corresponding thermites also affected the energy release characteristics. These conclusions can guide the design of metal/fluoropolymer-based RMs [2,9,10].

Sun et al. [11] investigated the mechanical properties, constitutive behaviors, and failure criteria of aluminum–polytetrafluoroethylene–tungsten (Al–PTFE–W) reactive materials with W content from 20% to 80%. Based on the experimental results and numerical iteration, the J–C constitutive (A, B, n, C, and m) and failure parameters (D1~D5) were well-determined. The research results would be useful for the numerical studies, design, and application of reactive materials. Lu et al. [12] proposed an impact-initiated chemical reaction model to describe the ignition and energy release behavior of Al/PTFE RM. The hotspot formation mechanism of pore collapse was first introduced to describe the decomposition process of PTFE. Material fragmentation and PTFE decomposition were used as ignition criteria. Then, the reaction rate of the decomposition product with aluminum was calculated according to the gas–solid chemical reaction model. Finally, the reaction states of RM calculated by the model are compared and qualitatively consistent with the experimental results. The model provided insight into the thermal–mechanical–chemical responses and references for the numerical simulation of the impact ignition and energy release behavior of RMs.

6. Structural Damage and Material Failure Caused by High-Velocity Impact

The deformation and failure of materials under high pressure are related to structural optimization and equipment design [4,8,13,14]. The electronic components inside a main battle tank (MBT) are the key components by which the tank exerts its combat effectiveness. An explicit nonlinear dynamic analysis was performed to study the vibration features and fault mechanism under instantaneous shock load by Li et al. [13]. They obtained the curves of stress–frequency and strain–frequency of the CPU board under different harmonic loads, which were applied to further identify the peak response of the structure. Validation of the finite element model and simulation results are performed by comparing those obtained from the model with experiments. Based on dynamic simulation and experimental analysis, fault patterns of the CPU board were discussed, and some optimization suggestions were proposed.

The space harpoon is a rigid–flexible coupled debris capture method with a simple, reliable structure and high adaptability to the target. Zhao et al. [14] studied the process of impacting and embedding the harpoon into the target plate, the effect of friction at a low-velocity impact, and the criteria for the effective embedding of the harpoon. A simulation model of the dynamics of the harpoon and the target plate considering tangential friction was established, and the reliability of the numerical simulation model was verified by comparing the impact test. The results showed that the frictional effect in the low-velocity impact was more obvious for the kinetic energy consumption of the harpoon itself, and the effective embedding of the harpoon into the anchored target ranges from 50 to ~90 mm, corresponding to a theoretical launch initial velocity between 88.4 and ~92.5 m/s.

This Special Issue includes a small number of discrete case studies on the topic of "Materials under High Pressure". We hope that this Special Issue will be of interest to the academic community, and the valuable contributions from all the authors are highly appreciated.

Funding: The authors are grateful for the support from NSFC (NO. U2241285).

Acknowledgments: The contributions of all authors are gratefully acknowledged.

Conflicts of Interest: The authors declare no conflict of interest.

References

1. Wang, S.; Yan, H.; Zhang, D.; Hu, J.; Li, Y. The Microstructures and Deformation Mechanism of Hetero-Structured Pure Ti under High Strain Rates. *Materials* **2023**, *16*, 7059. [CrossRef] [PubMed]
2. Zhang, Z.; He, Y.; He, Y.; Guo, L.; Ge, C.; Wang, H.; Ma, Y.; Gao, H.; Tian, W.; Wang, C. Compressive Mechanical Properties and Shock-Induced Reaction Behavior of Zr/PTFE and Ti/PTFE Reactive Materials. *Materials* **2022**, *15*, 6524. [CrossRef] [PubMed]
3. Sun, B.; Bai, C.; Zhao, C.; Li, J.; Jia, X. Dispersal Characteristics Dependence on Mass Ratio for Explosively Driven Dry Powder Particle. *Materials* **2023**, *16*, 4537. [CrossRef] [PubMed]
4. He, Y.; Guo, L.; Wang, C.; Du, J.; Wang, H.; He, Y. Study on Axial Dispersion Characteristics of Double-Layer Prefabricated Fragments. *Materials* **2023**, *16*, 3966. [CrossRef] [PubMed]
5. Meng, Y.; He, Y.; Wang, C.; Ma, Y.; Guo, L.; Jiao, J.; He, Y. The Effect of Surface Electroplating on Fragment Deformation Behavior When Subjected to Contact Blasts. *Materials* **2023**, *16*, 5464. [CrossRef] [PubMed]
6. Ma, B.; Huang, Z.; Wu, Y.; Wang, Y.; Jia, X.; Gao, G. Research on the Formation Characteristics of the Shaped Charge Jet from the Shaped Charge with a Trapezoid Cross-Section. *Materials* **2022**, *15*, 8663. [CrossRef] [PubMed]
7. Wang, Y.; Yin, J.; Zhang, X.; Yi, J. Study on Penetration Mechanism of Shaped-Charge Jet under Dynamic Conditions. *Materials* **2022**, *15*, 7329. [CrossRef] [PubMed]
8. Zhang, Z.; Li, H.; Wang, L.; Zhang, G.; Zong, Z. Formation of Shaped Charge Projectile in Air and Water. *Materials* **2022**, *15*, 7848. [CrossRef] [PubMed]
9. Guo, M.; Wang, Y.; Wang, H.; Xiao, J. The Mechanical and Energy Release Performance of THV-Based Reactive Materials. *Materials* **2022**, *15*, 5975. [CrossRef] [PubMed]
10. Yuan, Y.; Cai, Y.; Shi, D.; Chen, P.; Liu, R.; Wang, H. Controlling Shock-Induced Energy Release Characteristics of PTFE/Al by Adding Oxides. *Materials* **2022**, *15*, 5502. [CrossRef] [PubMed]
11. Sun, T.; Liu, A.; Ge, C.; Yuan, Y.; Wang, H. Mechanical Properties, Constitutive Behaviors and Failure Criteria of Al-PTFE-W Reactive Materials with Broad Density. *Materials* **2022**, *15*, 5167. [CrossRef] [PubMed]
12. Lu, G.; Li, P.; Liu, Z.; Xie, J.; Ge, C.; Wang, H. Theoretical Model for the Impact-Initiated Chemical Reaction of Al/PTFE Reactive Material. *Materials* **2022**, *15*, 5356. [CrossRef] [PubMed]
13. Li, X.; Wang, G.; Chen, Y.; Zhao, B.; Xiao, J. Failure Mechanism of the Fire Control Computer CPU Board inside the Tank under Transient Shock: Finite Element Simulations and Experimental Studies. *Materials* **2022**, *15*, 5070. [CrossRef]
14. Zhao, W.; Pang, Z.; Zhao, Z.; Du, Z.; Zhu, W. A Simulation and an Experimental Study of Space Harpoon Low-Velocity Impact, Anchored Debris. *Materials* **2022**, *15*, 5041. [CrossRef]

Article

The Microstructures and Deformation Mechanism of Hetero-Structured Pure Ti under High Strain Rates

Shuaizhuo Wang [1,2], Haotian Yan [1], Dongmei Zhang [1], Jiajun Hu [1] and Yusheng Li [1,2,*]

[1] National Key Laboratory of Transient Physics, Nanjing University of Science and Technology, Nanjing 210094, China; wsz@njust.edu.cn (S.W.); yanhaotian@njust.edu.cn (H.Y.); zbb@njust.edu.cn (D.Z.); hujiajun@njust.edu.cn (J.H.)
[2] School of Materials Science and Engineering, Nanjing University of Science and Technology, Nanjing 210094, China
* Correspondence: liyusheng@njust.edu.cn

Abstract: This study investigates the microstructures and deformation mechanism of hetero-structured pure Ti under different high strain rates ($500\ \mathrm{s}^{-1}$, $1000\ \mathrm{s}^{-1}$, $2000\ \mathrm{s}^{-1}$). It has been observed that, in samples subjected to deformation, the changes in texture are minimal and the rise in temperature is relatively low. Therefore, the influence of these two factors on the deformation mechanism can be disregarded. As the strain rate increases, the dominance of dislocation slip decreases while deformation twinning becomes more prominent. Notably, at a strain rate of $2000\ \mathrm{s}^{-1}$, nanoscale twin lamellae are activated within the grain with a size of 500 nm, which is a rarely observed phenomenon in pure Ti. Additionally, martensitic phase transformation has also been identified. In order to establish a correlation between the stress required for twinning and the grain size, a modified Hall–Petch model is proposed, with the obtained value of K_{twin} serving as an effective metric for this relationship. These findings greatly enhance our understanding of the mechanical responses of Ti and broaden the potential applications of Ti in dynamic deformation scenarios.

Keywords: pure Ti; hetero-structure; deformation mechanism; twinning; high strain rate

Citation: Wang, S.; Yan, H.; Zhang, D.; Hu, J.; Li, Y. The Microstructures and Deformation Mechanism of Hetero-Structured Pure Ti under High Strain Rates. *Materials* **2023**, *16*, 7059. https://doi.org/10.3390/ma16217059

Academic Editor: Matthias Bönisch

Received: 13 October 2023
Revised: 31 October 2023
Accepted: 2 November 2023
Published: 6 November 2023

1. Introduction

The usage of pure Ti has been steadily increasing in the fields of chemical, biomedical, and aerospace engineering. This is primarily due to its exceptional corrosion resistance, excellent biological compatibility, and impressive high specific strength [1,2]. As a structural material, pure Ti not only encounters static loads, but also confronts challenges from high-speed impact loads. Structural components in industries such as aerospace and defense industries may inevitably be involved in high-speed collision events [3]. Compared to static loading, the mechanical properties and deformation mechanisms of titanium and its alloys undergo significant changes under dynamic loading conditions, especially at high strain rates [4]. Therefore, conducting a thorough investigation of the deformation mechanism of pure Ti at high strain rates is of great significance for promoting the extensive utilization of pure Ti.

Previous studies have indicated that the deformation mechanism of pure Ti are affected by various factors, including its inherent characteristics (grain size, alloying elements, texture, etc.) and experimental conditions (loading mode, strain rate, temperature, strain level, etc.) [5–9]. In particular, strain rate plays a significant role and greatly influences the deformation behavior of materials. It has been found that higher strain rate facilitates the activation of twinning [10,11]. This can be attributed to the fact that a high strain rate promotes the rapid accumulation of dislocations in localized regions, leading to severe stress concentration. These regions of stress concentration provide effective nucleation sites for twinning [12]. It is worth noting that at high strain rates, the heat dissipation process in titanium is relatively slower compared to heat generation. This results in thermal softening during the accumulation of plastic deformation [4]. Consequently, instabilities in plastic flow and the formation of shear bands are observed, exerting a significant impact on the deformation mechanism. In addition, the influence of crystallographic texture on the mechanical behavior and deformation mechanism of pure Ti cannot be underestimated, owing to its hexagonal close-packed (HCP) structure. Extensive research has revealed that the presence of two distinctly different initial textures results in varying mechanical responses [13].

In addition to strain rate, the deformation mechanism of pure Ti also has a strong grain size dependency [14]. Studies have revealed that the occurrence of twinning becomes increasingly challenging as the grain size decreases, reaching a critical threshold at 750 nm. Below this grain size, deformation twinning is completely replaced by dislocation slip. Nevertheless, it is important to note that this conclusion solely takes into account the influence of static loads [15]. Currently, high-speed deformation studies of pure Ti primarily focus on samples with large grain sizes, where deformation are mainly dominated by dislocation slips and various types of twinning [13,16]. Unfortunately, there has been limited experimental research conducted on the high-speed deformation of pure Ti with hierarchical grain size, thereby impeding an in-depth investigation into grain size effect. The reason for this predicament is the difficulty in fabricating cross-scale structures with varying grain sizes. In recent years, an innovative approach, microstructural heterogenization, offering a promising avenue to overcome this challenge. This strategic paradigm has demonstrated a remarkable synergy in enhancing mechanical properties in pure Ti [17]. By incorporating heterogenization in grain size, ranging from coarse to ultra-fine regime, this structure can be served as an ideal model for studying the mechanism change during high-speed deformation processes.

Therefore, this study focuses on hetero-structured pure Ti samples with a hierarchical grain size. A room temperature split Hopkinson pressure bar test was conducted to examine the dynamic mechanical responses of the samples at different strain rates. Microstructural evolution during deformation was characterized using electron backscattering diffraction (EBSD) and transmission electron microscopy (TEM) techniques. Microstructures and deformation mechanism of hetero-structured pure Ti were systematically investigated, with a particular focus on exploring the influence of grain size on twinning under high strain rates.

2. Experiment

In this study, a split Hopkinson pressure bar (ALT1000) compression test was conducted to analyze the mechanical responses of the sample. The compression was performed along the ND (normal direction) axis of Ti samples at room temperature, and the compression rate was controlled by adjusting the gas pressure. This allowed for data collection on the dynamic mechanical responses of the samples at different strain rates ($500\ s^{-1}$, $1000\ s^{-1}$, $2000\ s^{-1}$). Figure 1(a-1) illustrates a schematic diagram of the Hopkinson impact test, while Figure 1(a-2) displays the observation surface of the sample in the transverse direction (TD) plane. Commercially pure Ti (grade 1) was selected as the material for this investigation, and the compositions is as follows (wt%): 0.001 C, 0.005 N, 0.0015 H, 0.085 O, 0.045 Fe, and balanced Ti. To prepare the initial hetero-structured pure Ti samples with hierarchical grain sizes, a combination of rolling and annealing techniques was employed. The Ti samples underwent cold rolling to an 80% reduction in thickness, followed by annealing at 480 °C for 5 min under vacuum conditions. EBSD observations were carried out using a Zeiss Auriga scanning electron microscope (SEM) with an acceleration voltage of 15 kV and a step size of 50 nm. TEM characterization was performed using an FEI TEM at 200 kV. Figure 1(b-1) presents the EBSD results of the hetero-structured pure Ti sample. The grain size exhibits a bimodal distribution, with an average grain size of 580 nm for the ultrafine grains and 2.14 μm for the coarse grains (Figure 1(b-3)). No twinning was observed in the sample. The grain boundary orientation distribution map (Figure 1(b-2)) reveals that the proportion of low angle grain boundaries (LAGBs) in the sample is approximately 19.5%.

This indicates that the material has undergone significant dislocation recovery during the annealing process.

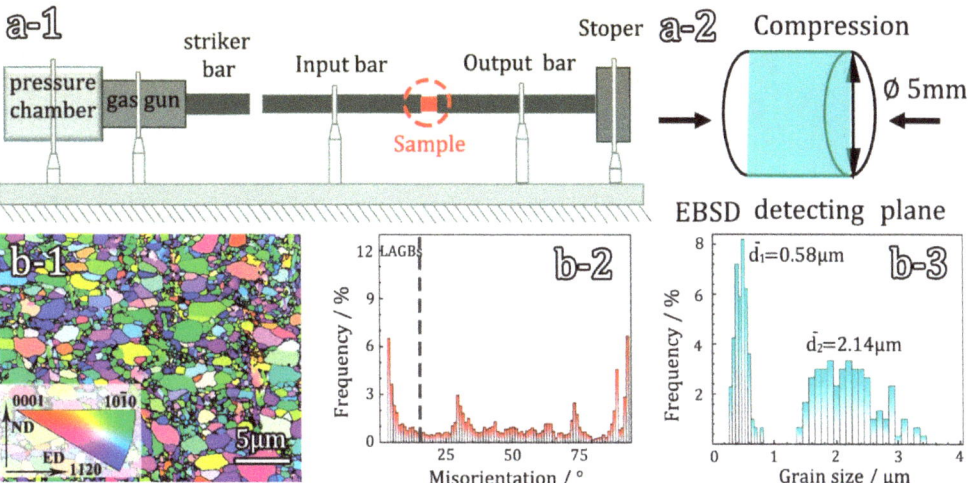

Figure 1. (**a-1**) The schematics of the split Hopkinson pressure bar system; (**a-2**) specimen; (**b-1**–**b-3**) EBSD characterization of the initial hetero-structured sample. (**b-1**) Inverse polar figure (IPF) map; (**b-2**) distribution of grain boundary orientation; (**b-3**) distributions of grain size.

3. Results and Discussion

Figure 2a presents the Hopkinson impact mechanical curves of the hetero-structured pure Ti samples deformed under strain rates of 500 s^{-1}, 1000 s^{-1}, 2000 s^{-1}. It can be observed that the dynamic compressive stress gradually increases as the strain increases, indicating that the sample absorbs impact energy during the impact process. It is important to note that the rapid decline in flow stress at the end of the curve is solely attributed to the termination of the applied load on the Hopkinson bar and not due to material failure. To obtain the true stress-strain curve, it can be calculated based on the compressive engineering stress-strain curve, as depicted in Figure 2b. This relationship can be described as follows:

$$\sigma_t = \sigma_e(1 - \varepsilon_e) \tag{1}$$

$$\varepsilon_t = -In(1 - \varepsilon_e) \tag{2}$$

where σ_t and ε_t represent the true stress and strain, respectively, σ_e and ε_e represent engineering stress and strain, respectively. It can be observed that the increase in true stress is not significant with increasing strain, but there is a noticeable oscillatory pattern. This oscillatory pattern arises from stress vibrations, which occur when an elastic wave, caused by an impact from a hammer in the test machine, propagates through the test sample and is detected by the load cell [18]. Figure 2c shows the strain hardening capacity of hetero-structured sample under different strain rates, and this is closely related to the deformation mechanism. In Figure 2d, the variation trends of different mechanical performance indicators with strain rate are illustrated. It can be seen that the corresponding yield strength (YS), ultimate compressive strength (UCS), and uniform elongation (UE) increase with the increase in strain rate.

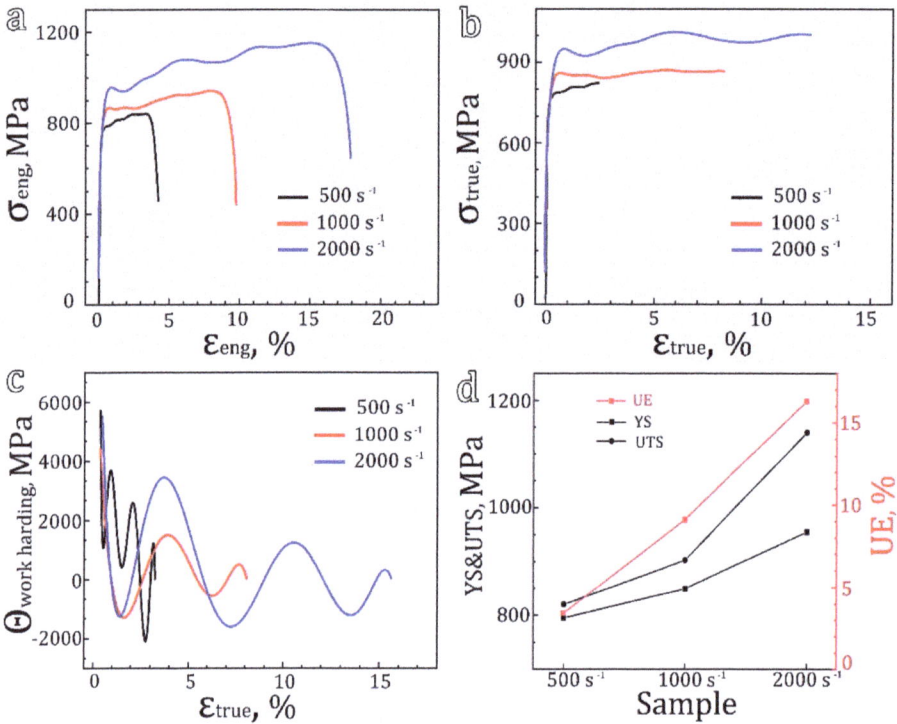

Figure 2. Mechanical properties of hetero-structured samples under different strain rate: (**a**) compressive engineering stress-strain curves; (**b**) true stress-strain curves from compressive tests; (**c**) strain hardening rate curves; (**d**) corresponding yield strength (YS), ultimate compressive strength (UCS) and compression strain of the samples under different strain rate.

During high-speed deformation, a substantial amount of the energy involved in plastic deformation is converted into heat energy. Due to insufficient heat dissipation, a noticeable local temperature rise occurs, leading to thermal softening of the material. The influence of temperature on the deformation mechanism of the material is crucial, thus necessitating a thorough examination of the impact of temperature rise during high-speed deformation. If we assume that all the energy generated from plastic deformation is completely transformed into heat energy and not dissipated, the resulting adiabatic temperature rise can be expressed as:

$$\Delta T = \frac{\beta}{\rho C_v} \int_0^{\varepsilon_f} \sigma_T d\varepsilon_T \tag{3}$$

where ρ represents density and C_v refers to specific heat capacity, σ_t and ε_t represent the true stress and strain, respectively, and ε_f is the max strain. β is a constant known as the Taylor–Quinney factor, which denotes the coefficient of plastic deformation work converted into heat. The values of ρ, C_v, β are $\rho = 4.51$ g/cm^3, $C_v = 527$ J/(kg·°C), $\beta = 0.9$ [19]. The results calculated from Equation (3) are shown in Figure 3. At a strain rate of 500 s^{-1}, the theoretical temperature rise is approximately 10 K. As the impact rate increases, the total strain also increases, resulting in an increase in the plastic deformation energy of the corresponding sample and a subsequent increase in adiabatic temperature rise. When the strain rate reaches 2000 s^{-1}, the corresponding theoretical temperature rise is approximately 45 K. Although the temperature rise increases obviously with the strain rate, it is still far from reaching the recrystallization temperature. Therefore, we believe that the temperature rise here has little effect on the deformation mechanism, and the emphasis should be placed on the effect of stress.

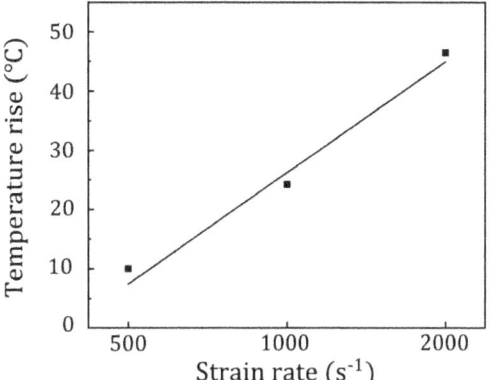

Figure 3. Adiabatic temperature rises values for overall hetero-structured samples under different strain rate.

Figure 4 presents the EBSD results of hetero-structured pure Ti subjected to various strain rates. From the IPF maps depicted in Figure 4(a-1–c-1), it is evident that the microstructure remains its hetero-structured nature, with no signs of recrystallization or significant grain refinement within the samples. However, after the impact tests, a noticeable occurrence of twinning is observed. Through analyzing the orientation differences (Figure 4(a-3–c-3), it can be observed that the predominant twinning type in the three deformed samples is $\{11\bar{2}2\}$ compression twins [20]. For pure Ti with an HCP structure, $\{11\bar{2}2\}$ twinning is more readily activated when the lattice experiences stress along the c-axis [21]. As depicted in the twin boundary map (Figure 4(a-2–c-2)), an increase in the percentage of twin boundaries, from 5.4% to 6.2% and 8.6%, is observed when the strain rate increases from $500\ s^{-1}$ to $1000\ s^{-1}$ and $2000\ s^{-1}$. The percentage of twin boundaries is determined by calculating the areal fractions of twin boundaries among all interfaces within the EBSD maps. It is noteworthy that at relatively low strain rates ($500\ s^{-1}$), twinning only occurs in coarse grains due to the notable influence of grain size on twinning behavior in pure Ti [22,23]. When the grain size is refined to the ultrafine range, twinning is nearly absent. However, when the strain rate is increased to $2000\ s^{-1}$, twinning appears in some smaller grains, as shown in Figure 4(c-2). This occurrence could be attributed to the unique microstructure of hetero-structured pure Ti and the high stress induced by the ultra-high strain rate impact, which triggers twinning [24]. Additionally, the proportion of LAGBs does not vary significantly under different strain rates, implying a decreased dominance of dislocations in the impact deformation process [25]. Figure 4(a-4–c-4) illustrate the statistical distribution plots of local misorientation corresponding to impact deformation at different strain rates. By analyzing these plots, the average Local Misorientation difference ($\bar{K}$) can be calculated. For strain rates of $500\ s^{-1}$, $1000\ s^{-1}$, and $2000\ s^{-1}$, the $\bar{K}$ values are 0.38, 0.54, and 0.59, respectively. An increase in $\bar{K}$ indicates a higher level of plastic deformation or a higher density of defects in the sample [26]. Specifically, the sample subjected to a strain rate of $2000\ s^{-1}$ shows a similar $\bar{K}$ to the sample at a strain rate of $1000\ s^{-1}$, suggesting insignificant dislocation accumulation as the strain rate increases from $1000\ s^{-1}$ to $2000\ s^{-1}$. This observation is consistent with the minimal increase in the proportion of LAGBs shown in Figure 4(b-2,c-2).

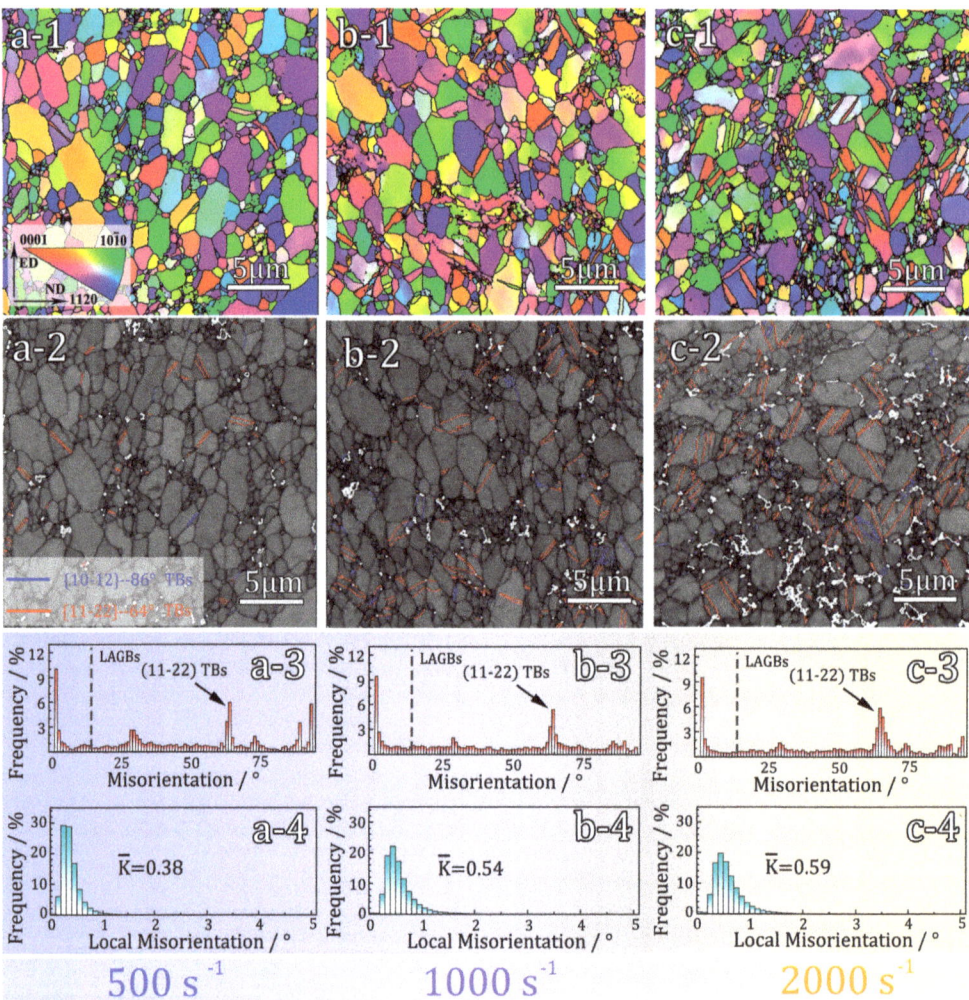

Figure 4. EBSD observations of deformed hetero-structured sample under a strain rate of 500 s^{-1}, 1000 s^{-1}, 2000 s^{-1}, respectively: (**a-1–c-1**) IPF map; (**a-2–c-2**) twin boundary map; (**a-3–c-3**) distribution of grain boundary orientation; (**a-4–c-4**) statistical distributions of Local Misorientation.

Figure 5a–c display the pole figures that correspond to the samples deformed under strain rates of 500 s^{-1}, 1000 s^{-1}, and 2000 s^{-1}, respectively. The pole figures reveal a bimodal basal texture, with the highest texture strengths measured at 13.5, 12.8, and 14.6, respectively. There is no significant change in the texture type when compared to the original sample. These results indicate that the strain rate has a minimal effect on the texture. In other words, the influence of texture on the change in deformation mechanism can be ignored.

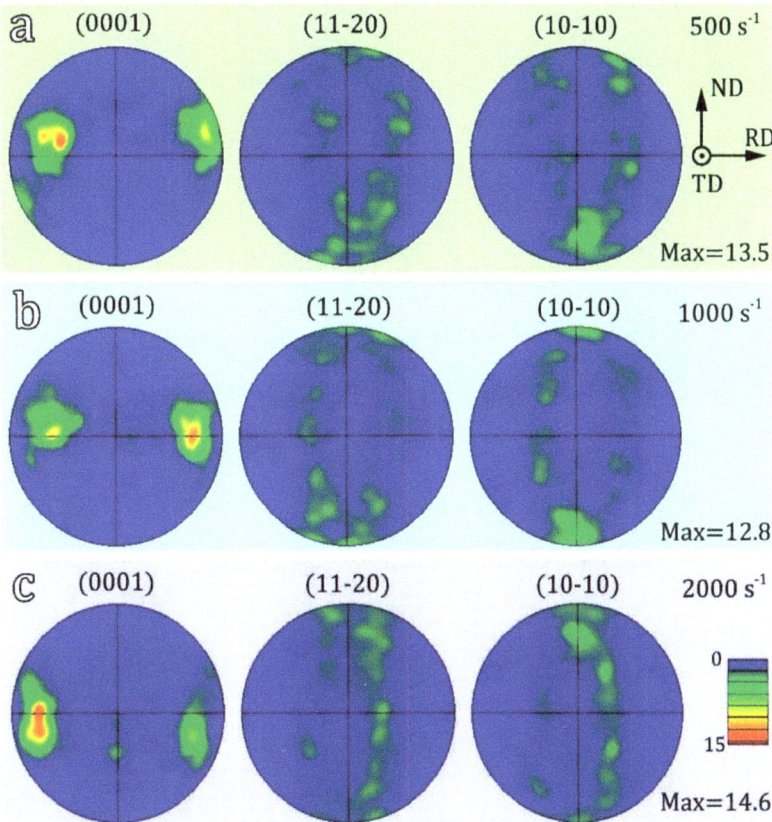

Figure 5. (0001), (11$\bar{2}$0) and (10$\bar{1}$0) pole figures of deformed hetero-structured samples under different strain rate: (**a**) 500 s^{-1}, (**b**) 1000s^{-1}, (**c**) 2000 s^{-1}.

Figure 6(a-1,a-2) shows a bright-field TEM image of Ti sample deformed at a strain rate of 500 s^{-1}. It can be observed that the deformed sample contains a high density of dislocations, which is a typical characteristic of plastic deformation. Twin boundaries are commonly observed within coarse grains, as illustrated by the yellow dotted lines in Figure 6(a-1,a-2). In contrast, no twinning is observed in the ultrafine grains, where dislocation slip serves as the primary mode of deformation [27]. When the impact strain rate increases to 1000 s^{-1}, the microstructure of the sample is shown in Figure 6(b-1,b-2), revealing a typical deformed structure characterized by a significant presence of defects such as dislocations and twins. Similarly, no twinning is observed in the ultrafine grain region of the sample. This implies that even at a strain rate of 1000 s^{-1}, twinning remains suppressed in the ultrafine grains. On the other hand, within the coarse grains, a prominent twin boundary is present, accompanied by a significant distribution of dislocations that form a ring-like pattern. In the magnified region (Figure 6(b-2)), a large number of dislocations can be seen near the twin boundary, indicating that the newly formed twin boundary inhibits the movement of dislocations [28]. Additionally, dislocations are also observed within the twin lamellae.

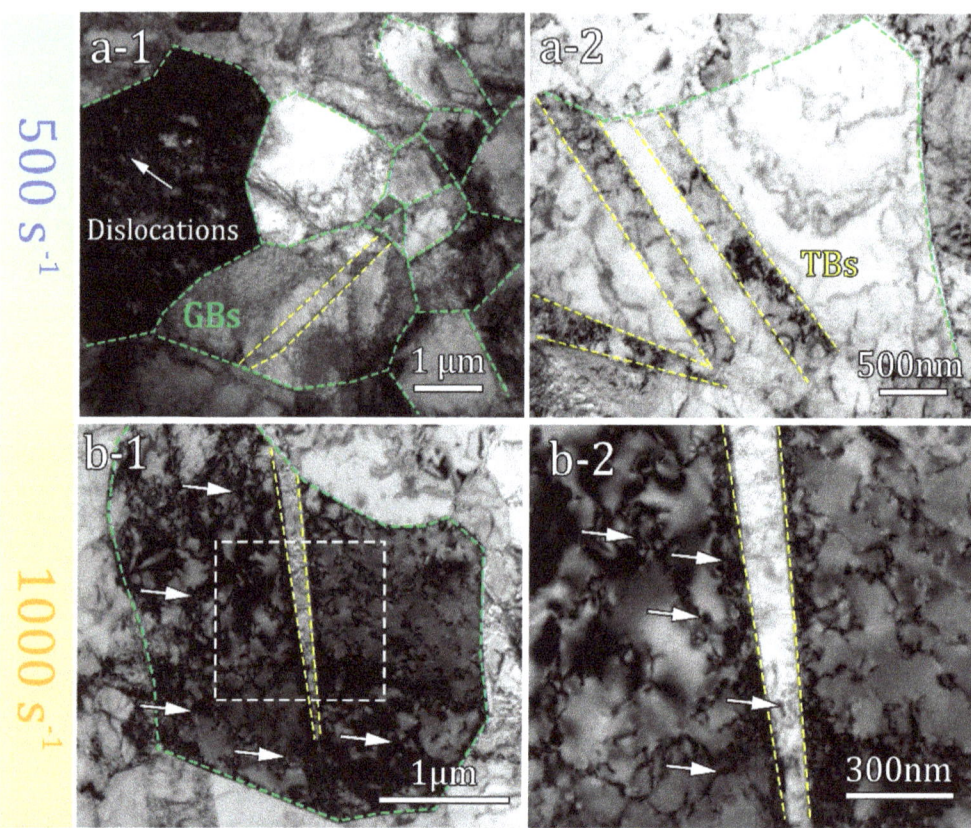

Figure 6. TEM images of deformed hetero-structured sample with different strain rates. (**a-1,a-2**) 500 s^{-1}; (**b-1,b-2**) 1000 s^{-1}, The white arrows indicate dislocations.

When the impact strain rate reaches 2000 s^{-1}; the microstructure undergoes severe plastic deformation, leading to significant disordering within the grains, as shown in Figure 7. Consequently, the grains exhibit a remarkably high density of defects, making it difficult to clearly distinguish grain boundaries and twin boundaries based solely on morphology (Figure 7a,b). Although it has been verified by EBSD observation (Figure 4(c-2)) that the areal twin percentage is 8.6%, TEM bright-field images at high magnification reveal the formation of dislocation cell structures in certain regions, as depicted by the solid rectangle in Figure 7b. Additionally, Figure 7(c-1,c-2) show the presence of numerous fine needle-like structures, which were not found in the deformed samples at strain rates of 500 s^{-1} and 1000 s^{-1}. These structures resemble the needle-like martensitic phases found in titanium alloys. Previous studies have indicated that under extreme deformation conditions, pure Ti can undergo stress-induced martensitic phase transformation, transitioning from the HCP phase to the face-centered cubic (FCC) phase, ultimately resulting in the formation of fine needle-like martensitic structures [29,30].

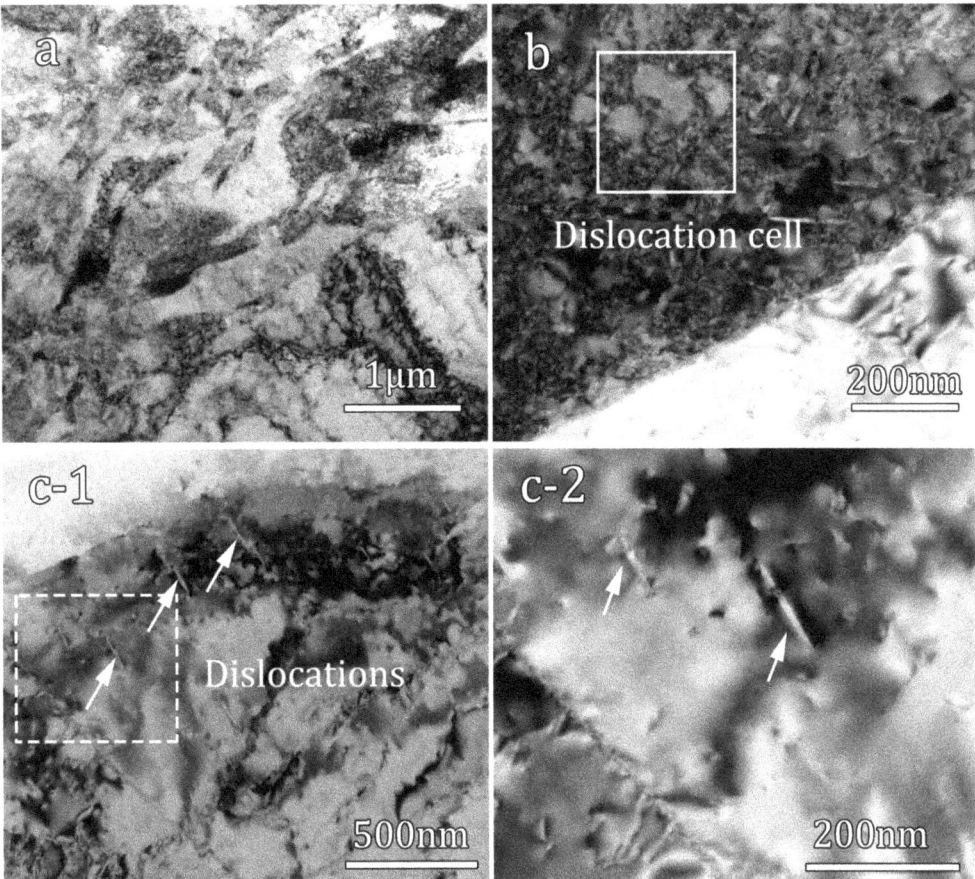

Figure 7. Microstructure of deformed hetero-structured sample with a strain rate of 2000 s^{-1}: (**a**,**b**) bright field TEM images; (**c-1**) The TEM image of martensitic phases; (**c-2**) close-up view of the area marked by the white dash line box in (**c-1**), the white arrows indicate needle-like martensitic phases.

For the hetero-structured pure Ti investigated in this study, the grain sizes are generally below 3 μm and contain a significant number of ultrafine grains, resulting in a relatively small grain size (Figure 1(b-3)). Considering the influence of grain size on deformation mechanism, dislocation slip is the most prevalent deformation carrier under this grain size [31,32]. On one hand, higher strain rates lead to the activation of more dislocation sources, resulting in the entanglement of dislocations and impeding their motion. This leads to the accumulation and pile-up of a significant number of dislocations, as confirmed by TEM images of the deformed samples. On the other hand, the hierarchical structure contains numerous hetero-structure interfaces. These interfaces give rise to hetero-deformation-induced (HDI) stress, which in turn leads to a significant accumulation of geometrically necessary dislocations [33,34].

It is generally believed that twinning is unlikely to occur in ultrafine grains [22,31]. However, in the deformed samples under a strain rate of 2000 s^{-1}, TEM bright-field images reveal the presence of small nanoscale twin lamellae within the ultrafine grains with an equivalent size of approximately 500 nm, as shown in Figure 8. The thickness of these twin lamellae is approximately 10 nm. The occurrence of twinning within the ultrafine grains is not widespread, but rather localized, observed in specific individual grains. This localized twinning may be due to the inhomogeneous deformation during impact, which leads to highly localized stress levels that satisfy the nucleation conditions for twinning within the ultrafine grains [35,36]. The significance of this discovery lies in gaining a profound understanding of the fundamental reasons behind the inhibitory effect of grain refinement on twinning. It uncovers that twinning is not completely absent within ultrafine grains, but rather occurs under extremely stringent conditions.

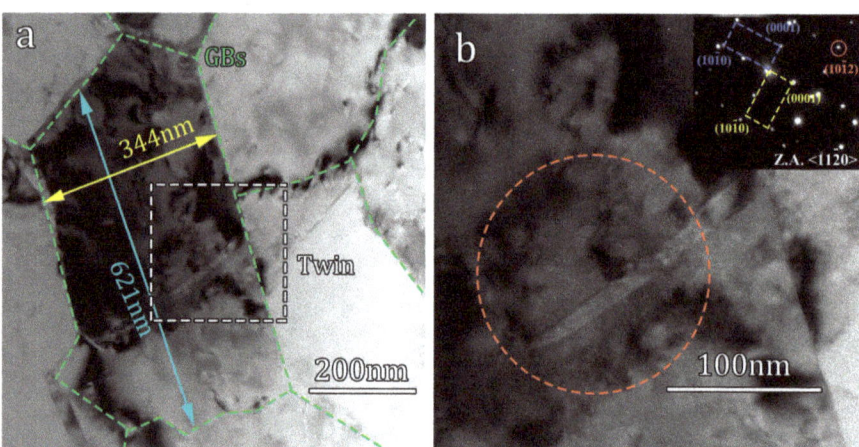

Figure 8. TEM images of ultra-fine grain of deformed hetero-structured sample under a strain rate of 2000 s^{-1}: (**a**) bright field TEM image of ultra-fine grain; (**b**) close-up view of the area marked by the white dash line box in (**a**), and the corresponding selected area diffraction pattern.

Numerous studies have demonstrated that dislocations play a pivotal role in the initiation of twinning in pure Ti [37–39]. Furthermore, the Hall–Petch relationship, which is based on the principle of dislocation obstacle at grain boundaries, has effectively explained the origin of strength [15]. Hence, we hypothesize that the correlation between the required stress for twin activation and the grain size is similar to the Hall–Petch relationship. Consequently, in order to gain a better understanding of the emergence of nanoscale twin lamellae in ultrafine grain, we propose a modified Hall–Petch relationship, which takes into account the fact that the stress required to initiate deformation twinning, σ_{twin}, is inversely proportional to the square root of grain size. This the relationship can be described as:

$$\sigma_{twin} = K_{twin} \times d^{-\frac{1}{2}} \tag{4}$$

where d represent the average grain size and K_{twin} is a constant that represents the sensitivity of twin activation to grain size. A higher K_{twin} value indicates a stronger grain size effect on twinning. The experimental results in this study demonstrate that twinning is activated at an equivalent grain size of 500 nm (the finest observable grain size where twinning occurs) when the strain rate reached 2000 s^{-1}. Based on the stress–strain curve (Figure 2), the corresponding range of flow stress is between 0.92 GPa and 1.04 GPa. According to Equation (4), the K_{twin} can be calculated to range from $0.651 \text{ MNm}^{-3/2}$ to $0.735 \text{ MNm}^{-3/2}$, which is significantly larger than the K value obtained from the traditional Hall–Petch relationship ($0.24 \text{ MNm}^{-3/2}$) [40]. Extensive research has indicated that the critical shear stress for compression twinning in pure Ti falls within the range of 125 MPa to 255 MPa when the grain size is between 10 μm and 50 μm [41–43]. To validate the equation, by assuming values of K_{twin} as 0.651 and 0.735, when K_{twin} is set to 0.651, the critical shear stress for twinning at grain sizes ranging from 10 μm to 50 μm can be calculated as 92 MPa to 203 MPa. On the other hand, when K_{twin} is set to 0.735, the critical shear stress can be calculated as 105 MPa to 232 MPa. These values are in good agreement with previous studies, confirming the scientific validity of this equation. The value of K_{twin} reported in this study can be used to predict the stress required to initiate deformation twinning across a wide range of grain sizes spanning from coarse to ultrafine.

In summary, our experimental results and empirical model greatly enhance our understanding of the twinning mechanism at ultrafine grain level, as well as establish a correlation between grain size and critical shear stress. These findings also offer potential avenues for strengthening mechanisms through twinning in HCP structured materials with ultrafine grains.

4. Conclusions

This study investigates the influence of deformation rate on the microstructures and deformation mechanism in hetero-structured pure Ti. The results can be summarized as follows:

1. The mechanical responses and deformation mechanism of hetero-structured pure Ti samples are closely related to the strain rate. As the strain rate increases from 500 s^{-1} to 2000 s^{-1}, dislocation activities are the primary deformation carrier, but the dominance of dislocation slipping reduces and the dislocation configurations undergo changes. Conversely, there is an increase in the percentage of deformation twinning with higher strain rates.

2. It has been found that under different high deformation rates (500 s^{-1}, 1000 s^{-1}, 2000 s^{-1}), the changes in texture are relatively minimal, and the degree of temperature rise is low (far lower than the recrystallization temperature). Consequently, the alterations in these two influential factors are insufficient to induce changes in the deformation mechanism.

3. When subjected to a strain rate of 2000 s^{-1}, martensitic phase transformation is identified in the deformed Ti sample. Moreover, nanoscale twin lamellae are observed within the ultrafine grain, which can be attributed to the high flow stress. A modified Hall–Petch model is proposed, and the obtained K_{twin} can be used to effectively establish the correlation between the stress required for twinning and the grain size.

Author Contributions: S.W. and Y.L. conceived and designed the experiments; S.W., H.Y., D.Z. and J.H. performed the experiments; Y.L. and S.W., analyzed the data; S.W. and Y.L. wrote the paper. Each contributor was essential to the production of this work. All authors have read and agreed to the published version of the manuscript.

Funding: This research was funded by the National Key Laboratory of Transient Physics Foundation (6142604220103), the Key Program of National Natural Science Foundation of China (No. 51931003), the National Natural Science Foundation of China (No. 52071180), and the Projects in Science and Technique Plans of Ningbo City (No. 2019B10083).

Institutional Review Board Statement: Not applicable.

Informed Consent Statement: Not applicable.

Data Availability Statement: The data that support the findings of this study are available from the corresponding author, upon reasonable request.

Conflicts of Interest: The authors declare that they have no known competing financial interest or personal relationship that could have appeared to influence the work reported in this paper.

References

1. Özcan, M.; Hämmerle, C. Titanium as a Reconstruction and Implant Material in Dentistry: Advantages and Pitfalls. *Materials* **2012**, *5*, 1528–1545. [CrossRef]
2. Konstantinov, A.S.; Bazhin, P.M.; Stolin, A.M.; Kostitsyna, E.V.; Ignatov, A.S. Ti-B-Based Composite Materials: Properties, Basic Fabrication Methods, and Fields of Application (Review). *Compos. Part A Appl. Sci. Manuf.* **2018**, *108*, 79–88. [CrossRef]
3. Zochowski, P.; Bajkowski, M.; Grygoruk, R.; Magier, M.; Burian, W.; Pyka, D.; Bocian, M.; Jamroziak, K. Ballistic Impact Resistance of Bulletproof Vest Inserts Containing Printed Titanium Structures. *Metals* **2021**, *11*, 225. [CrossRef]

4. Yan, N.; Li, Z.; Xu, Y.; Meyers, M.A. Shear Localization in Metallic Materials at High Strain Rates. *Prog. Mater. Sci.* **2021**, *199*, 100755. [CrossRef]
5. Lei, L.; Zhao, Q.; Zhao, Y.; Wu, C.; Huang, S.; Jia, W.; Zeng, W. Gradient nanostructure, phase transformation, amorphization and enhanced strength-plasticity synergy of pure titanium manufactured by ultrasonic surface rolling. *J. Am. Acad. Dermatol.* **2022**, *299*, 117322. [CrossRef]
6. Wang, M.; Wang, Y.; He, Q.; Wei, W.; Guo, F.; Su, W.; Huang, C. A Strong and Ductile Pure Titanium. *Mater. Sci. Eng. A* **2022**, *833*, 142534. [CrossRef]
7. Won, J.W.; Lee, S.; Kim, W.C.; Hyun, Y.-T.; Lee, D.W. Significantly increased twinning activity of pure titanium during room-temperature tensile deformation by cryogenic-deformation treatment. *Mater. Sci. Eng. A* **2023**, *862*, 144453. [CrossRef]
8. Zhao, S.; Zhang, R.; Yu, Q.; Ell, J.; Ritchie, R.O.; Minor, A.M. Cryoforged Nanotwinned Titanium with Ultrahigh Strength and Ductility. *Science* **2021**, *373*, 1363–1368. [CrossRef]
9. Huang, Z.; Cao, Y.; Nie, J.; Zhou, H.; Li, Y. Microstructures and Mechanical Properties of Commercially Pure Ti Processed by Rotationally Accelerated Shot Peening. *Materials* **2018**, *11*, 366. [CrossRef]
10. Zhou, P.; Xiao, D.; Jiang, C.; Sang, G.; Zou, D. Twin Interactions in Pure Ti Under High Strain Rate Compression. *Met. Mater. Trans. A* **2016**, *48*, 126–138. [CrossRef]
11. Jia, H.; Marthinsen, K.; Li, Y. Revealing Abnormal {112$^-$1} Twins in Commercial Purity Ti Subjected to Split Hopkinson Pressure Bar. *J. Alloys Compd.* **2019**, *783*, 513–523. [CrossRef]
12. Wang, T.B.; Li, B.L.; Li, M.; Li, Y.C.; Nie, Z.R. The Dynamic Mechanical Behavior and Microstructural Evolution of Commercial Pure Titanium. *Adv. Mater. Res.* **2014**, *968*, 7–11. [CrossRef]
13. Gurao, N.; Kapoor, R.; Suwas, S. Deformation behaviour of commercially pure titanium at extreme strain rates. *Acta Mater.* **2011**, *59*, 3431–3446. [CrossRef]
14. Deguchi, M.; Yamasaki, S.; Mitsuhara, M.; Nakashima, H.; Tsukamoto, G.; Kunieda, T. Tensile Deformation Behaviors of Pure Ti with Different Grain Sizes under Wide-Range of Strain Rate. *Materials* **2023**, *16*, 529. [CrossRef]
15. Ovid'Ko, I.; Valiev, R.; Zhu, Y. Review on superior strength and enhanced ductility of metallic nanomaterials. *Prog. Mater. Sci.* **2018**, *94*, 462–540.A. [CrossRef]
16. Xu, F.; Zhang, X.; Ni, H.; Liu, Q. Deformation twinning in pure Ti during dynamic plastic deformation. *Mater. Sci. Eng. A* **2012**, *541*, 190–195. [CrossRef]
17. Wu, X.; Yang, M.; Yuan, F.; Wu, G.; Wei, Y.; Huang, X.; Zhu, Y. Heterogeneous lamella structure unites ultrafine-grain strength with coarse-grain ductility. *Proc. Natl. Acad. Sci. USA* **2015**, *112*, 14501–14505. [CrossRef]
18. Tanaka, Y.; Kondo, M.; Miyazaki, N.; Ueji, R. Deformation Behavior of Pure Titanium at a Wide Range of Strain Rates. *J. Phys. Conf. Ser.* **2010**, *240*, 012021. [CrossRef]
19. Li, Q.; Xu, Y.B.; Bassim, M.N. Dynamic Mechanical Behavior of Pure Titanium. *J. Mater. Process. Technol.* **2004**, *155–156*, 1889–1892. [CrossRef]
20. Xu, S.; Wang, J. Deformation Twins Stimulated by {112$\bar{2}$} Twinning in Adjacent Grain in Titanium. *Acta Mater.* **2022**, *229*, 117805. [CrossRef]
21. Huang, Z.; Jin, S.; Zhou, H.; Li, Y.; Cao, Y.; Zhu, Y. Evolution of twinning systems and variants during sequential twinning in cryo-rolled titanium. *Int. J. Plast.* **2019**, *112*, 52–67. [CrossRef]
22. Yu, Q.; Mishra, R.K.; Minor, A.M. The Effect of Size on the Deformation Twinning Behavior in Hexagonal Close-Packed Ti and Mg. *JOM* **2012**, *64*, 1235–1240. [CrossRef]
23. Li, L.; Zhang, Z.; Shen, G. Effect of Grain Size on the Tensile Deformation Mechanisms of Commercial Pure Titanium as Revealed by Acoustic Emission. *J. Mater. Eng. Perform.* **2015**, *24*, 1975–1986. [CrossRef]
24. Yu, K.; Wang, X.; Mahajan, S.; Beyerlein, I.J.; Cao, P.; Rupert, T.J.; Schoenung, J.M.; Lavernia, E.J. Twin Nucleation from Disconnection-Dense Sites between Stacking Fault Pairs in a Random Defect Network. *Materialia* **2023**, *30*, 101835. [CrossRef]
25. Mittemeijer, E.J. The Crystal Imperfection; Structure Defects. In *Fundamentals of Materials Science*; Springer: Berlin/Heidelberg, Germany, 2021.
26. Wei, K.; Hu, R.; Yin, D.; Xiao, L.; Pang, S.; Cao, Y.; Zhou, H.; Zhao, Y.; Zhu, Y. Grain size effect on tensile properties and slip systems of pure magnesium. *Acta Mater.* **2021**, *206*, 116604. [CrossRef]
27. Huang, Z.; Wen, D.; Hou, X.; Li, Y.; Wang, B.; Wang, A. Grain size and temperature mediated twinning ability and strength-ductility correlation in pure titanium. *Mater. Sci. Eng. A* **2022**, *849*, 143461. [CrossRef]
28. Ahmadikia, B.; Wang, L.; Kumar, M.A.; Beyerlein, I.J. Grain boundary slip—Twin transmission in titanium. *Acta Mater.* **2023**, *244*, 118556. [CrossRef]
29. Zhao, H.; Ding, N.; Ren, Y.; Xie, H.; Yang, B.; Qin, G. Shear-induced hexagonal close-packed to face-centered cubic phase transition in pure titanium processed by equal channel angular drawing. *J. Mater. Sci.* **2019**, *54*, 7953–7960. [CrossRef]
30. Zheng, X.; Gong, M.; Xiong, T.; Ge, H.; Yang, L.; Zhou, Y.; Zheng, S.; Wang, J.; Ma, X. Deformation Induced Fcc Lamellae and Their Interaction in Commercial Pure Ti. *Scr. Mater.* **2019**, *162*, 326–330. [CrossRef]
31. Sun, J.; Trimby, P.; Yan, F.; Liao, X.; Tao, N.; Wang, J. Grain size effect on deformation twinning propensity in ultrafine-grained hexagonal close-packed titanium. *Scr. Mater.* **2013**, *69*, 428–431. [CrossRef]
32. Palán, J.; Procházka, R.; Džugan, J.; Nacházel, J.; Duchek, M.; Németh, G.; Máthis, K.; Minárik, P.; Horváth, K. Comprehensive Evaluation of the Properties of Ultrafine to Nanocrystalline Grade 2 Titanium Wires. *Materials* **2018**, *11*, 2522. [CrossRef]

33. Zhu, Y.; Wu, X. Heterostructured Materials. *Prog. Mater. Sci.* **2023**, *131*, 101019. [CrossRef]
34. Zhu, Y.; Ameyama, K.; Anderson, P.M.; Beyerlein, I.J.; Gao, H.; Kim, H.S.; Lavernia, E.; Mathaudhu, S.; Mughrabi, H.; Ritchie, R.O.; et al. Heterostructured materials: Superior properties from hetero-zone interaction. *Mater. Res. Lett.* **2020**, *9*, 1–31. [CrossRef]
35. Xu, S.; Zhou, P.; Liu, G.; Xiao, D.; Gong, M.; Wang, J. Shock-Induced Two Types of {10$\bar{1}$2} Sequential Twinning in Titanium. *Acta Mater.* **2019**, *165*, 547–560. [CrossRef]
36. He, Y.; Li, B.; Wang, C.; Mao, S.X. Direct observation of dual-step twinning nucleation in hexagonal close-packed crystals. *Nat. Commun.* **2020**, *11*, 2483. [CrossRef]
37. Kou, Z.; Yang, Y.; Huang, B.; Luo, X.; Li, P.; Zhao, G.; Zhang, W. Observing the Dynamic {10$\bar{1}$1} twining Process in Pure Ti at Atomic Resolution. *Scr. Mater.* **2017**, *139*, 139–143. [CrossRef]
38. Liao, X.; Wang, J.; Nie, J.; Jiang, Y.; Wu, P. Deformation twinning in hexagonal materials. *MRS Bull.* **2016**, *41*, 314–319. [CrossRef]
39. Paudel, Y.; Giri, D.; Priddy, M.W.; Barrett, C.D.; Inal, K.; Tschopp, M.A.; Rhee, H.; El Kadiri, H. A Review on Capturing Twin Nucleation in Crystal Plasticity for Hexagonal Metals. *Metals* **2021**, *11*, 1373. [CrossRef]
40. Zherebtsov, S.; Dyakonov, G.; Salem, A.; Sokolenko, V.; Salishchev, G.; Semiatin, S. Formation of nanostructures in commercial-purity titanium via cryorolling. *Acta Mater.* **2013**, *61*, 1167–1178. [CrossRef]
41. Wang, L.; Zheng, Z.; Phukan, H.; Kenesei, P.; Park, J.-S.; Lind, J.; Suter, R.; Bieler, T. Direct measurement of critical resolved shear stress of prismatic and basal slip in polycrystalline Ti using high energy X-ray diffraction microscopy. *Acta Mater.* **2017**, *132*, 598–610. [CrossRef]
42. Cao, Y.; Ni, S.; Liao, X.; Song, M.; Zhu, Y. Structural evolutions of metallic materials processed by severe plastic deformation. *Mater. Sci. Eng. R Rep.* **2018**, *133*, 1–59. [CrossRef]
43. Yang, H.; Li, H.; Ma, J.; Wei, D.; Chen, J.; Fu, M. Temperature dependent evolution of anisotropy and asymmetry of α-Ti in thermomechanical working: Characterization and modeling. *Int. J. Plast.* **2020**, *127*, 102650. [CrossRef]

Article

Compressive Mechanical Properties and Shock-Induced Reaction Behavior of Zr/PTFE and Ti/PTFE Reactive Materials

Zhenwei Zhang [1], Yong He [1,*], Yuan He [1], Lei Guo [1], Chao Ge [2], Haifu Wang [2], Yue Ma [1], Hongyin Gao [1], Weixi Tian [1] and Chuanting Wang [1,*]

[1] School of Mechanical Engineering, Nanjing University of Science and Technology, Nanjing 210094, China
[2] State Key Laboratory of Explosion Science and Technology, Beijing Institute of Technology, Beijing 100081, China
* Correspondence: yonghe1964@163.com (Y.H.); ctwang@njust.edu.cn (C.W.)

Abstract: Existing research on PTFE-based reactive materials (RMs) mostly focuses on Al/PTFE RMs. To explore further possibilities of formulation, the reactive metal components in the RMs can be replaced. In this paper, Zr/PTFE and Ti/PTFE RMs were prepared by cold isostatic pressing and vacuum sintering. The static and dynamic compressive mechanical properties of Zr/PTFE and Ti/PTFE RMs were investigated at different strain rates. The results show that the introduction of zirconium powder and titanium powder can increase the strength of the material under dynamic loading. Meanwhile, a modified J-C model considering strain and strain rate coupling was proposed. The parameters of the modified J-C model of Zr/PTFE and Ti/PTFE RMs were determined, which can describe and predict plastic flow stress. To characterize the impact-induced reaction behavior of Zr/PTFE and Ti/PTFE RMs, a quasi-sealed test chamber was used to measure the over-pressure induced by the exothermic reaction. The energy release characteristics of both materials were more intense under the higher impact.

Keywords: reactive materials (RMs); mechanical properties; constitutive model; quasi-sealed chamber test; shock-induced reaction

Citation: Zhang, Z.; He, Y.; He, Y.; Guo, L.; Ge, C.; Wang, H.; Ma, Y.; Gao, H.; Tian, W.; Wang, C. Compressive Mechanical Properties and Shock-Induced Reaction Behavior of Zr/PTFE and Ti/PTFE Reactive Materials. *Materials* **2022**, *15*, 6524. https://doi.org/10.3390/ma15196524

Academic Editors: Antonio Gil Bravo and Daniela Iannazzo

Received: 27 July 2022
Accepted: 15 September 2022
Published: 20 September 2022

Publisher's Note: MDPI stays neutral with regard to jurisdictional claims in published maps and institutional affiliations.

1. Introduction

Metal/polytetrafluoroethylene (PTFE) reactive materials (RMs) are usually composed of PTFE and single or multiple reactive metal elements. It is a new type of high-efficiency damage material different from traditional energetic materials such as explosives, propellants, and pyrotechnics. Compared with traditional energetic materials, metal/PTFE RMs have high energy density and remain inert under normal conditions [1]. When reactive fragments hit the target, a violent redox reaction occurs under the action of high-velocity impact, releasing a large amount of chemical energy and even deflagration or explosion, which can cause multiple damages to the target [2,3].

Among the existing metal/PTFE RMs, Al/PTFE RMs have attracted extensive attention from scholars. In the 1970s, M. J. Willis firstly reported that the Al/PTFE combination can produce reactions under high-velocity impact. D. B. Nielson and V. S. Joshi et al. [4,5] proposed the preparation method and sintering process curve of Al/PTFE RMs. W. Mock et al. [6] studied the reaction threshold of Al/PTFE RMs by Taylor test. M. N. Raftenberg et al. [7] determined the parameters of the Johnson–Cook constitutive model and pressure-shear damage model of Al/PTFE RMs via the Hopkinson compression bar and universal testing machine, which can describe the dynamic response of the materials under various strain rates. From the perspective of energy release, R. G. Ames [8,9] conducted a secondary impact experiment with a vented chamber device and established the relationship between the pressure change in the quasi-sealed chamber, the impact velocity, and the energy release efficiency combined with theoretical analysis.

In recent years, metal/PTFE RMs have been mainly used in the manufacturing of reactive fragments and reactive liners. However, Al/PTFE RMs have the disadvantages of low density and strength, which restricts their further development. To explore more possibilities of formulation, the reactive metal components in the RMs can be replaced. Both zirconium and titanium have excellent physical and mechanical properties, and their strength and density are higher than aluminum. They are widely used as functional or structural materials in the industry, biomedical, aerospace, and other fields [10–12]. It is worth noting that the energy release characteristics of zirconium-based materials and titanium-based materials under impact have also attracted more and more attention, such as W/Zr fragments [13,14], Zr/THV fragments [8] (the energy release is better than Al/PTFE of the same volume), and TNTZ alloys fragments [15]. Therefore, Zr/PTFE and Ti/PTFE RMs have potential applications as energetic materials [16,17]. It is very important to explore the dynamic response and shock response characteristics of Zr/PTFE and Ti/PTFE RMs for the design of their multifunctional properties. The main objectives of the work were to investigate the compressive property under various strain rates and the impact-initiated reaction behavior of Zr/PTFE and Ti/PTFE RMs.

2. Materials and Methods

2.1. Material Types

In this study, Zr/PTFE and Ti/PTFE RMs with mass ratios of 47.6:52.4 and 32:68 were prepared. The mass ratio is calculated by assuming that Zr and Ti are rapidly oxidized by fluorine in PTFE under impact conditions and the reactions have zero oxygen balance. The reaction chemical equation is as follows:

$$Zr_{(s)} + (-C_2F_4-)_{(s)} \rightarrow ZrF_{4(s)} + 2C_{(s)} \ \Delta H = -5769J/g, \tag{1}$$

$$Ti_{(s)} + (-C_2F_4-)_{(s)} \rightarrow TiF_{4(s)} + 2C_{(s)} \ \Delta H = -5683J/g. \tag{2}$$

The particle size of the Zr powder and Ti powder was 48 μm (Zhuzhou Runfeng New Material Co., Ltd., Zhuzhou, China) and the particle size of the PTFE was 100 μm (DuPont, Wilmington, DE, USA). The purity of Zr, Ti, and PTFE were all above 99%.

2.2. Preparation of Specimens

The preparation process of Zr/PTFE and Ti/PTFE RMs refers to the molding and sintering method for preparing PTFE-based RMs [4,5]. The preparation process can be described as follows:

(a) The dried powder of Zr, Ti, and PTFE was filled into a chrome steel tank by mass ratio and ground at room temperature;
(b) The grounded powder was placed into a uniaxially pressed mold, the pressure of 30 MPa was held for 5 min, and a cylindrical sample of Φ10 mm × 10 mm was prepared;
(c) The pressed sample was left at room temperature for 4 h to remove any residual stress. The sample was put into the vacuum sintering furnace for sintering. The sintering temperature control curve has been reported by D. B. Nielson and J. Zhou [1,4].

The theoretical maximum density (TMD) of the sintered Zr/PTFE RMs is 3.21 g/cm³, the actual density is 2.97 ± 0.08 g/cm³, and the porosity is 7.5%. The TMD of the Ti/PTFE RMs is 2.63 g/cm³, the actual density is 2.47 ± 0.03 g/cm³, and the porosity is 6.1%.

2.3. Mechanical Property Experiment

Quasi-static compression tests were carried out on Zr/PTFE and Ti/PTFE RMs samples (Φ10 mm × 10 mm) at a strain rate of 10^{-3} s^{-1} to 10^{-2} s^{-1}, and the moving speed of the cross-head is 0.6 mm/min to 6 mm/min. During the experiment, Vaseline was coated between the specimen and the cross-head to reduce the friction. All experiments were performed at least three times so as to get repeatable results.

The dynamics compression behavior of Zr/PTFE and Ti/PTFE RMs specimens were studied via a split Hopkinson pressure bar (SHPB). The SHPB tests were conducted under 298 K and 373 K to evaluate the role of temperature softening effect. Due to the low wave impedance of PTFE-based RMs, in order to obtain clear transmitted wave signals, aluminum rods were used for the incident rod and transmission rod in the test. The incident rod and transmission rod were both 1500 mm in length, the rod diameter was 14.5 mm, and the bullet length was 300 mm. The sizes of the specimens were Φ10 mm × 10 mm and Φ10 mm × 5 mm. The specimens of Φ10 mm × 5 mm were used for the tests at 373 K to prevent the material from creeping at high temperature. During the test, Φ5 mm × 0.3 mm copper discs were used to adjust the incident pulse and slow down the rising edge of the incident wave. All experiments were performed at least three times so as to get repeatable results. The detailed composition and layout of the SHPB device have been reported by W. Li [18] and R. G. Zhao [19].

2.4. Quasi-Sealed Chamber Experiment

A hemispherical quasi-sealed chamber was employed in the experiment. This device was improved according to R. G. Ames' "Vented chamber calorimetry" [9]. The test device is shown in Figure 1. The hemispherical quasi-sealed chamber is made of a thick steel plate with a volume of 23 L, and the front end is sealed with a 0.5 mm thickness iron lid. An observation window and pressure sensors are installed on the outer wall of the chamber to monitor the entire reaction process.

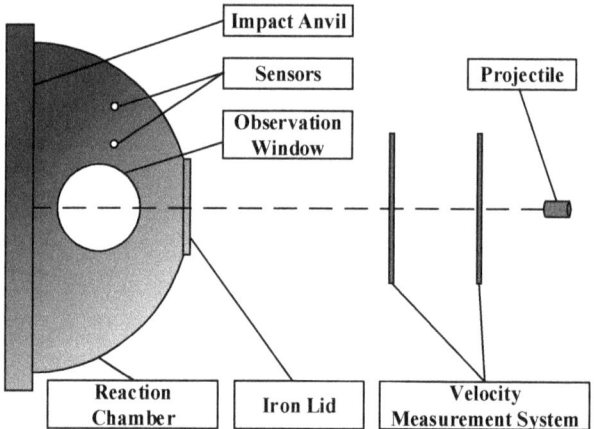

Figure 1. Schematic illustration of the quasi-sealed chamber device.

The Zr/PTFE and Ti/PTFE RMs were packed into steel shells to ensure that the materials would not be broken when launching or hitting the iron lid. The size of Zr/PTFE and Ti/PTFE RMs was Φ10 mm × 9 mm, the mass of cylindrical steel shells was 4.6 ± 0.1 g, and the thickness of the steel shells was 1 mm. A ballistic gun was used to launch the projectiles, and the velocity of the projectiles was adjusted by changing the charge quantity. The velocity of the projectiles was measured through the speed measurement system in front of the ballistic gun. The projectiles passed through the iron lid and hit the impact anvil, causing a strong initial impact inside the RMs, triggering a violent chemical reaction and releasing energy.

3. Results

3.1. Microstructure of the Composites

The mechanical property and dynamic responses of sintered materials are closely related to their microstructure. The cross-sections of Zr/PTFE and Ti/PTFE RMs were scanned using a scanning electron microscope (JSM-IT500HR, JEOL Ltd., Tokyo, Japan) under a voltage of 20 kV. It is shown in Figure 2 that the PTFE is dark gray, and the metal particles are bright silver. The sintered PTFE forms a continuous matrix, and the two metal particles are relatively uniformly distributed in the matrix. Both metal particles are irregular in shape, and there are pores between metal particles and PTFE. When the pores are under pressure, cracks will occur due to stress concentration, which will accelerate the failure of the material.

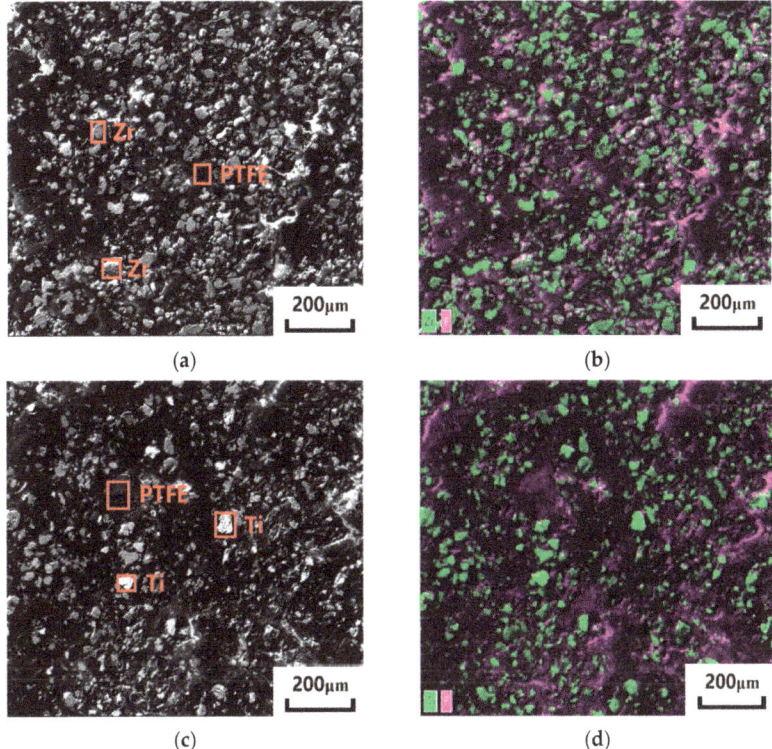

Figure 2. Microstructure of the Zr/PTFE, Ti/PTFE section. (**a**) SEM of Zr/PTFE, (**b**) EDS of Zr/PTFE, (**c**) SEM of Ti/PTFE, (**d**) EDS of Ti/PTFE.

3.2. The Mechanical Property under Quasi-Static Compression

Deformed specimens and true stress-strain curves of Zr/PTFE and Ti/PTFE RMs under quasi-static compression are shown in Figure 3. The static mechanical properties of Zr/PTFE and Ti/PTFE RMs show obvious plastic properties, with the linear elastic response. After reaching the yield stress, the specimen will produce lateral plastic deformation, and the specimen after compression is in a drum shape. The test results show that when the strain rate is 10^{-3} s^{-1} and 10^{-2} s^{-1}, the true stress of Zr/PTFE RMs is 32.1 ± 0.2 MPa and 37.0 ± 0.2 MPa, respectively, while the true stress of Ti/PTFE RMs is 27.4 ± 0.8 MPa and 30.4 ± 0.8 MPa, respectively. Due to the good plasticity of the specimens, an obvious upward setting effect appeared during the compression tests. Therefore, the tests were terminated when the engineering strain reached around 0.4.

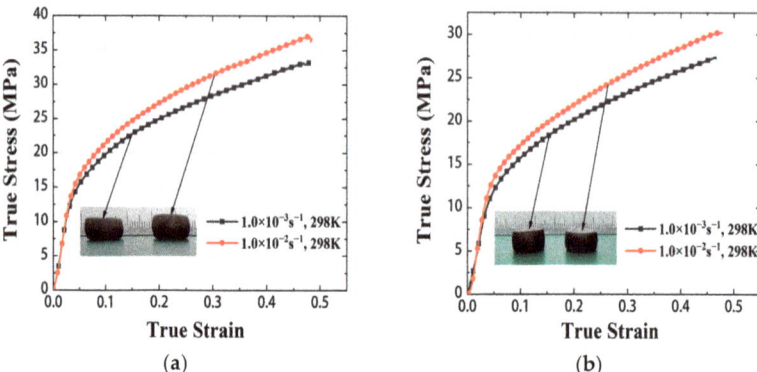

Figure 3. Stress-strain curves of Zr/PTFE, Ti/PTFE under quasi-static compression. (**a**) Zr/PTFE, (**b**) Ti/PTFE.

3.3. The Mechanical Property under Dynamic Compression

During the SHPB test, the high-pressure air chamber drove the bullet to hit the incident rod, and the generated stress wave propagated in the incident rod, the transmission rod, and the specimen. The strain-time signal was collected by the strain gauges on the incident rod and the transmission rod, then adjusted by the Wheatstone bridge to convert it into a voltage pulse-time signal. Figure 4 shows the voltage pulse-time curves under various strain rates at 298 K.

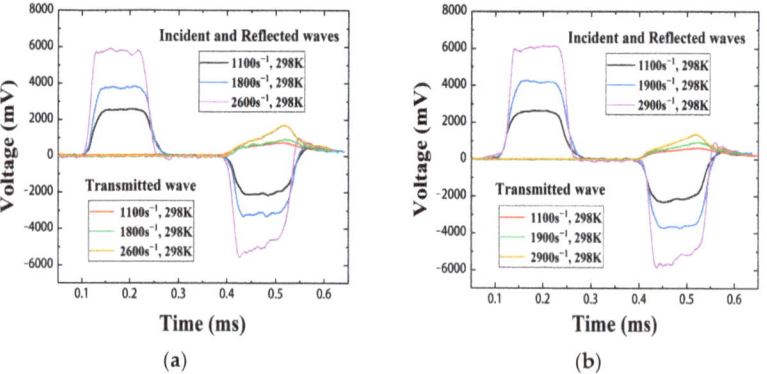

Figure 4. Strain circuit pulse signal-time curves of Zr/PTFE, Ti/PTFE at 298 K. (**a**) Zr/PTFE, (**b**) Ti/PTFE.

Based on mass conservation and momentum conservation, using the one-dimensional stress wave theory and the assumption of stress uniformity, the true stress $\sigma(t)$, true strain $\varepsilon(t)$, and strain rate $\dot{\varepsilon}(t)$ formulas in the SHPB test can be expressed as follows [20]:

$$\begin{cases} \dot{\varepsilon}(t) = \frac{-2C_B}{L_s}\varepsilon_R(t) \\ \sigma(t) = \frac{E_B A_B}{A_S}\varepsilon_T(t) \\ \varepsilon(t) = \frac{2C_B}{L_s}\int_0^t \varepsilon_R(t)dt \end{cases}, \tag{3}$$

where $\varepsilon_R(t)$ and $\varepsilon_T(t)$ are the reflection and transmission strains generated after the loading pulse, A_B is the cross-sectional area of the compression bar, A_S is the cross-sectional area of the specimen, C_B is the longitudinal propagation velocity of the stress wave in the compression bar, L_S is the length of the specimen, and E_B is the elastic modulus of the compression rod.

Combined with the pulse signal-time curve of the strain gauge and Equation (3), the true stress and strain curves of Zr/PTFE and Ti/PTFE RMs in the range of strain rate 1100–2900 s^{-1} under 298 K and 373 K were obtained. As shown in Figure 5, the material exhibits typical elastic-plastic mechanical behavior under dynamic loading, and the fluctuation of elastic modulus and yield strength is small in the range of strain rate 1100–2900 s^{-1}. In a certain strain range, the stress of the material increases approximately linearly. Meanwhile, both materials showed an obvious temperature softening effect.

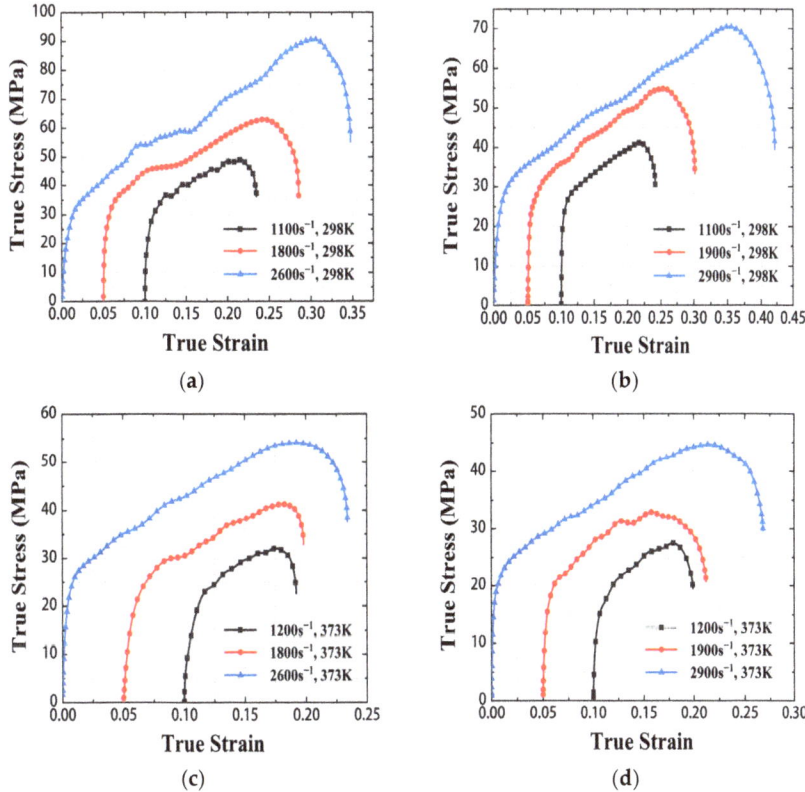

Figure 5. The stress-strain curves of Zr/PTFE, Ti/PTFE under dynamic loading. (**a**) Zr/PTFE, 298 K, (**b**) Ti/PTFE, 298 K, (**c**) Zr/PTFE, 373 K, (**d**) Ti/PTFE, 373 K.

According to the true stress-strain curve, the strength of Zr/PTFE is 49 ± 1 MPa, 63 ± 1 MPa, 90 ± 3 MPa at 1100 s^{-1}, 1800 s^{-1}, and 2600 s^{-1} strain rates under 298 K, respectively. The strength of Ti/PTFE is 41.3 ± 1 MPa, 55 ± 2 MPa, and 70 ± 1 MPa at 1100 s^{-1}, 1900 s^{-1}, and 2900 s^{-1} strain rates under 298 K, respectively. Since both Zr/PTFE and Ti/PTFE RMs were fabricated by the same sintering process, the distribution of metal particles in the PTFE matrix was similar. When the metal particle detaches from the PTFE matrix, stress concentrates on the defects in the material, which can lead to material failure [16]. The failure behavior and mechanical properties of the two materials are similar. The Zr/PTFE has slightly higher strength compared to the Ti/PTFE, and the reason may be the different strengths of Zr particles and Ti particles. Under the same conditions, the static mechanical strength of Zr is higher than those of Ti [21]. It is also not excluded that the binding force between the two metal particles and the PTFE matrix might also be different. Similarly, the ultimate strength of the sintered Al/PTFE specimen at a strain rate of about 3000 s^{-1}

was 40–62 MPa [18,22], which shows that replacing aluminum with zirconium or titanium improves the strength of the material under dynamic loads.

3.4. Shock-Induced Reaction Characteristics

Figure 6 shows the high-velocity photographic frames of Zr/PTFE and Ti/PTFE RMs impacting the impact anvil at 970–1340 m/s. As shown in the figure, the flash on the observation window is bright, and the gas and reaction products in the chamber are sprayed out from the perforation of the iron lid due to high temperature and high pressure, forming a special venting phenomenon. At the same time, the flash intensity and venting effect increase with the increase of the impact velocity. This shows that both materials undergo a violent chemical reaction under high-velocity impact, and the energy release characteristics will increase with the increase of the impact velocity. The results are consistent with an earlier investigation on Al/PTFE [1,8].

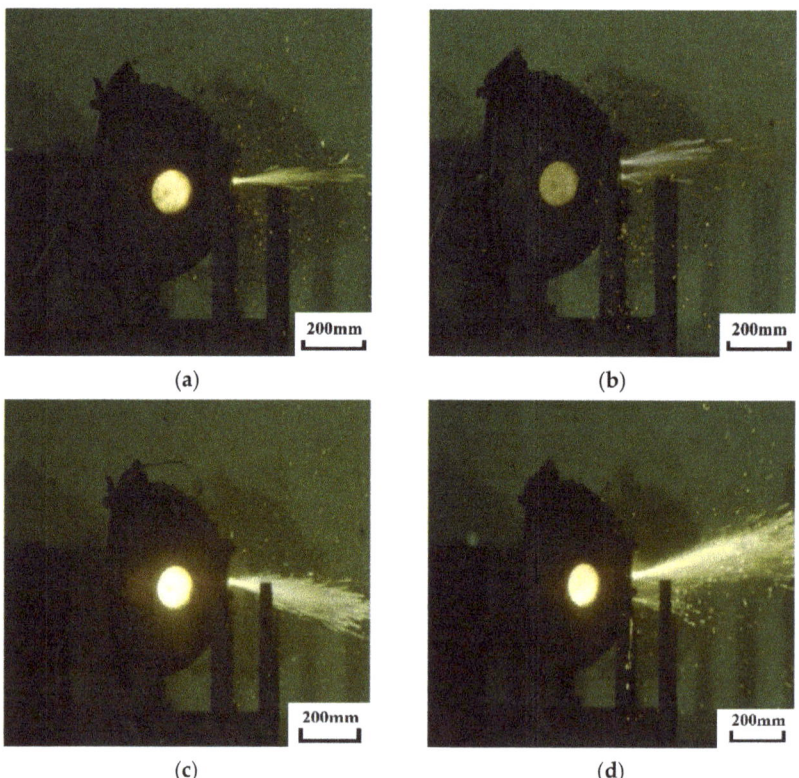

Figure 6. Video frames of Zr/PTFE and Ti/PTFE hitting the impact anvil. (**a**) Zr/PTFE: v = 970 m/s, t = 11 ms, (**b**) Ti/PTFE: v = 990 m/s, t = 16 ms, (**c**) Zr/PTFE: v = 1250 m/s, t = 10 ms, (**d**) Ti/PTFE: v = 1340 m/s, t = 18 ms.

4. Discussion

Through the quasi-static compression test and SHPB test, it is shown that Zr/PTFE and Ti/PTFE RMs have obvious plastic properties, and the strength under dynamic loading is higher than that of Al/PTFE RMs. Meanwhile, similar energy release phenomenon to Al/PTFE RMs also occurred when Zr/PTFE and Ti/PTFE RMs impacted the quasi-sealed chamber device under high-velocity.

4.1. Validation of SHPB Test Results

In order to ensure the validity of the SHPB test results, it is necessary to ensure that the specimen reaches the dynamic balance of stress—that is, the stress on the front and rear surfaces of the specimen is equal during loading. Since the material of the incident rod and the transmission rod are the same material and have the same diameter, the stress can be directly expressed by the pulse signal of the strain gauge. The front-end stress signal is represented by the absolute value of the incident signal, and the back-end stress signal is represented by the sum of the absolute value of the transmitted signal and the absolute value of the reflected signal. Figure 7 shows the front and rear stress signals of the Ti/PTFE RMs at $1100\,\mathrm{s}^{-1}$ strain rates under 298 K. It is shown that the stress signal curve completely coincides with the loading rising stage and the final unloading stage. There is a difference at the peak of the initial load rise, but the stress difference decreases with time. It shows that the specimen reaches dynamic stress balance and deforms at a constant strain rate, and the test results are valid.

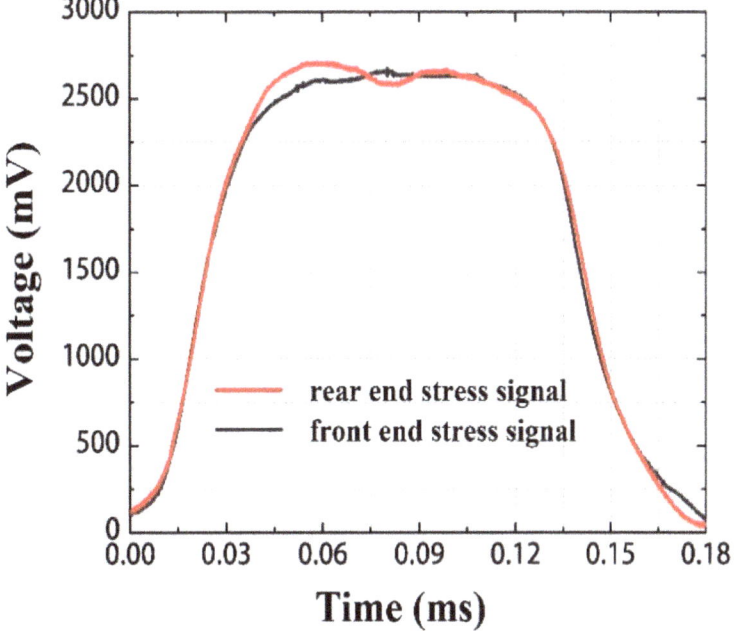

Figure 7. Front and rear stress signals of the specimen under dynamic compression.

4.2. Constitutive Model Building

4.2.1. Johnson–Cook Constitutive Model

The constitutive model of the material can describe the mechanical properties and predict the plastic flow stress of the material under various strain rates. The Johnson–Cook model is a purely empirical viscoplastic constitutive model, which has the characteristics of easy fitting of parameters and simple form. It is often used to describe the dynamic mechanical properties of materials under impact loads. Johnson–Cook model considers the strain hardening effect, strain rate effect, and temperature softening effect, and its mathematical form is [23]:

$$\sigma = [A + B\varepsilon^{n}][1 + C\ln(\dot{\varepsilon}^{*})][1 - T^{*m}], \tag{4}$$

where σ is the von Mises stress, ε is the equivalent plastic strain, $\dot{\varepsilon}^{*} = \dot{\varepsilon}/\dot{\varepsilon}_{0}$ is the plastic strain rate, $\dot{\varepsilon}$ is the equivalent plastic strain rate, and $\dot{\varepsilon}_{0}$ is the reference strain rate of the Johnson–

Cook model, generally the quasi-static strain rate of 10^{-3} s^{-1}, $T^* = (T - T_r)/(T_m - T_r)$ is the dimensionless temperature parameter, T_r and T_m are the reference temperature, and the melting point of the material, respectively, and A, B, C, n, and m are undetermined parameters: A is the yield strength of the material at the reference strain rate and reference temperature, B and n are the strain hardening parameters, C is the strain rate sensitive parameter, and m is the temperature softening parameter.

4.2.2. Determination of Parameters in Constitutive Models

Since the strain, strain rate, and temperature in the Johnson–Cook model are not coupled with each other, the parameters of the Zr/PTFE RMs constitutive model can be obtained by using the variable separation method and the least-squares method.

Under quasi-static conditions at room temperature, the effects of strain rate strengthening and thermal softening on the flow stress of the materials are ignored, taking the Zr/PTFE RMs as an example: Figure 4 shows that when $\dot{\varepsilon}_0 = 1 \times 10^{-3}$ s^{-1}, the static yield stress of the Zr/PTFE RMs is 11.596 MPa, that is, A = 11.596. So, Equation (4) can be transformed as:

$$\lg(\sigma - 11.596) = \lg B + n \lg \varepsilon. \tag{5}$$

Using the least-squares fitting, B = 37.0508, n = 0.70609.

The parameter C in the Johnson–Cook model can be obtained from the dynamic compression test data at different strain rates under 298 K in Figure 5a [24]. Because the strain hardening effect of the material in the plastic stage has little effect on the flow stress, only the strain rate strengthening effect is considered. Equation (4) can be simplified as:

$$\sigma = 11.596(1 + C \ln(\dot{\varepsilon}/\dot{\varepsilon}_0)). \tag{6}$$

Fitted using least squares, C = 0.11457.

The parameter m can be obtained from the dynamic compression test data under 298 K and 373 K in Figure 5a,c. At high temperature and high strain rate, Equation (4) can be simplified as:

$$\sigma(T^*) = \sigma(T_r)\left[1 - \left(\frac{T - T_r}{T_m - T_r}\right)^m\right], \tag{7}$$

where $\sigma(T^*)$ and $\sigma(T_r)$ represent the yield stress of the material at high temperature (373 K) and the yield stress of the material at the reference temperature (298 K), respectively. T_r is 298 K and T_m is 598 K.

Fitted using least squares, m = 0.9536.

In summary, the Johnson–Cook constitutive model of Zr/PTFE RMs is expressed as:

$$\sigma = (11.596 + 37.0508\varepsilon^{0.7061})(1 + 0.1146 \ln(\dot{\varepsilon}/\dot{\varepsilon}_0))[1 - T^{*0.9536}]. \tag{8}$$

Similarly, the Johnson–Cook constitutive model of Ti/PTFE RMs is expressed as:

$$\sigma = (10.227 + 35.7976\varepsilon^{0.8525})(1 + 0.1134 \ln(\dot{\varepsilon}/\dot{\varepsilon}_0))[1 - T^{*0.937}]. \tag{9}$$

4.2.3. Fitting and Modification of Johnson–Cook Constitutive Model

According to the Johnson–Cook constitutive model, the corresponding stress-strain curves are fitted and compared with the existing test curves under dynamic loading. As shown in Figure 8, in the initial plastic stage, the model-predicted curves of the two materials fit well with the experimental curves. However, with the increase of strain rate and strain, the prediction error of the Johnson–Cook constitutive model gradually increases.

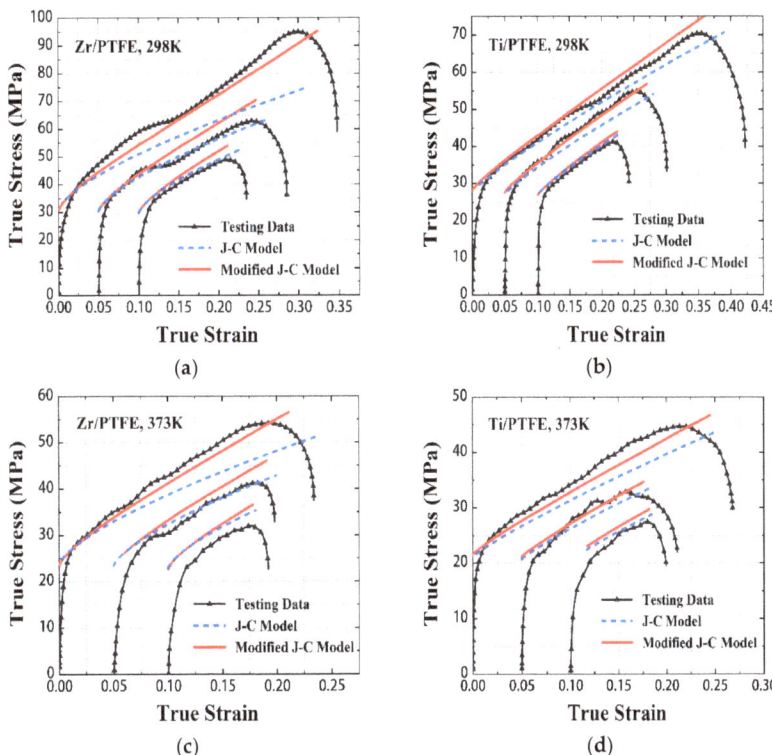

Figure 8. Comparisons of the true stress under dynamic loading for Zr/PTFE, Ti/PTFE between two model predictions and testing data. (**a**) Zr/PTFE, 298 K. (**b**) Ti/PTFE, 298 K. (**c**) Zr/PTFE, 373 K. (**d**) Ti/PTFE, 373 K.

According to the research results of Nagy, A [25], the strain and strain rate will have a coupling effect on stress. Therefore, in this paper, a modification of strain and strain rate coupling is added to the Johnson–Cook constitutive model to reduce the error. The modified J-C model is expressed as:

$$\sigma = [A + B\varepsilon^n][1 + C\ln(\dot{\varepsilon}^*)][1 - T^{*m}][(\dot{\varepsilon}^*)^{a\varepsilon+b}]. \tag{10}$$

The parameter a, b can be obtained from the dynamic compression test data under 298 K and 373 K in Figure 5. The modified Zr/PTFE and Ti/PTFE RMs constitutive models can be expressed as:

$$\text{Zr/PTFE RMs}: \sigma = (11.596 + 37.0508\varepsilon^{0.7061})(1 + 0.1146\ln(\dot{\varepsilon}/\dot{\varepsilon}_0))[1 - T^{*0.9536}][(\dot{\varepsilon}/\dot{\varepsilon}_0)^{0.0485\varepsilon-0.00061}], \tag{11}$$

$$\text{Ti/PTFE RMs}: \sigma = (10.227 + 35.7976\varepsilon^{0.8525})(1 + 0.1134\ln(\dot{\varepsilon}/\dot{\varepsilon}_0))[1 - T^{*0.937}][(\dot{\varepsilon}/\dot{\varepsilon}_0)^{0.0147\varepsilon-0.00157}]. \tag{12}$$

As shown in Figure 8, the fitting accuracy of the modified J-C model at high strain rates is improved. At the strain rate of 2600 s^{-1}, the fitting errors of Zr/PTFE RMs at 298 K and 373 K decreased from 21.8% to 4.5% and 14% to 4.3%, respectively. At the strain rate of 2900 s^{-1}, the fitting errors of Ti/PTFE RMs at 298 K and 373 K decreased from 5.7% to 4.6% and 8.7% to 6.4%, respectively.

4.2.4. Comparison of the Existing Johnson–Cook Constitutive Model

As shown in Table 1, the Johnson–Cook constitutive model parameters of pure PTFE and some PTFE-based RMs are summarized. It is shown that the parameters are close to the range of the parameters of the above constitutive model, which proves the feasibility and universality of the model proposed in this paper.

Table 1. Summary of the Johnson–Cook constitutive model parameters for PTFE-based RMs.

Material	A	B	n	C	m	a	b
PTFE [26]	11	44	1	0.12	1	/	/
Al/PTFE [7,27] (26.5/73.5)	8.044~17	45.403~250.6	0.86659~1.8	0.4~0.4873	1	/	/
AL/PTFE/W [28] (12/38/50)	23	20.26	0.67604	0.19707	/	/	/
Zr/PTFE (47.6/52.4)	11.596	37.0508	0.7061	0.1146	0.9536	0.0485	0.00061
Ti/PTFE (32/68)	10.227	35.7976	0.8525	0.1134	0.937	0.0147	0.00157

4.3. Shock-Induced Reaction Behavior

Figure 9a shows the quasi-static pressure-time curve of Zr/PTFE RMs in the velocity range 640–1250 m/s. Figure 9b shows the quasi-static pressure-time of the Ti/PTFE RMs in the velocity range 840–1340 m/s. It is shown that the pressure in the chamber rises sharply at the initial stage of the reaction, resulting in a pressure peak, and then gradually decreases with the ejection of reaction products and gases. At the same time, the pressure peak will increase with the increase of the impact velocity.

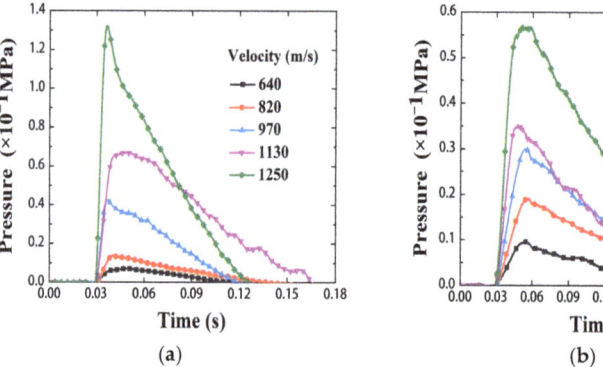

(a)   (b)

Figure 9. Quasi-static pressure-time curves in the chamber for the Zr/PTFE, Ti/PTFE. (**a**) Zr/PTFE. (**b**) Ti/PTFE.

The test chamber can be regarded as a closed system before the pressure peak. The relationship between the quasi-static pressure peak and the energy increase in the chamber can be written as [8]:

$$\Delta P = \frac{\gamma - 1}{V} \Delta E, \tag{13}$$

where γ is the air specific heat ratio and the value 1.4. V is the volume of the test chamber. ΔP is the peak pressure value and ΔE is the heat increase of the gas in the chamber, that is, the energy release value of the RMs. Table 2 shows the energy release values of the Zr/PTFE and Ti/PTFE RMs at different velocities. Under the same impact velocity, the release energy of Zr/PTFE RMs is higher than that of Ti/PTFE RMs.

Table 2. Calculation results of the energy released by RMs.

Material	Velocity (m/s)	ΔP (10^{-1} MPa)	ΔE (J)
Zr/PTFE	640	0.076 ± 0.02	437
	820	0.135 ± 0.03	776.25
	970	0.428 ± 0.03	2461
	1130	0.67 ± 0.05	3852.5
	1250	1.317 ± 0.03	7572.75
Ti/PTFE	840	0.094 ± 0.02	540.5
	990	0.19 ± 0.02	1092.5
	1090	0.29 ± 0.03	1667.5
	1140	0.347 ± 0.03	1995.25
	1340	0.566 ± 0.05	3254.5

The reaction heat of Zr/PTFE RMs is -5769 J/g and the reaction heat of Ti/PTFE RMs is -5683 J/g. The density of Zr/PTFE RMs is higher than those of Ti/PTFE RMs, so the mass of the Zr/PTFE sample is higher. Besides, the difference in energy release between Zr/PTFE and Ti/PTFE RMs is also closely related to the reaction threshold and the mechanical/chemical properties of the oxide films on the two metal particles [29]. All the above reasons could lead to difference of the reaction intensity between the two composites under impact.

5. Conclusions

In this paper, Zr/PTFE and Ti/PTFE RMs were prepared by replacing the metal components in RMs. The compressive mechanical properties of the two materials at different strain rates were studied by quasi-static compression test and SHPB test. In addition, the shock-induced reaction behavior of Zr/PTFE and Ti/PTFE RMs were investigated via the quasi-sealed chamber test. The following conclusions can be drawn:

1. The addition of zirconium and titanium powders improves the strengths of the materials under dynamic load compared to Al/PTFE RMs. The strengths of Zr/PTFE RMs are higher than Ti/PTFE RMs;
2. A modified J-C model considering the strain and strain rate coupling was proposed. The parameters of the modified J-C model of Zr/PTFE and Ti/PTFE RMs were determined, which can describe and predict plastic flow stress;
3. The test results of the quasi-sealed chamber show that the two materials have violent chemical reactions under the high-velocity impact, and the energy release characteristics will increase with the increase of the impact velocity. At the same impact velocity, the released energy of Zr/PTFE RMs is higher than that of Ti/PTFE RMs.

Author Contributions: Conceptualization, Z.Z., Y.H. (Yuan He) and C.W.; Data curation, Z.Z. and C.W.; Formal analysis, Z.Z., L.G. and C.W.; Investigation, Z.Z. and C.W.; Methodology, Z.Z. and C.W.; Project administration, Y.H. (Yong He); Resources, Y.H. (Yong He) and H.W.; Supervision, Y.H. (Yong He); Validation, Z.Z. and C.W.; Writing—original draft, Z.Z.; Writing—review & editing, Z.Z., Y.H. Y.H. (Yuan He), L.G., C.G., H.W., Y.M., H.G., W.T. and C.W. All authors have read and agreed to the published version of the manuscript.

Funding: This research was funded by the Fundamental Research Funds for the National Natural Science Foundation of Jiangsu China [BK20160832] and the National Natural Science Foundation of China [NSFC51601095].

Institutional Review Board Statement: Not applicable.

Informed Consent Statement: Not applicable.

Data Availability Statement: Not applicable.

Acknowledgments: The authors thank Z. R. Zhu, H. Gao, and C. L. Gao of NJUST for assistance with the mechanical property test.

Conflicts of Interest: The authors declare no conflict of interest.

References

1. Zhou, J.; He, Y.; He, Y.; Wang, C.T. Investigation on impact initiation characteristics of fluoropolymer-matrix reactive materials. *Prop. Explos. Pyrotech.* **2017**, *42*, 603–615. [CrossRef]
2. Wang, C.T.; He, Y.; Ji, C.; He, Y.; Han, W.; Pan, X.C. Investigation on shock-induced reaction characteristics of a Zr-based metallic glass. *Intermetallics* **2018**, *93*, 383–388. [CrossRef]
3. Wang, C.T.; He, Y.; Ji, C.; He, Y.; Guo, L.; Meng, Y.P. Dynamic fragmentation of a Zr-based metallic glass under various impact velocities. *J. Mater. Sci.* **2021**, *56*, 2900–2911. [CrossRef]
4. Nielson, D.B.; Tanner, R.L.; Lund, G.K. High Strength Reactive Materials and Methods of Making. U.S. Patent 2004/0116576 A1, 17 June 2004.
5. Joshi, V.S. Process for Making Polytetrafluoroethylene-Aluminum Composite and Product Made. U.S. Patent US 6547993 B1, 15 April 2003.
6. Mock, W.; Holt, W.H. Impact initiation of rods of pressed polytetrafluoroethylene (PTFE) and aluminum powders. In Proceedings of the Conference of the American Physical Society Topical Group on Shock Compression of Condensed Matter, Baltimore, MD, USA, 31 July–5 August 2005.
7. Raftenberg, M.N.; Mock, W.; Kirby, G.C. Modeling the impact deformation of rods of a pressed PTFE/Al composite mixture. *Int. J. Impact Eng.* **2008**, *35*, 1735–1744. [CrossRef]
8. Ames, R.G. Energy release characteristics of impact-initiated energy materials. In Proceedings of the Materials Research Society Symposium, Boston, MA, USA, 28 November–2 December 2005.
9. Ames, R.G. Vented chamber calorimetry for impact-initiated energetic materials. In Proceedings of the 43rd AIAA Aerospace Sciences Meeting and Exhibit, Reno, NV, USA, 10–13 January 2005.
10. Rack, H.J.; Qazi, J.I. Titanium alloys for biomedical applications. *Mater. Sci. Eng. C* **2006**, *26*, 1269–1277. [CrossRef]
11. Suyalatu; Nomura, N.; Oya, K.; Tanaka, Y.; Kondo, R.; Doi, H.; Tsutsumi, Y.; Hanawa, T. Microstructure and magnetic susceptibility of as-cast Zr–Mo alloys. *Acta Biomater.* **2010**, *6*, 1033–1038. [CrossRef] [PubMed]
12. Feng, Z.H.; Xia, C.Q.; Zhang, X.Y.; Ma, M.Z.; Liu, R.P. Development and applications of zirconium alloys with high strength and toughness. *Mater. Sci. Technol.* **2018**, *26*, 1–8.
13. Wang, L.Y.; Jiang, J.W.; Li, M.; Ma, Y.Y. Experimental research on energy release behavior of W/Zr/Hf Alloy Fragment. *Acta Armamentarii* **2019**, *40*, 1603–1608.
14. Chen, W.; Zhao, W.T.; Wang, J.; Liang, D.; Ge, W.Y.; Xiong, X.S. Study of damaging process for W-Zr alloy fragment. *Ordnance Mater. Sci. Eng.* **2009**, *32*, 108–111. [CrossRef]
15. Guo, Z.P.; Liu, R.; Wang, C.T.; He, Y.; He, Y.; Ma, Y.; Hu, X.B. Compressive Mechanical Properties and Shock-Induced Reaction Behavior of a Ti–29Nb–13Ta–4.6Zr Alloy. *Met. Mater. Int.* **2019**, *26*, 1498–1505. [CrossRef]
16. Li, W.; Ren, H.L.; Ning, J.G.; Hao, L. Dynamic compression characteristics of zirconium/polytetrafluoroethylene reactive material. *Acta Armamentarii* **2020**, *41*, 56–62.
17. Tong, Y.; Wang, Z.C.; Cai, S.Y.; Hu, W.X.; Li, S.Y. Damage Effects of Metal Plates by Ti/W/PTFE Energetic Fragment Impact. *J. Ordnance Equip. Eng.* **2019**, *40*, 1–6.
18. Li, W.; Ren, H.L.; Ning, J.G.; Liu, Y.B. Dynamic mechanical behavior and impact ignition characteristics of Al/PTFE reactive materials. *Chin. J. Energ. Mater.* **2020**, *28*, 38–45.
19. Zhao, R.G.; Chen, C.Z.; Luo, W.B.; Luo, X.Y.; Li, H.C.; Tan, D.H. Key issues in the SHPB experiment of polymer materials. *Chin. J. Solid Mech.* **2011**, *32*, 134–144.
20. Wang, L.L. *Stress Wave Basics*, 2nd ed.; National Defense Industry Press: Beijing, China, 1985; pp. 52–60.
21. Tang, W.H.; Zhang, R.Q. *Introduction to Theory and Computation of Equations of State*, 2nd ed.; Higher Education Press: Beijing, China, 2008; pp. 301–302.
22. Yang, X.L.; He, Y.; He, Y.; Wang, C.T.; Qi, L.; Guo, Z.P. Study of the effect of interface properties on the dynamic behavior of Al/PTFE composites using experiment and 3D meso-scale modelling. *Compos. Interfaces* **2020**, *27*, 401–418. [CrossRef]
23. Johnson, G.R.; Cook, W.H. A constitutive model and data for materials subjected to large strains, high strain rates, and high temperatures. In Proceedings of the 7th International Symposium on Ballistics, Hague, The Netherlands, 19–21 April 1983.
24. Wang, Y.; Zhang, C.M.; Zhang, Y. Study on static and dynamic mechanical properties of aviation Al7050 alloy and construction of JC constitutive model. *Mater. Rep.* **2021**, *35*, 10096–10102.
25. Nagy, A.; Ko, W.L.; Lindholm, U.S. Mechanical behavior of foamed materials under dynamic compression. *J. Cell. Plast.* **1974**, *10*, 127–134. [CrossRef]
26. Herbold, E.B.; Cai, J.; Benson, D.J.; Nesterenko, V.F. Simulation of particle size effect on dynamic properties and fracture of PTFE-Al-W composite. *AIP Conf. Proc.* **2007**, *3*, 161–173.
27. Wang, Z. Parameter Calibration of PTFE/Al Reactive Materials and Numerical Simulation Research of Its Impact-Induced Deflagration Behavior. Master's Thesis, North University of China, Shanxi, China, 2021.
28. Zhang, J. Study on Dynamic Mechanical Properties of Multifunctional Energetic Structural Materials. Master's Thesis, Nanjing University of Science and Technology, Nanjing, China, 2013.
29. Dreizin, E.L. Metal-based reactive nanomaterials. *Prog. Energy Combust. Sci.* **2009**, *35*, 141–167. [CrossRef]

Article

Dispersal Characteristics Dependence on Mass Ratio for Explosively Driven Dry Powder Particle

Binfeng Sun [1,*], Chunhua Bai [1], Caihui Zhao [2], Jianping Li [1] and Xiaoliang Jia [3]

1   State Key Laboratory of Explosion Science and Technology, Beijing Institute of Technology, Beijing 100081, China
2   Jinxi Industries Group Co., Ltd., Taiyuan 030027, China
3   Liaoning Jinhua Electromechanical Co., Ltd., Huludao 125000, China
*   Correspondence: sunbinfeng@bit.edu.cn

Abstract: An investigation on the dispersal characteristics of the cylindrically packed material of dry powder particles driven by explosive load is presented. By establishing a controllable experimental system under laboratory conditions and combining with near-field simulation, the particle dispersal process is described. Additionally, Kelvin–Helmholtz instability is observed during the process of jet deceleration dispersal. The characteristic parameters of radially propagated particles are explored under different mass ratio of particle-to-charge (M/C). Results indicate that, when the charge mass remains constant, an increase in M/C leads to a decrease in dispersed jet number, void radius and maximum velocity, wherein the maximum velocity correlates with calculations by the porous Gurney model. The case of the smaller M/C always has a higher outer-boundary radius and area expansion factor. Findings indicate that when particles detach from the jet upon reaching minimum acceleration and entering low-speed far-field stage from high-speed near-field stage, the outer-boundary radius is 30~36 times the initial particles' body radius under different M/C. In addition, particle concentration distribution over time and distance is qualitatively analyzed by the grayscale image method. This research can be referential for improving the fire-extinguishing capacity of extinguishing bombs and the damage property of fuel air explosive (FAE).

Keywords: explosive dispersal; dry powder; mass ratio; particle jet

Citation: Sun, B.; Bai, C.; Zhao, C.; Li, J.; Jia, X. Dispersal Characteristics Dependence on Mass Ratio for Explosively Driven Dry Powder Particle. *Materials* **2023**, 16, 4537. https://doi.org/10.3390/ma16134537

Academic Editor: Giovanni Polacco

Received: 17 May 2023
Revised: 14 June 2023
Accepted: 19 June 2023
Published: 23 June 2023

1. Introduction

Explosive dispersal of the granular materials is extensively applied in industrial security and military engineering [1–5]. Dry powder fire-extinguishing bombs explode and disperse the dry powder particles to form stable aerosols, serving for fire control and suppression [6]. Fuel air explosive (FAE) brings about severe damage to the target by detonating the dispersed cloud driven by explosive load [7,8]. Among these application fields, it is of great significance to investigate the evolution law in the process of the particle dispersal, involving relevant parameters such as dispersal velocity and cloud area, etc. Such parameters can determine the fire-extinguishing capacity of the dry powder fire-extinguishing bomb, as well as the secondary initiation time, dropping position, and damage power of FAE.

In typical explosive dispersal, it is often a research challenge to obtain the detailed temporal visualization of the interface for high-pressure gas/particle contact and the particle/air contact, because these experiments are usually conducted in outdoor environments where the lighting and environment-related parameters are difficult to be controlled [9,10]. Zhang et al. [11] conducted a large-scale spay detonation using gasoline explosive load. Such an outdoor dispersal experiment allows only the outer-trajectories of the clouds to be observed, while the particles' concentration distribution in the cloud region and the morphology of the inner-void are difficult to be accurately observed by optical shooting. To clearly obtain the dispersed cloud trajectory, small-scale dispersal experiments based on the laboratory environment have been carried out, mainly involving the use of the shock tubes or detonation tubes to generate shockwaves and drive particle dispersal [12–14]. The shockwave generated by shock tubes or detonation tubes belongs to weak shock load on the order of 10^{-1}~10^0 MPa, while the blast wave driven by explosive load is on the order of 10^1 GPa. There is significant difference in load strength between the two [15]. Therefore, the dispersal experiments conducted using weak shock load are difficult to simulate the real particles' dispersal driven by explosive loads in real situations.

The initial stage of explosive dispersal involves research problems such as the interaction between shockwaves and particles, detonation products and particles, as well as the interaction among the particles. However, this process occurs inside the particles' body within the microsecond level, making it hard to be recorded with high-speed shooting. To better explore the evolution of particle dispersal driven by explosive load, it is necessary to conduct studies in a synchronous method by a combination of experimental and numerical simulation.

This paper focuses on the acquisition and research of the characteristic parameters in a dry powder material dispersal system driven by explosive load. A particle dispersal test system under controllable laboratory conditions is established, and the dispersal process of the annular dry powder material driven by a central detonator is experimentally investigated. By using a surface light source as the background light, the complete particle dispersal morphology, including inner- and outer-trajectories, as well as the grayscale distribution are captured more visibly compared to previous tests [3] carried out under natural light. To obtain the initial-stage particle dispersal process that cannot be measured experimentally, a numerical simulation method based on the smoothed particle hydrodynamics finite element (SPH–FEM) coupling algorithm is proposed, followed by a comparison of the results with the porous Gurney models. The characteristic parameters, including the number of jets and the maximum velocity in the dispersal process, are investigated under different mass ratio (the ratio of dry powder mass-to-charge mass (M/C)). Finally, the time-dependent laws of the characteristic parameters, including the dispersal radius, cloud area, the radius and the area of the explosive void, are explored. Particle concentration distribution is qualitatively analyzed by the grayscale image.

2. Materials and Methods

2.1. Dry Powder and Central Charge

Dry powder extinguishant (75% $NH_4H_2PO_4$, 15% $(NH_4)_2SO_4$, 9% $Mg_3[Si_4O_{10}](OH)_2$, purchased from Zhengzhou Haichao fire agent Co., Ltd., Zhengzhou, China) is used for dispersal experiments, with a density of 1.8 g/cm^3. The particle size and sphericity distribution, as well as the SEM (scanning electron microscope) image are shown in Figure 1. In the selection of central charge, factors including indoor test safety, space limitation, and the strength for fixing the device are comprehensively considered. The explosive equivalent should be minimized on the premise of ensuring the experimental effect. Accordingly, No. 8 electric detonator with a diameter of 7 mm is adopted as the central charge. It contains 1.074 g RDX, 30 mm behind the tail of the detonator.

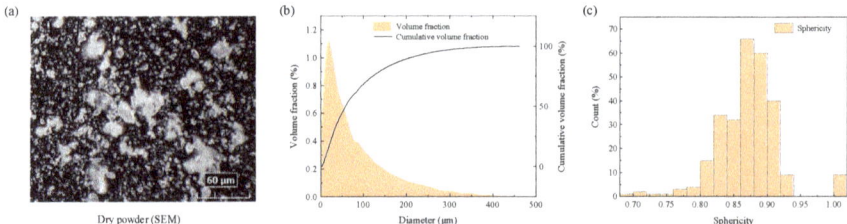

Figure 1. Dry powder material properties. (**a**) SEM image, (**b**) particle size (the median diameter D50 of the whole particles is 58.4 μm) and (**c**) sphericity distribution.

2.2. Experimental Setup

Considering the experimental requirements for environment stability (no wind or natural light) and a large space, an enclosed darkroom with an inner-dimension of 20 m × 10 m × 5 m and safety protection measures is selected for the test. A cylindrical explosive dispersal device is designed with a central electric detonator surrounded by annular dry powder particles. To eliminate the impact of the shell debris on the particle dispersal, A4 paper is used for the packaging of the dry powder. Each end of the shell configures a plate with a diameter slightly larger than that of the particles' body, so as to constrain the axial movement of the particles at the initial explosive stage while not affecting the optical shooting. The radial movement is research-oriented here. Specific dimension of the dispersal device is shown in Figure 2a. The height of the particles' body is set to be 30 mm. To clearly observe the morphology of the dispersed particle cloud and capture the grayscale images, a uniform, flicker-free, and brightness-adjustable surface light source is selected as the backlight. The dimension of the surface light source is set to 2 m × 2 m, considering the particle dispersal radius.

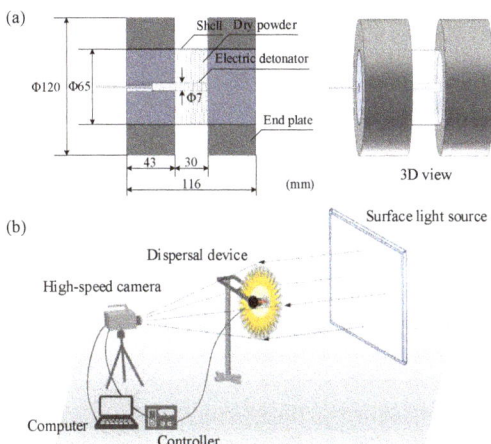

Figure 2. Diagram of (**a**) the dispersal device and (**b**) experimental setup.

FASTCAM NOVA S16 high-speed camera is used to record the dispersal process driven by explosive, with a frame rate of 10,000 fps and a resolution of 1024 × 1024. A controller is configured for the synchronous trigger of the high-speed camera and the detonator. An absolute pixel length scale is established according to the dimension of the surface light source.

2.3. Numerical Simulation Methodology

Explosively driven particle dispersal involves matters including large-scale displacement and the deformation of the particles' body. As for the dispersed particles, separate finite element (FEM)-based method is not applicable due to its disadvantages of requiring complex computation and a reduced reliability when referring to mesh deformation. While for the SPH method, it employs the space-independent particle as the computational domain as a Lagrange particle algorithm [16,17]. Particle motion is used to describe the deformation of materials, which solves the problem of calculating large deformation in FEM. However, compared with FEM, the SPH method is less efficient because it requires information of other surrounding particles to calculate the physical quantity of a single particle. FEM method has advantages such as simple structural modeling and a mature dynamic response analysis technology. Hence, smoothed particle hydrodynamics (SPH) algorithm coupled with FEM is adopted for numerical simulation. Thereinto, the end plate used to constrain the axial movement of particles is simulated by FEM and defined as

SHELL element. The paper shell that has little effect on the movement is ignored. For charge and dry powder with large deformation, SPH-based simulation is performed in ANSYS/LS-DYNA software. Thus, both computational efficiency and reliability of simulation results are reconciled. The coupling between SPH particles and FEM elements is achieved through automatic nodes-to-surface contact algorithm, where SPH particles are defined as slave bodies and FEM units are defined as main bodies. The physical information transfer among structures is realized through the coupling algorithm of SPH and FEM.

The material model chosen to describe the RDX charge is MAT_HIGH_EXPLOSIVE_BURN with a Jones–Wilkins–Lee (JWL) equation of state expressed as [18]:

$$P_c = A\left(1 - \frac{\omega}{R_1 V}\right)e^{-R_1 V} + B\left(1 - \frac{\omega}{R_2 V}\right)e^{-R_2 V} + \frac{\omega E}{V} \tag{1}$$

where P_c denotes the pressure of the detonation product, V is the relative specific volume, and E is the internal energy per unit volume of the charge. A, B, R_1, R_2 and ω denote the parameters of the JWL equation of state, as shown in Table 1 [19]. ρ_c is the charge density and D is the detonation velocity.

Table 1. Detonation properties of the RDX charge and JWL equation-of-state parameters.

ρ_c (g/cm³)	D (m/s)	P_c (GPa)	A (GPa)	B (GPa)	R_1	R_2	ω	E (MJ/kg)
1.436	7500	17.5	100.3	22.2	6.028	1.8519	0.48	9

Dry powder is described by the MAT_SOIL_AND_FOAM model [20], whose ideal plastic yield function can be expressed as:

$$\phi = \frac{S_{ij}S_{ij}}{2} - (a_0 + a_1 P_m + a_2 P_m{}^2) \tag{2}$$

where a_0, a_1 and a_2 are user-defined constants. S_{ij} denotes the deviatoric stress component and P_m ($m = 1\sim10$) denotes the pressure value at the corresponding points where volume strain occurs. The main calculation parameters are listed in Table 2, where ρ_0 is the density, G denotes the shear modulus and KUN denotes the bulk modulus for unloading. PC and VCR denote the pressure cutoff for tensile fracture and volumetric crushing option, respectively. EPS_m denotes the volume strain values.

Table 2. Dry powder material parameters.

ρ_0 (g/cm³)	G (GPa)	KUN (GPa)	a_0	a_1	a_2	PC	VCR		
1.8	1.601×10^{-2}	13,280	0.0033	1.31×10^{-7}	0.1232	0	0		
EPS_1	EPS_2	EPS_3	EPS_4	EPS_5	EPS_6	EPS_7	EPS_8	EPS_9	EPS_{10}
0	0.05	0.09	0.11	0.15	0.19	0.21	0.22	0.25	0.3
P_1 (GPa)	P_2 (GPa)	P_3 (GPa)	P_4 (GPa)	P_5 (GPa)	P_6 (GPa)	P_7 (GPa)	P_8 (GPa)	P_9 (GPa)	P_{10} (GPa)
0	3.42	4.53	6.76	12.7	20.8	27.1	39.2	56.6	123

Here, the deformation of the end plate is not considered in the simulation. In order to improve the calculation efficiency, MAT_RIGID model is used for description of the end plate, having a density of 7.8 g/cm³, Young's modulus of 210 GPa and Poisson's ratio of 0.3. The displacement and rotation in all directions are constrained. In the initial stage of the explosive dispersal, the shockwave generated by the explosion is much greater than the atmospheric pressure, thus ignoring the air resistance. The numerical simulation model ($M/C = 102.4$) of dry powder particles driven by charge is shown in Figure 3, where dry powder contains 787,140 SPH particles and charge contains 11,700 SPH particles. R_{in0} and R_{out0} denote the initial radius of the central charge and dry powder particles, respectively.

Figure 3. SPH–FEM coupling simulation model (M/C = 102.4).

3. Results and Discussion

The dispersal evolution law of the dry powder particles driven by RDX charge under different mass ratio (ratio of the dry powder material mass to charge mass, M/C) was studied experimentally and numerically. M/C was changed by adjusting R_{out0}, while the bulk density, height of the dry powder particles and the central charge-related parameters remained constant. Five operation conditions (M/C = 14.1, 28.8, 48.5, 73.0 and 102.4, respectively) were selected for subsequent analysis. Specific parameter setup is shown in Table 3.

Table 3. Parameters setup for the experiment and simulation.

No.	Charge Mass C (g)	Dry Power Mass M (g)	Mass Ratio (M/C)	Inner-Radius R_{in0} (mm)	Outer-Radius R_{out0} (mm)
1	1.074	15.17	14.1	3.5	12.5
2	1.074	30.98	28.8	3.5	17.5
3	1.074	52.05	48.5	3.5	22.5
4	1.074	78.39	73.0	3.5	27.5
5	1.074	110.00	102.4	3.5	32.5

3.1. Particle Dispersal Process and Morphology

Previous studies by Loiseau et al. [9] have revealed the spherical materials' dispersal driven by a C-4 charge. The spherical materials' dispersal is omni-directional. Only the morphology of the outer-layer particles' cloud can be obtained while the inner-layer particles' cloud is not visible due to being surrounded by outer-layer particles. In this study, the dry powder particles' body in a cylindrical profile with a constrained axial movement can clearly exhibit the particles' morphology of both the inner- and outer-trajectories in a radial direction. Moreover, the detachment process of the jet tip fragments at the later dispersal process is clearly observed, which is not realized in previous studies.

A diagram of the particle dispersal driven by explosive charge is depicted in Figure 4. The cross-section of the initial dispersal structure is shown in Figure 4a, appearing as a cylinder with central charge surrounded by dry powder particles. After the explosive detonation, the aroused shockwave propagates radially outward and compresses the porous particles to be compact. Simultaneously, the detonation product drives particle dispersal to form an inner-void (Figure 4b,c). The blast wave is then transmitted into the surrounding air, and the rarefaction wave travels inward through the compacted particles. The compacted particles expand under tension and then break into fragments (Figure 4d). A radial jet structure forms as the fragments propagate radially outward and the unconsolidated loose particles then fall off (Figure 4e,f). The velocity of the jet rapidly decreases and begins to slowly disperse as the consolidated fragments detach from the jet at a high speed. The lower part of Figure 4 presents the simulation (at the initial dispersal stage) and experimental images (at the later dispersal stage) corresponding to the different stages of the dispersal process under M/C = 102.4.

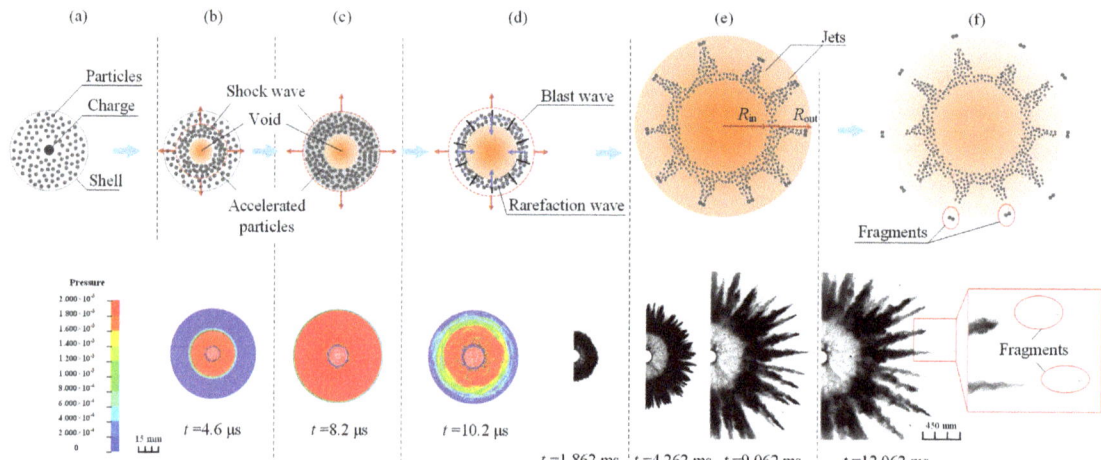

Figure 4. Dry powder particle dispersal process driven by cylindrical charge. Top (**a~f**) depict the diagram of the whole dispersal process. Bottom left presents the simulated process at the initial dispersal stage that is unavailable to be experimentally measured (M/C = 102.4). Bottom right displays the experimental images taken at the later dispersal stage (M/C = 102.4).

During the deceleration dispersal stage of particles jet, Kelvin–Helmholtz instability [21] is observed, as shown in Figure 5. It is caused by the tangential velocity difference at the jet/air contact interface, exhibiting obvious vortex profile. The emergence of Kelvin–Helmholtz instability increases the mixing velocity and degree of the particles and air near the jet/air contact interface.

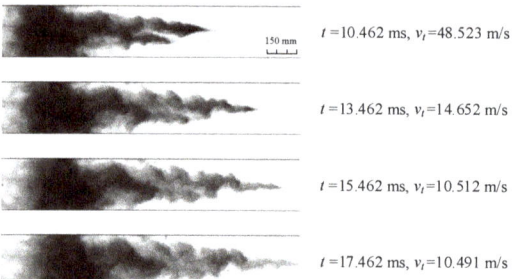

Figure 5. Kelvin–Helmholtz instability observed during the process of jet deceleration dispersal (M/C = 102.4 and v_t denotes the real-time velocity).

Reynolds number R_e, i.e., the ratio of inertial force to the frictional force in the particles system, is used to characterize the critical conditions for the jet formation. R_e is expressed as [22]:

$$R_e = (\rho v L)/(\gamma_s c_s d_s) \tag{3}$$

where ρ, v and L denote the loading density of the particles, maximum dispersal velocity and the thickness ($L = R_{ou0} - R_{in0}$) of the particles' body. γ_s, c_s and d_s denote the mass density, sound velocity and the average diameter of the particles. Frost et al. [22] found that the jet number N increases with increasing Reynolds number. In this study, where the mass ratio M/C is taken as an independent variable, v and L are two variables that change accordingly. The experimental dependence of normalized product term vL and the jet number N on the mass ratio M/C is shown in Figure 6. It can be seen that vL is positively correlated with the jet flow number N, i.e., R_e is positively correlated with the jet number N, which is consistent with Frost's theory.

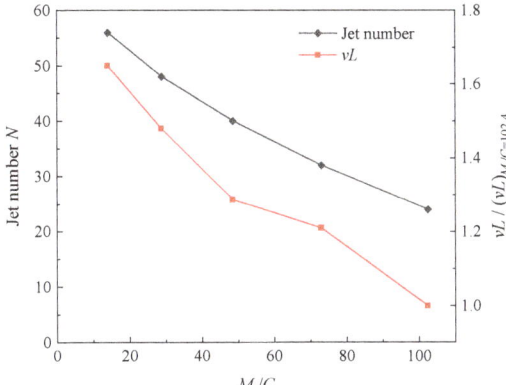

Figure 6. Dependence of normalized vL and the jet number N on mass ratio M/C.

To obtain the trajectory radius and velocity of the dispersed particles, the experimental image is converted into a grayscale image, followed by a binarized operation with a self-defined grayscale threshold. The inner- and outer-boundary trajectories of the image are accordingly extracted. Then, a ray is drawn in each image from the charge center toward the direction of jet movement. The intersection points between the ray and the inner-/outer-boundary of the particles are defined as the inner-void radius R_{in}/outer-radius R_{out} of the particles cloud, as marked in Figure 4e. The cloud area is extracted by the grayscale threshold on each image, and the inner-void area is calculated by R_{in}. Following this, the characteristic parameters of the particle dispersal dependence on mass ratio M/C are presented, including maximum velocity, dispersal radius and area, the characteristics of the dispersed void, as well as the grayscale distribution of the dispersed particles.

3.2. Particle Dispersal Maximum Velocity

Previously, Gurney [23] proposed a model to predict the initial/maximum velocity of spherical and cylindrical homogeneous shells driven by high explosive charge. However, for the heterogeneous particle system in this study, there are pores among the particles. Under high explosive load, the compaction and deformation of the particle body can collapse and heat the pores, leading to a significant entropy dissipation of the explosive energy in the interstitial air and particles. Therefore, the standard Gurney model is no longer applicable for the valid prediction of the maximum velocity of granular material. Taking into account the porosity and bulk density effect, Milne [24] empirically modified the Gurney model, which is now expressed as:

$$V_{\text{Gurney}}(M/C) = \sqrt{2E}(M/C + 0.5)^{-0.5} \tag{4}$$

$$V(M/C, \rho_0, \varphi) = V_{\text{Gurney}} \left[\frac{M/C}{a(\rho_0)} \right] \times F(\varphi, M/C) \tag{5}$$

$$\alpha(\rho_0) = 0.2\rho_0^{0.18} \tag{6}$$

$$F(\varphi, M/C) = 1 + \left(0.162e^{1.127\varphi} - 0.5\right) \log_{10}(M/C) \tag{7}$$

where M is the mass of the powder, C is the mass of the charge, $\sqrt{2E}$ is the Gurney velocity coefficient (2.93 km/s for RDX [25,26]). ρ_0 is material density and φ is the loading density ($\varphi = \rho/\rho_0 = 1.1/1.8 \approx 0.6$, where ρ denotes the bulk density).

The maximum velocity of the particles under different M/C is given in Figure 7, including the results by experiments, numerical simulation and the porous Gurney model. It can be seen that there is good consistency among the three, verifying the reliability of the simulation model. In the early expansion stage of the explosive products, the process can be considered as isentropic adiabatic expansion because there is no heat exchange between the products and the particles due to the rapid action [27]. When M/C is increased by increasing the particle mass M while the charge mass C remains constant, the kinetic energy obtained by the whole particles system under different M/C is approximate. According to the kinetic energy theorem, the maximum velocity of the particles tends to decrease.

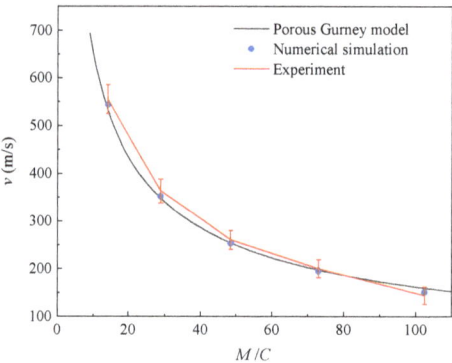

Figure 7. Dependence of maximum dispersal velocity on mass ratio M/C.

3.3. Particle Dispersal Radius and Area

To accurately obtain the variation of the dispersal radius and area over time, the starting time of the dispersal process need to be firstly determined. Although the high-speed camera shooting is controlled synchronously with the charging of the detonator, there are detonation time differences among the electric detonators, making it difficult to determine the starting time for every measurement. To address this problem, the simulated R_{out}-time data under different M/C is referred. Taking the case of $M/C = 102.4$ as an example, the starting time when the first image is taken can be determined according to the simulated R_{out}-time data. As shown in Figure 8a, the outer-boundary radius R_{out} extracted from the first image is 152.3 mm, then the corresponding time $t = 0.862$ ms is obtained by simulation. The simulated data are then concatenated between 0 and 0.862 ms with the experimental data after 0.862 ms, and a complete R_{out}–time curve for describing the dispersal process is obtained. Accordingly, the time-dependent velocity and acceleration are given by performing first and second derivative operations on the R_{out}–time curve, as shown in Figure 8b.

Previously, the process of explosive dispersal and cloud formation is divided into three stages [28]. As shown in Figure 8b, it is the acceleration at the near-field stage during 0~0.3 ms where the pressure from detonation products is greater than the air resistance; it is the uniform transition stage during 0.3~5 ms where the two are roughly equal; and after 5 ms, it is the deceleration at the far-field stage where the air resistance plays a dominant role. At the moment of minimum acceleration, the consolidated particle fragments detach from the cloud. Prior to this, the particle jet exhibits approximate high-speed ballistic movement; afterward, the movement velocity of the particles decreases rapidly and approaches zero. The moment ($t = 9.262$ ms) of minimum acceleration is marked in Figure 8b. According to the turning point of acceleration, the whole process can be re-divided into two stages: the near-field stage with high-speed motion and the far-field stage with low-speed motion.

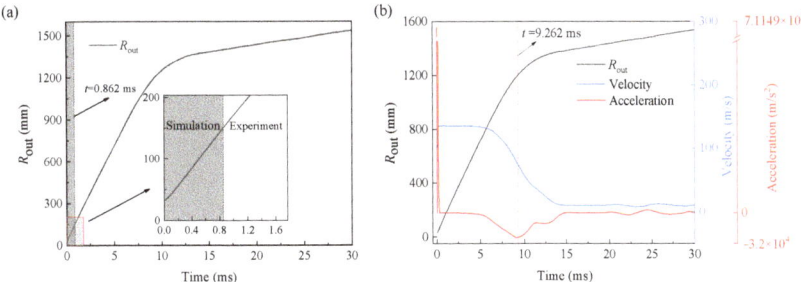

Figure 8. Time-dependent curves of (**a**) the particles' outer-boundary radius, (**b**) the velocity, and the acceleration ($M/C = 102.4$).

The time-dependent curves of the particles' outer-boundary radius increment $\Delta R_{\text{out}} = R_{\text{out}} - \Delta R_{\text{out0}}$ under different M/C are shown in Figure 9a. In the case of smaller M/C, ΔR_{out} increases faster at the initial stage, but the high-growth-rate duration is shorter than that of a larger M/C. Moreover, the initial ΔR_{out} is larger under a smaller M/C, but later, it is surpassed by the case with a larger M/C. This can be explained by the kinetic energy theorem. When the total kinetic energy obtained by the system remains constant, the particle body with a smaller mass can obtain a higher velocity at the initially explosive stage. However, the particles' body is thinner, leading to the unconsolidated particles detaching earlier from the consolidated fragments. The loose particles have a smaller density and are subject to greater air resistance. Therefore, the particles' body with smaller mass will ultimately achieve a smaller radius and displacement. The opposite accounts for particles' body with a larger mass.

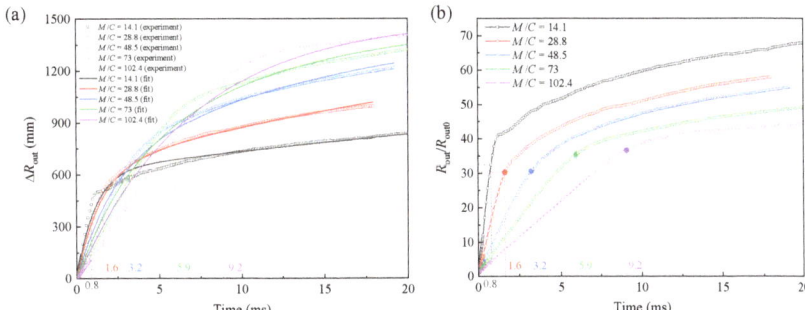

Figure 9. Time-dependent curves of (**a**) the outer-radius increment ΔR_{out} (experimental and fitting results, variance $R^2 = 0.997$) and (**b**) radius expansion factor (experimental results) of the dispersed particles under different M/C. The moment when the minimum acceleration occurs is marked.

As a result, to determine the outer-boundary radius ΔR_{out}, not only the initial velocity but also the duration of high-speed movement should be comprehensively considered. In order to quantitatively reveal the dependence of ΔR_{out} and mass ratio M/C on time, the experimental results are fitted, as shown in Figure 9a. The fitting expression (units: g-mm-ms) is given as follows:

$$\Delta R_{\text{out}} = 1500 - 620 \exp\left\{ -t \,\middle/\, \left[0.334 e^{(M/C)/35} \right] \right\} - 880 \exp\left\{ -t \,\middle/\, \left[177 e^{-(M/C)/14} \right] \right\} \quad (8)$$

The radius expansion factor, defined as $R_{\text{out}}/R_{\text{out0}}$, is time-dependent under different mass ratio M/C, as shown in Figure 9b. It shows a trend of rapid growth in the initial stage and gentle growth in the later stage. Under a smaller M/C, $R_{\text{out}}/R_{\text{out0}}$ increases faster at the initial stage but does not last long. With the increase in M/C, $R_{\text{out}}/R_{\text{out0}}$ has a smaller initial growth rate but lasts longer. Unlike the ΔR_{out}–time curves, the R_{out}/R_{out0}–time curves always have a larger R_{out}/R_{out0} at low M/C than at high M/C. Moreover, the particles under a smaller M/C possess a larger radius expansion factor than that of a larger M/C at the same moment. Provided that the particles' body mass remains constant, a larger expansion factor, i.e., a larger outer-boundary dispersal distance, can be obtained by increasing the charge mass within a certain range. When the minimum acceleration is reached, R_{out} is 30~36 times that of R_{out0} as is marked in the vertical coordinate in Figure 9b.

Particle dispersal radius determines the dispersal distance. However, due to the structural morphology of particle jets, it is necessary to study the variation of cloud area to obtain the dispersal coverage range. The initial area of the particles' body is defined as S_{out0}, and the area S_{out} formed after dispersal can be obtained by $S_{out} = S_{out0} + \Delta S_{out}$ (ΔS_{out} denotes the area increment). The time-dependent curves of the dispersal cloud area under different M/C are given in Figure 10a. The expansion factor of the cloud area, defined as S_{out}/S_{out0}, is time-dependent under different mass ratio M/C, as shown in Figure 10b. S_{out} and S_{out}/S_{out0} exhibit a similar trend to that of the particles' outer-boundary radius increment ΔR_{out} and radius expansion factor R_{out}/R_{out0}, respectively. Provided that the particles' body mass remain constant, a larger expansion factor, i.e., a larger cloud coverage area, can be obtained by increasing the charge mass within a certain range.

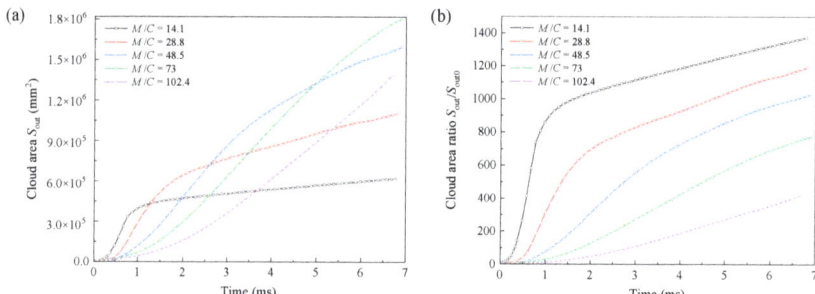

Figure 10. Time-dependent curves of the (**a**) cloud area and (**b**) area expansion factor.

3.4. Characteristics of the Dispersed Void

Previous studies [29,30] are unable to clearly obtain the internal explosive void formed by the particle dispersal. Through the experimental method here, a circular explosion void was clearly observed as demonstrated in Figure 4, which is defined as the region within the interface of the cloud, with a radius of R_{in}.

The initial radius R_{in0} of the void is equal to the radius of the central charge, and the real-time radius R_{in} can be obtained by $R_{in} = R_{in0} + \Delta R_{in}$ (ΔR_{in} denotes the radius increment). The time-dependent curves of ΔR_{in} and void area are shown in Figure 11. The results indicate that the variation of ΔR_{in} over time appears smooth without a significant turning point. The variation of void area over time tends to be more linear. As the pressure of the detonation products in the void gradually decreases, the resistance gradually dominates. Accordingly, the particles disperse slower, exhibiting a decreased slope of ΔR_{in}–time curves. The void is generated by the expansion of the detonation products to promote particles dispersal. Under a larger M/C, i.e., a larger particle mass M, the resistance is larger, leading to a slower expansion of the detonation products. Accordingly, the generated void has a smaller radius.

The fitted results of the time-dependent ΔR_{in} under different M/C are shown in Figure 11a, and the expression is given as follows:

$$\Delta R_{in} = 380\{1 - \exp[-t/(1.44856 + 0.02127 \times (M/C))]\} \tag{9}$$

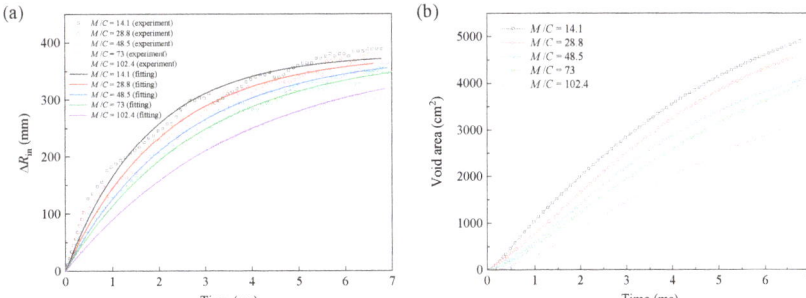

Figure 11. Time-dependent curves of (**a**) the void radius increment ΔR_{in} (experimental and fitting results, variance $R^2 = 0.991$) and (**b**) the void area (experimental results) under different M/C.

3.5. Particle Dispersal Grayscale Distribution

Particles' concentration distribution can be qualitatively analyzed by the grayscale image. The grayscale image is obtained by subtracting the initial background image from the cloud image, as shown in Figure 12a, with a grayscale value between 0 and 255. The black pixels with a grayscale value of 0 represent the background, and the pixels with grayscale values between 1 and 255 represent the particles. The larger the grayscale value, the higher the concentration. Figure 12b–f depict the particles' grayscale distribution over time and distance under different M/C. Compared with research conducted by Gao et al. [31], where an ultrasonic–electric hybrid detection method was adopted for partial-area particle concentration acquisition, here, the overall grayscale distribution in the cloud area can be obtained to analyze the overall concentration distribution law. The peak near the coordinate origin of distance axis is due to the frictional force between the particles and the end plate, causing a small portion of the particles to remain near the end plate. The inner-boundary radius R_{in} and outer-boundary radius R_{out} of the main part of the cloud are marked in Figure 12f. Over time, the cloud width ($R_{in} - R_{out}$) increases and the concentration decreases. Radial bimodal fluctuation is presented in the main part of the cloud from the grayscale value variation over distance. Under a larger M/C, the cloud width and concentration are larger.

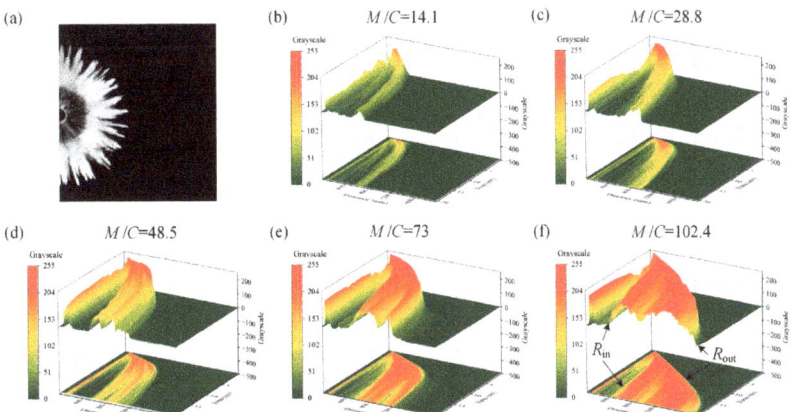

Figure 12. Particles' grayscale extraction method and distribution. (**a**) Grayscale extraction. (**b–f**): Grayscale distribution of the dispersed particles under different M/C.

4. Conclusions

A controllable explosively driven particles dispersal system has been established under laboratory conditions for investigating the dispersal process. By conducting dispersal experiments on cylindrical particles' body with a charge placed in the center, both the inner- and outer-dispersal trajectories in radial direction were clearly obtained. Through the combination of experimental and numerical simulation, the characteristic parameters evolution of dry powder particles dispersal under different mass ratio M/C was studied. Additionally, the particle concentration distribution over time and distance was qualitatively analyzed under different M/C by the grayscale image.

The research results indicate that when the charge mass remains constant, the jet number decreases as the mass ratio increases. Additionally, the outer-boundary radius and area of the cloud show a tendency of rapid growth in the early stage and gentle growth in the later stage. Moreover, the initial outer-radius and area of particles under low mass ratio conditions are larger, but later, they are surpassed by particles under high mass ratio conditions. The expansion factors of the outer-boundary radius and area are always higher under the small mass ratio in the whole process. When the outer-boundary reaches the minimum acceleration, the radius of the outer-boundary is 30~36 times the initial radius of the particle body. From this moment on, the particles move from a high-speed near-field stage to low-speed far-field stage. In addition, the radius of the void decreases with increasing mass ratio.

The proposed experimental system is not only applicable to the research of explosive dispersal for cylindrically packed granular materials, but can also be extended to the research of liquid or solid–liquid mixed materials with other shapes. It can be referential for improving the fire-extinguishing capacity of the extinguishing bombs and the damage property of FAE, as well as the performance prediction of the dispersal materials. SPH or an SPH-coupled simulation method can also be extended to other research fields involving thr prediction of non-military explosive characteristics such as mines, tunnel and structural demolition.

Author Contributions: Conceptualization, B.S. and C.B.; methodology, B.S.; software, B.S. and C.Z.; validation, J.L.; investigation, C.Z. and X.J.; data curation, C.B.; writing—original draft preparation, B.S.; writing—review and editing, B.S., C.B. and J.L.; supervision, C.B.; project administration, C.B. All authors have read and agreed to the published version of the manuscript.

Funding: This research received no external funding.

Institutional Review Board Statement: Not applicable.

Informed Consent Statement: Not applicable.

Data Availability Statement: Not applicable.

Conflicts of Interest: The authors declare no conflict of interest.

References

1. Yang, L.; Zhao, Q.; Chai, X.; Ma, X.Y. Research on the Regular Between Concentration of superfine powder extinguishing agent explosion scatter and fire-extinguishing ability. *Adv. Mater. Res.* **2014**, *877–888*, 1017–1023. [CrossRef]
2. Bai, C.; Wang, Y.; Li, J.; Chen, M. Influences of the Cloud Shape of Fuel-Air Mixtures on the Overpressure Field. *Shock Vib.* **2016**, *2016*, 9748536. [CrossRef]
3. Apparao, A.; Rao, C.R.; Tewari, S.P. Studies on Formation of Unconfined Detonable Vapor Cloud Using Explosive Means. *J. Hazard. Mater.* **2013**, *254*, 214–220. [CrossRef] [PubMed]
4. Singh, S.K.; Singh, V.P. Extended near-Field Modelling and Droplet Size Distribution for Fuel-Air Explosive Warhead. *Def. Sci. J.* **2001**, *51*, 303–314. [CrossRef]
5. Guo, X.; Zhang, H.; Pan, X.; Zhang, L.; Hua, M.; Zhang, C.; Zhou, J.; Yan, C.; Jiang, J. Experimental and Numerical Simulation Research on Fire Suppression Efficiency of Dry Powder Mediums Containing Molybdenum Flame Retardant Additive. *Process Saf. Environ. Prot.* **2022**, *159*, 294–308. [CrossRef]
6. Guan, X.; Liu, S. Research on Dispersion Mechanism of Extinguishing Agent Scattered by Grenade Explosion. *Adv. Mater. Res.* **2014**, *900*, 738–741. [CrossRef]

7. Zhang, F.; Findlay, R.; Anderson, J.; Ripley, R. Large Scale Unconfined Gasoline Spray Detonation. In Proceedings of the 24th International Colloquium on the Dynamics of Explosion and Reactive Systems, Taipei, Taiwan, 28 July–2 August 2013.

8. Wang, Y.; Liu, Y.; Xu, Q.; Li, B.; Xie, L. Effect of Metal Powders on Explosion of Fuel-Air Explosives with Delayed Secondary Igniters. *Def. Technol.* **2021**, *17*, 785–791. [CrossRef]

9. Loiseau, J.; Pontalier, Q.; Milne, A.M.; Goroshin, S.; Frost, D.L. Terminal Velocity of Liquids and Granular Materials Dispersed by a High Explosive. *Shock Waves* **2018**, *28*, 473–487. [CrossRef]

10. Apparao, A.; Rao, C.R. Performance of Unconfined Detonable Fuel Aerosols of Different Height to Diameter Ratios. *Propellants Explos. Pyrotech.* **2013**, *38*, 818–824. [CrossRef]

11. Zhang, F.; Ripley, R.; Yoshinaka, A.; Findlay, C.R.; Anderson, J.; von Rosen, B. Large-scale Spray Detonation and Related Particle Jetting Instability Phenomenon. *Shock Waves* **2015**, *25*, 239–254. [CrossRef]

12. Rodriguez, V.; Saurel, R.; Jourdan, G.; Houas, L. Solid-particle Jet Formation under Shock-wave Acceleration. *Phys. Rev. E* **2013**, *88*, 063011. [CrossRef] [PubMed]

13. Rodriguez, V.; Saurel, R.; Jourdan, G.; Houas, L. Impulsive Dispersion of a Granular Layer by A Weak Blast Wave. *Shock Waves* **2017**, *27*, 187–198. [CrossRef]

14. Xiong, X.; Gao, K.; Zhang, J.; Li, B.; Xie, L.; Zhang, D.; Mensah, R.A. Interaction between Shock Wave and Solid Particles: Establishing A Model for the Change of Cloud's Expansion Rate. *Powder Technol.* **2021**, *381*, 632–641. [CrossRef]

15. Frost, D.L. Heterogeneous/particle-laden blast waves. *Shock Waves* **2018**, *28*, 439–449. [CrossRef]

16. Hou, Q.; Liu, J.; Lian, J.; Lu, W. A Lagrangian Particle Algorithm (SPH) for an Autocatalytic Reaction Model with Multicomponent Reactants. *Processes* **2019**, *7*, 421. [CrossRef]

17. Salis, N.; Franci, A.; Idelsohn, S.; Reali, A.; Manenti, S. Lagrangian Particle-Based Simulation of Waves: A Comparison of SPH and PFEM Approaches. *Eng. Comput.* **2023**. [CrossRef]

18. Chen, M.; Bai, C.; Li, J. Simulation on Initial Velocity and Structure Dynamic Response for Fuel Dispersion. *Chin. J. Energetic Mater.* **2015**, *23*, 323–329.

19. Feng, Z. Experiment and Numerical Simulation of Critical Diameter of RDX Compound Explosive. Master's Thesis, North University of China, Taiyuan, China, 2007.

20. Wang, Z. The Design and Simulation of a New Type of Forest Fire Extinguishing Bomb. Master's Thesis, North University of China, Taiyuan, China, 2016.

21. Mansouri, Z.; Boushaki, T. Investigation of Large-Scale Structures of Annular Swirling Jet in a Non-Premixed Burner Using Delayed Detached Eddy Simulation. *Int. J. Heat Fluid Flow* **2019**, *77*, 217–231. [CrossRef]

22. Frost, D.L.; Gregoire, Y.; Goroshin, S.; Zhang, F. Interfacial Instabilities in Explosive Gas-particle Flows. In Proceedings of the 23rd International Colloquium on the Dynamics of Explosion and Reactive Systems, Irvine, CA, USA, 24–29 July 2011.

23. Gurney, R.W. *The Initial Velocities of Fragments from Bombs, Shell and Grenades*; Technical Report; Army Ballistic Research Laboratory: Aberdeen, MD, USA, 1943.

24. Milne, A.M. Gurney Analysis of Porous Shells. *Propellants Explos. Pyrotech.* **2016**, *41*, 665–671. [CrossRef]

25. Frem, D. Estimating the Metal Acceleration Ability of High Explosives. *Def. Technol.* **2020**, *16*, 225–231. [CrossRef]

26. Frem, D. A Mathematical Model for Estimating the Gurney Velocity of Chemical High Explosives *FirePhysChem* 2022, *in press*.

27. Qi, Z.; Chunhua, B.; Qingming, L.; Zhongqi, W.; Huimin, L.; Shaoqing, X. Study on near Field Dispersal of Fuel Air Explosive. *J. Beijing Inst. Technol.* **1999**, *8*, 113–118.

28. Gardner, D.R. *Near-Field Dispersal Modeling for Liquid Fuel-Air Explosives*; Technical Report; Sandia National Laboratory: Kekaha, HI, USA, 1990.

29. Li, L.; Lu, X.; Ren, X.; Ren, Y.J.; Zhao, S.T.; Yan, X.F. The Mechanism of Liquid Dispersing from a Cylinder Driven by Central Dynamic Shock Loading. *Def. Technol.* **2020**, *17*, 1313–1325. [CrossRef]

30. Song, X.; Zhang, J.; Zhang, D.; Xie, L.; Li, B. Dispersion and Explosion Characteristics of Unconfined Detonable Aerosol and Its Consequence Analysis to Humans and Buildings. *Process Saf. Environ. Prot.* **2021**, *152*, 66–82. [CrossRef]

31. Zhang, Y.; Lou, W.; Wang, H.; Guo, M.; Fu, S. Experimental Research on Dynamic Concentration Distribution for Combustible Dust Based on Ultrasonic-Electric Hybrid Detection. *Heat Mass Transf.* **2020**, *56*, 1673–1684. [CrossRef]

Article

Study on Axial Dispersion Characteristics of Double-Layer Prefabricated Fragments

Yuan He [1], Lei Guo [1], Chuanting Wang [1], Jinyi Du [1], Heng Wang [2] and Yong He [1,*]

1 School of Mechanical Engineering, Nanjing University of Science & Technology, Nanjing 210094, China
2 Xi'an Institute of Modern Control Technology, Xi'an 710065, China
* Correspondence: yonghe1964@163.com

Abstract: The axial distribution of initial velocity and direction angle of double-layer prefabricated fragments after an explosion were investigated via an explosion detonation test. A three-stage detonation driving model of double-layer prefabricated fragments was proposed. In the three-stage driving model, the acceleration process of double-layer prefabricated fragments is divided into three stages: "detonation wave acceleration stage", "metal–medium interaction stage" and "detonation products acceleration stage". The initial parameters of each layer of prefabricated fragments calculated by the three-stage detonation driving model of double-layer prefabricated fragments fit well with the test results. It was shown that the energy utilization rate of detonation products acting on the inner-layer and outer-layer fragments were 69% and 56%, respectively. The deceleration effect of sparse waves on the outer layer of fragments was weaker than that on the inner layer. The maximum initial velocity of fragments was located near the center of the warhead where the sparse waves intersected, located at around 0.66 times of the full length of warhead. This model can provide theoretical support and a design scheme for the initial parameter design of double-layer prefabricated fragment warheads.

Keywords: three-stage detonation driving model; double-layer prefabricated fragments; fragment initial parameters; explosion detonation

Citation: He, Y.; Guo, L.; Wang, C.; Du, J.; Wang, H.; He, Y. Study on Axial Dispersion Characteristics of Double-Layer Prefabricated Fragments. *Materials* **2023**, *16*, 3966. https://doi.org/10.3390/ma16113966

Academic Editor: Enrique Casarejos

Received: 17 March 2023
Revised: 21 May 2023
Accepted: 22 May 2023
Published: 25 May 2023

1. Introduction

Blasting-fragmentation warheads destroy targets by the shock wave and fragments formed during the detonation. With the development of blast-fragmentation warhead technology, prefabricated fragments are often used to increase the power of the warhead. It is found that the prefabricated fragments have a high energy utilization rate and a strong killing ability towards long-distance targets. To improve the energy utilization rate of explosives, double-layer prefabricated fragments are designed to increase the number of fragments. Therefore, it is important to carry out the theoretical study of the fragments' initial parameters, including the initial velocity and the initial direction angle [1–3].

The research on the initial state of fragments mainly focuses on the initial velocity and flying direction angle of fragments. Currently, studies are mainly focused on three aspects: experimental research, finite element simulation calculation, and theoretical calculation.

In experimental research, various technical means are often used to record the spatiotemporal parameters of prefabricated fragments, such as fragment initial velocity, fragment dispersion direction angle, fragment mass, etc. The commonly used technical methods include X-ray photography, high-speed photography, etc. These methods can simultaneously record data such as initial velocity of fragmentation, and direction of fragmentation. However, as research progresses, researchers have gradually discovered that for the study of initial parameters of fragments, a large amount of experimental data is required to obtain universal rules. Shortcomings such as high experimental research costs, long cycle, a lack of clearly targeted research objectives, and poor applicability of research methods have gradually emerged [4,5].

With the development of finite element simulation technology, a large number of researchers have carried out finite element simulation studies on initial parameters of fragments that can generate a large amount of data in a relatively short period of time. However, researchers have found that finite element simulation methods have very strict requirements for material constitutive models and calculation parameter settings, and even slight differences in parameters can cause differences in results. Moreover, if there are minor changes in the structure of the warhead, the finite element simulation results are unreliable, resulting in a significant waste of time and cost. These issues have led researchers to shift their focus towards more universal and efficient calculation methods [6,7].

The earliest theoretical calculation methods for the initial parameters of fragments were proposed by Gunny and Taylor, who proposed the Gunny formula for calculating the initial velocity of fragments and the Taylor formula for calculating the dispersion direction angle of fragments, respectively [8]. Based on this initial work, later scholars conducted much theoretical, experimental, and simulation work and devised various modification formulas based on the above two formulas. Some started with the axial position of fragments and proposed improved methods for calculating the axial distribution of initial velocity of fragments. Other researchers started with lateral sparse waves and added a correction term to the Taylor angle [9,10].

When the warhead contains a large number of prefabricated fragments placed in multiple layers, there are strong collisions between these prefabricated fragments during the detonation drive process, ultimately leading to changes in the fragmentation behavior. However, such a theoretical model that considers multi-layer prefabricated fragments has not yet been well developed.

Therefore, the main objective of the work was to introduce the "metal–medium interaction stage" between the "detonation wave acceleration stage" and the "detonation products acceleration stage".

2. Experimental Procedures

The structure of the warhead for the static explosion test is shown in Figure 1. A double-layer of prefabricated fragments was placed outside the steel cylinder, the fragments were installed individually to ensure that each fragment of the outer layer was on the same radial direction as that of the inner layer. Figure 1 shows the symmetrical structure of the warhead, with the yellow part being the charge and the outer lining of the charge. The inner lining separates two layers of spherical prefabricated fragments from the explosive, with air outside the prefabricated fragments.

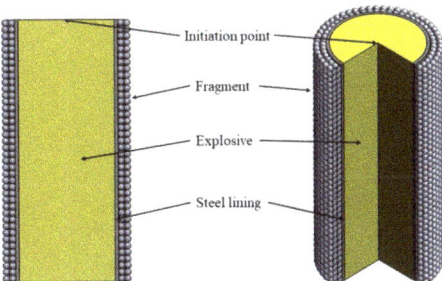

Figure 1. The scheme of the warhead used in the test.

In Figure 1, the charge radius of the warhead is 38 mm, the charge length is 100 mm, the mass of the explosive is around 760 g, the thickness of steel lining is 1.3 mm, the diameter of prefabricated tungsten fragments is 2.5 mm, and the total mass of prefabricated fragments is 806 g. The test layout is shown in Figure 2.

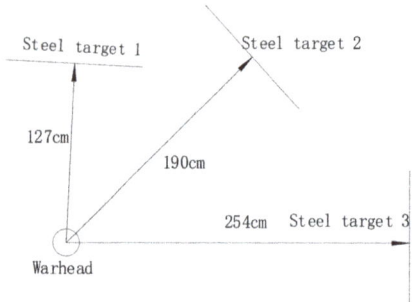

(**a**) Layout diagram   (**b**) Site layout

Figure 2. The layout of the explosion experiment.

Prior to ignition, the warhead was placed vertically on a stand with height of 67 cm. Three steel targets were placed at various distances perpendicular to the direction of the warhead. A detonator was connected to the upper surface of the warhead charge. Each steel target center was arranged at a horizontal distance of 127 cm, 190 cm, and 254 cm. Each steel target was 150 cm high, 40 cm wide, and 1 cm thick and fixed on the target frame. In the detonation test, the fragments impacted the steel targets and formed dimples on the targets. The position of each dimple was recorded carefully to study the dispersion distribution behavior of the fragments. The diameter of each dimple was recorded carefully to calculate the corresponding depth, and these depth values were then employed to calculate the impact velocity of the fragments.

The relationship between the penetration depth on a steel target and the velocity of a tungsten fragment was established by ballistic gun test. The test layout is shown in Figure 3. A ballistic gun was employed to shoot the tungsten fragments with diameter of 2.5 mm towards a steel target with thickness of 10 mm at various velocities. The velocities of the tungsten fragments were controlled by adjusting the amount of gun powder, and the velocity was measured by velocity measuring systems positioned in front of the steel target.

Figure 3. The set-up of the ballistic gun test.

3. Experimental Results

The warhead explosion process is shown in Figure 4. After ignition, the explosion expanded inside the charge and drove the fragments with high velocity. The fragments in the middle area of the cylinder warhead had the highest velocity, while the fragments at the end area had lower velocity due to the leakage effect.

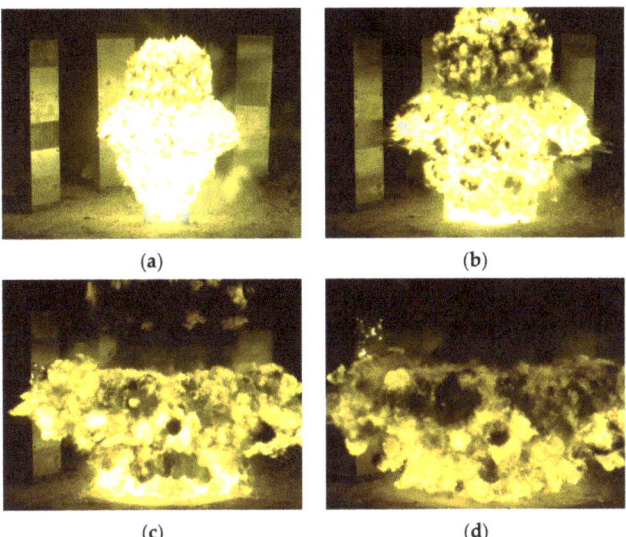

Figure 4. High-speed photography images of the explosion process, at (**a**) 166 μs, (**b**) 332 μs, (**c**) 664 μs and (**d**) 830 μs after ignition.

The landing position of fragments on the steel targets after the explosion is shown in Figure 5. On the steel target 1 which was closest to the warhead, most of the fragments were located in the center area of the target. As the targets moved further away to target 2 and 3, the distribution of fragments became more dispersed. It was also observed that the dimples in the center area of the steel target had greater depth compared with dimples at both ends, which demonstrates that the fragments striking the center area had higher impact velocities compared with those striking the end areas. This is consistent with the observations made with high-speed photography.

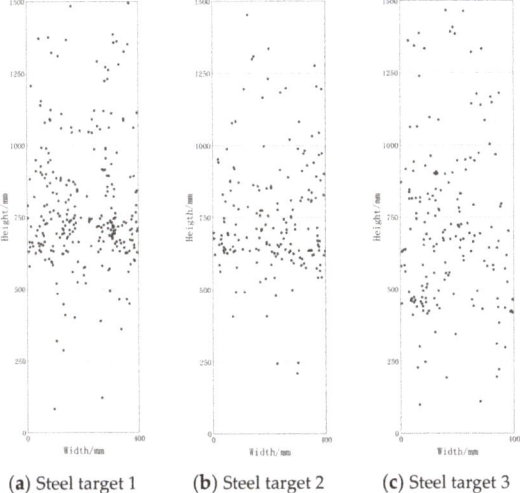

(**a**) Steel target 1   (**b**) Steel target 2   (**c**) Steel target 3

Figure 5. Landing position of fragments on steel targets after explosion.

In the ballistic gun test, the tungsten fragments with various impact velocities formed dimples of differing sizes on the steel target. The diameter of the dimples was carefully measured and used to calculate the depth of the dimple. The results are shown in Table 1.

Table 1. Penetration depth of tungsten fragments at various impact velocities.

Test	h_0 (dm)	v_b (m/s)
1	0.0113	870
2	0.0198	1199
3	0.0262	1394
4	0.0321	1570

In Table 1, v_b is the velocity of a prefabricated fragment, and h_0 is the penetration depth. Using the test results, the De Marre formula is modified to show the relationship between the fragment velocity and the penetration depth:

$$v_b = 4135 \frac{d^{0.937} \cdot h_0^{0.563}}{m_f^{0.5} \cos \omega} \tag{1}$$

where d is the prefabricated fragment diameter, and ω is the angle between the movement direction of the prefabricated fragment and the normal direction of the target plate. The formula also includes the mass in kilograms, length in decimeters, and velocity in meters per second.

4. Discussion

4.1. Classic Models for Fragment Velocity and Dispersion Calculation

Gurney assumed that the axial fragments fly out at the same velocity and proposed a famous formula based on energy conservation law [8], which can be expressed as:

$$v_0 = \sqrt{2E}\sqrt{\frac{C}{M + 0.5C}} \tag{2}$$

where $\sqrt{2E}$ is the Gurney constant, C is the charge mass, M is the shell mass, and v_0 is the Gurney velocity.

However, the Gurney formula does not consider the effect of the sparse waves incoming from the explosion initiation end and the explosive termination end on the initial velocity of the fragment. Therefore, researchers established modified formulas according to the Gurney formula. Based on experimental data, Huang [11] introduced the correction factor into the Gurney formula to describe the axial distribution of cylindrical warhead fragments at the endpoint, which can be expressed as:

$$v_f(x) = v_0[1 - 0.361e^{-1.111x/(2r_0)}][1 - 0.192e^{-3.03(L-x)/(2r_0)}] \tag{3}$$

where x is the axial distance between the fragment and the explosion initiation end, r_0 is the charge radius, v_f is the initial velocity of the fragment, and L is the charge length.

Since the energy released by the explosive cannot be fully utilized by the fragments, to describe the law of distribution of explosive energy along the axial, Randers-Pehrson [12] proposed a method dependent on the axial proportional coefficient to correct the loading mass along the axial direction, which can be expressed as:

$$f(x) = 1 - \{1 - \min[x/(2r_0), 1, (L - x)/r_0]\}^2 \tag{4}$$

Taking the coefficient into Equation (1), the initial velocity of fragment is shown as:

$$v_f(x) = \sqrt{2E}\sqrt{\frac{Cf(x)}{M + 0.5Cf(x)}} \qquad (5)$$

To describe the initial direction angle of the fragments, Taylor [13] calculated the direction angle of the cylindrical warhead based on the action law of the detonation products on the shell, called the Taylor formula, which is shown as:

$$\alpha(x) = 90° - \arcsin\left[\frac{v_f(x)}{2D}\cos\theta_1\right] \qquad (6)$$

where θ_1 is the angle between the detonation front and the normal to the explosive/metal interface, D is the detonation velocity, and α is the initial direction angle of the fragment.

Neither the Gurney formula nor the Taylor formula considers the effect of the sparse waves incoming from the explosion initiation end and the explosive termination end on the initial velocity of the fragment. Therefore, researchers [14] introduced the warhead structure parameters into the formula to calculate the initial direction angle, which can be expressed as:

$$\alpha(x) = 90° - K\left[\frac{df(x)}{dx}\right]^2\sqrt{\frac{M}{C} + 0.5} - \arcsin\left[\frac{v_f(x)}{2D}\cos\theta_1\right] \qquad (7)$$

where the coefficient K is 1.295 at the explosion initiation end and -6.315 at the explosion termination end, respectively, of the blast-fragmentation warhead.

The above method only discusses the acceleration process of the fragment in the "detonation wave acceleration stage" and the "detonation wave product acceleration stage" [15], while the effect of the interaction between the layers on the fragments initial parameters is ignored [16]. Meanwhile, the difference in the fragments' initial parameters along the axis [17] and the applicability to the initial parameters of the double-layer prefabricated fragments [18–21] is not considered.

4.2. Three-Stage Detonation Drive Model along the Axial Distribution

In this study, the acceleration process of double-layer prefabricated fragments is divided into three stages. During the "detonation wave acceleration stage", the detonation wave spreads in the detonation product after the explosion. The detonation wave obliquely incident to the interface of the lining medium forms the reflection wave, and the transmission wave travels from the detonation product to the lining medium. The liner collides with the inner layer of prefabricated fragments at high velocity. The process of the inner layer of prefabricated fragments acting on the outer layer of prefabricated fragments is called the "metal–medium interaction stage". This paper only discusses the first interaction between the various metal mediums. The "detonation product acceleration stage" is regarded as the acceleration process of the detonation product pressure on the prefabricated fragment based on the first two acceleration stages.

It is believed that the "detonation wave acceleration stage" and "metal–medium interaction stage" are completed instantly as the detonation wave and shock wave transmit in the medium with high velocity. Therefore, the "detonation wave acceleration stage" and the "metal–medium interaction stage" are independent of each other,

The "detonation product acceleration stage" is the main stage when prefabricated fragments are accelerated. It is believed that the "detonation product acceleration stage" occurs after the above two acceleration stages, and the initial velocity of prefabricated fragments is the result of the above three stages acting together. A diagram of each acceleration stage is shown in Figure 6.

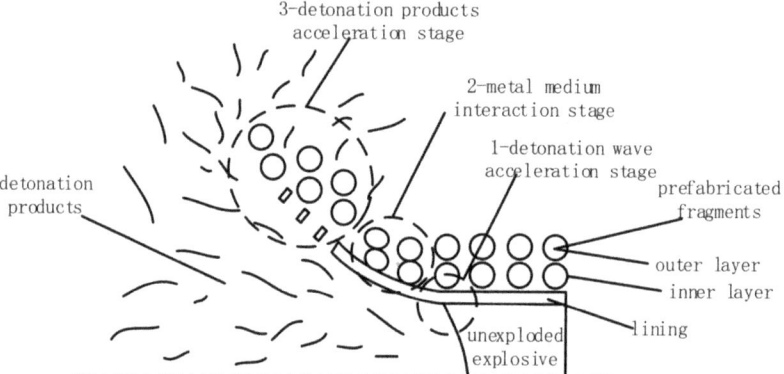

Figure 6. Three-stage detonation driving model of double-layer prefabricated fragments.

4.2.1. Detonation Wave Acceleration Stage

The detonation wave acceleration phase can be solved via the detonation wave oblique incidence theory. The detonation wave oblique incident model is established by taking the coordinate origin at the point where the detonation wave contacts the medium interface, as shown in Figure 7. U_0 is the velocity component of detonation velocity D along the direction of the metal wall; U_H is the velocity difference of medium on both sides of the OI interface; U_i is the flow velocity of medium in region (i); $U_i n$ is the velocity component of U_i normal to the oblique reflected shock wave front; $U_i t$ is the velocity component of U_i along the oblique reflected shock wave front; Ds is the velocity of the medium passing through the OT interface and u_m is the velocity component of Ds along the metal wall.

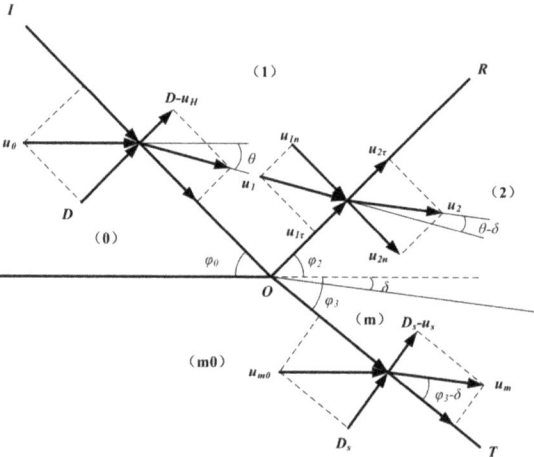

Figure 7. Normal oblique reflection of detonation wave on lining surface.

In Figure 7, Area (0) is unexploded explosive, Area (1) is the detonation product area behind the oblique detonation wave, Area (2) is the detonation product area behind the reflecting shock wave, Area (m0) is the initial media area, and Area (m) is the area where the medium is disturbed behind the oblique transmission of the shock wave. OI is an oblique detonation wave front. The angle between OI and the contact medium is φ_0. OR is the wavefront of the oblique reflected shock wave in the detonation product. The angle between OR and the initial interface of the media is φ_2. OT is the oblique transmission shock wave front in the medium. The angle between OT and the initial interface of the media is φ_3. The medium is deformed under the action of the detonation products. The angle between the interface and the initial boundary of the media is δ. The folding angle θ in the model and the Mach number of the region (1) can be shown as [22]:

$$\tan\theta = \frac{\tan\varphi_0}{1 + k(1 + \tan^2\varphi_0)} \tag{8}$$

$$M_1 = \sqrt{1 + \left(\frac{k+1}{k}\right)^2 \cot^2\varphi_0} \tag{9}$$

where k is the isentropic index. According to the geometric relationship of the model, C-J theory [23], the impact compression law of material is expressed as:

$$\sin^2(\varphi_2 + \theta) = \frac{1}{2kM_1^2}\left[\frac{\rho_{m0}(k+1)^2\sin\varphi_3}{\rho_0 b_m \sin\varphi_0}\left(\frac{\sin\varphi_3}{\sin\varphi_0} - \frac{a_m}{D}\right) + k - 1\right] \tag{10}$$

$$\frac{(k-1)\sin(\varphi_2+\theta)-(k+1)\cos(\varphi_2+\theta)\tan\varphi_2+\frac{2}{M_1^2\sin(\varphi_2+\theta)}}{(k-1)\sin(\varphi_2+\theta)\tan\varphi_2+(k+1)\cos(\varphi_2+\theta)+\frac{2\tan\varphi_2}{M_1^2\sin(\varphi_2+\theta)}}$$

$$= \frac{(1-\frac{a_m\sin\varphi_0}{D\sin\varphi_3})\tan\varphi_3}{b_m+(b_m-1+\frac{a_m\sin\varphi_0}{D\sin\varphi_3})\tan^2\varphi_3} \tag{11}$$

where ρ_0 is the charge density, ρ_{m0} is the initial density of the lining medium, and a_m and b_m are the Hugoniot parameters of the lining medium. The joint solution of the above two equations results in φ_2 and φ_3. The parameters of media behind the shock wave can be shown as:

$$\rho_m = \frac{\rho_{m0}}{1 - \frac{1}{b_m}(1 - \frac{a_m\sin\varphi_0}{D\sin\varphi_3})} \tag{12}$$

$$p_m = \frac{\rho_{m0}D^2\sin\varphi_3}{b_m\sin\varphi_0}\left(\frac{\sin\varphi_3}{\sin\varphi_0} - \frac{a_m}{D}\right) \tag{13}$$

$$tg(\varphi_3 - \delta) = \frac{\rho_{m0}}{\rho_m}tg\varphi_3 \tag{14}$$

$$u_s = \frac{D\sin\varphi_3}{\sin\varphi_0} - \frac{u_m}{\sin(\varphi_3 - \delta)} \tag{15}$$

where u_m is the tangential point velocity of the mass in the lining medium, ρ_m is the lining medium density behind the oblique transmission shock wave, p_m is the lining medium pressure behind the oblique transmission shock wave, and u_s is the lining medium velocity of the shock wave after the oblique transmission.

4.2.2. Metal–Medium Interaction Stage

Shock waves are generated in the lining medium and all layers of the prefabricated fragments to change the motion of the medium when the lining works with the double-layer prefabricated fragments. The equations are obtained according to the conservation equations of the medium before and after the collision, as well as the boundary conditions when considering the collision of the lining with the inner layer of prefabricated fragments:

$$p_{me} - p_m = \rho_m[a_m + b_m(u_s - u_{se})](u_s - u_{se}) \tag{16}$$

$$p_2 - p_{20} = \rho_{20}[a_2 + b_2(u_2 + u_{20})](u_2 - u_{20}) \tag{17}$$

$$p_{me} = p_2 \tag{18}$$

$$u_{se} = u_2 \qquad (19)$$

where u_{20} is the velocity of the inner layer of prefabricated fragments before the action of the shock wave, ρ_{20} is the density of the inner layer of prefabricated fragments before the action of the shock wave, p_{20} is the pressure of the inner layer of prefabricated fragments before the action of the shock wave, u_{se} is the velocity of the lining medium after the action of the shock wave, p_{me} is the pressure in the lining medium after the action of the shock wave, u_2 is the velocity of the inner layer of prefabricated fragments after the action of the shock wave, p_2 is the pressure of the inner layer of prefabricated fragments after the action of the shock wave, and a_2 and b_2 are the Hugoniot parameters of the inner layer of prefabricated fragments.

The action process of the inner-layer prefabricated fragments on the outer layer prefabricated fragments is discussed based on u_2. The velocity of each layer in the direction of collision becomes half the velocity of the inner layer based on the conservation of momentum, when the outer layer of prefabricated fragments is stationary before the collision and the two layers have the same quality and material. The velocity status of each layer before the collision is shown in Figure 8 when the prefabricated fragments are evenly arranged in the radius direction and the axis direction, where the x-direction is the axial direction of the warhead, the y-direction is the radial direction of the warhead, θ_1 is the angle between the velocity direction and the warhead axis direction after the "metal–medium interaction stage" of the inner layer of prefabricated fragments, the value of θ_1 is $\pi/2-\varphi_3$, and θ_1 is a function of x.

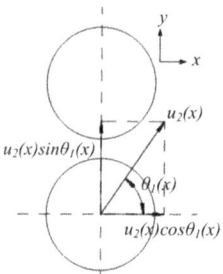

Figure 8. Velocity status before interaction of prefabricated fragments.

The velocity component of the inner layer along the axis of the warhead is $u_2(x)\cos\theta_1(x)$, and the velocity component along the radial of the warhead is $u_2(x)\sin\theta_1(x)$. The velocity of each layer is shown in Figure 9 after the collision. Therefore, after the "metal–medium interaction stage", the velocity of the prefabricated fragments is indicated by V_i, and the angle between V_i and the axial direction of the warhead is indicated by Φ_i.

$$V_i(x) = \begin{cases} \sqrt{[u_2(x)\cos\theta_1(x)]^2 + [0.5u_2(x)\sin\theta_1(x)]^2} & (i=1) \\ 0.5u_2(x)\sin\theta_1(x) & (i=2) \end{cases} \qquad (20)$$

$$\varphi_i(x) = \begin{cases} \operatorname{atan}\left[\frac{\tan\theta_1(x)}{2}\right] & (i=1) \\ \frac{\pi}{2} & (i=2) \end{cases} \qquad (21)$$

where i is the number of the prefabricated fragment layers.

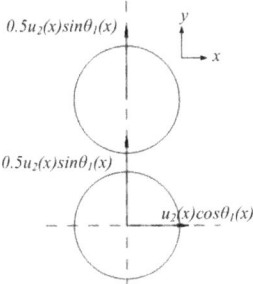

Figure 9. Velocity status after interaction of prefabricated fragments.

4.2.3. Detonation Products Acceleration Stage

The following assumptions are made to analyze the acceleration process of prefabricated fragments: The mass loss of prefabricated fragments is not considered. The explosive product expansion is isentropic. The change of detonation products density caused by the leakage along the axis is ignored. The detonation products are evenly distributed in space, and after the first sparse wave is introduced, the detonation products no longer work on the prefabricated fragments. The model of detonating product action on the double-layer spherical prefabricated fragments is shown in Figure 10, where OX is the symmetry axis of the explosive, the shaded part is the explosive, and the O point is the detonation point. The explosive has openings at both ends, and the outer layer of the explosive has a lining. There are two layers of spherical prefabricated fragments outside the lining, and outside the prefabricated fragments is air. There is no shell to block the outward movement of the fragments.

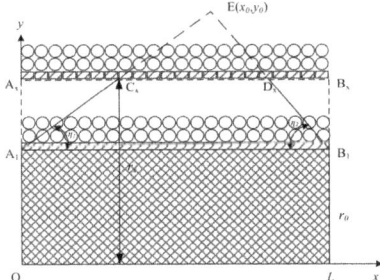

Figure 10. The effect of detonation products on double-layer prefabricated fragments.

After the explosive has exploded at point O, there are sparse wave afferent detonation products in the A_1C_x direction at the explosion initiation end, and sparse wave afferent detonation products in the B_1D_x direction at the explosion termination end. The angle between the A_1C_x direction and axial direction is η_1; the angle between the B_1D_x direction and axial direction is η_2; A_1C_x and B_1D_x intersect at the point E with the abscissa x_0 and the ordinate y_0. The detonation products in regions $A_1A_xC_x$ and $B_1B_xD_x$ no longer work on the prefabricated fragments when the detonation product expands to A_xB_x.

After the explosion, the detonation products diffuse outward in a circumferential direction [4,9]. The change in detonation products density caused by the leakage along the axis is ignored. The expression of the detonation product density ρ_x at any time is shown as:

$$\rho_x = \rho_0 \left(\frac{r_0}{r_x}\right)^2 \tag{22}$$

where r_x is the distance between the detonation product interface and the axis of the warhead, and according to the iso entropy equation, the detonation product pressure at A_xB_x can be shown as:

$$p_x = p_H \left(\frac{\rho_x}{\rho_H}\right)^k = \frac{\rho_0 D^2}{k+1}\left[\frac{k\rho_x}{(k+1)\rho_0}\right]^k = \frac{\rho_0 D^2}{k+1}\left[\frac{kr_0^2}{(k+1)r_x^2}\right]^k \tag{23}$$

where p_H and ρ_H are the C-J parameters. r_{max} is the working radius of the detonation product, which can be shown as:

$$r_{max} = \begin{cases} k_{1,i}x + k_{r1,i}r_0 & (x \le x_0) \\ k_{2,i}(L - x) + k_{r2,i}r_0 & (x > x_0) \end{cases} \tag{24}$$

where $k_{1,i}$ and $k_{2,i}$ represent the tangent value of the angles η_1 and η_2 of the sparse wave in layer i, and $k_{r1,i}$ and $k_{r2,i}$ are the correction coefficients of the initiation end radius and the termination end radius of the prefabricated fragments of layer i to correct the boundary explosive dimensions [9]. Equation (24) reflects the process of the sparse wave introducing detonation products in the three-stage mode, where coefficients $k_{1,I}$ and $k_{2,i}$ represent the slope of the sparse wave introducing detonation products, reflecting the influence of the sparse wave on the overall velocity of the prefabricated fragments. Their value increases with increasing work distance of the detonation products on the prefabricated fragments, which leads to a larger overall velocity of the prefabricated fragments. Therefore, the kinetic energy increment of a single prefabricated fragment can be expressed as:

$$e_k = \int_{r_0}^{r_{max}} p_x S_f dr_x \tag{25}$$

where e_k is the kinetic energy increment of a single prefabricated fragment, and S_f is the effective area of a single preformed fragment. Solving the integral, the kinetic energy increment can be shown as:

$$e_k(x) = \frac{\rho_0 D^2 S_f}{(1-2k)(k+1)}\left[\frac{r_0^2 k}{k+1}\right]^k \left[r_{max}^{1-2k}(x) - r_0^{1-2k}\right] \tag{26}$$

According to the laws of energy conservation, combining Equation (24) with Equation (26), the work performed by the detonation products on the prefabricated fragments can be obtained. Afterwards, by adding this work to the initial velocity obtained from the first two stages of the "three-stage model", the final initial velocity, v_i, of the prefabricated fragments can be obtained as:

$$v_i(x) = \sqrt{V_i^2(x) + \frac{\xi_i e_k(x)}{m_f}} \tag{27}$$

where m_f is the mass of the prefabricated fragment. In the model, the influence of explosive mass and energy leakage is taken into account in the correction coefficient ξ_i [24]. The axial distribution model of the initial velocity of the double-layer prefabricated fragments can be shown as:

$$v_i(x) = \begin{cases} \sqrt{V_i^2(x) + \frac{\xi_i}{m_f}\frac{\rho_0 D^2 S_f}{(1-2k)(k+1)}\left[\frac{r_0^2 k}{k+1}\right]^k \left[(k_{1,i}x + k_{r1,i}r_0)^{1-2k} - r_0^{1-2k}\right]} & (x \le x_0) \\ \sqrt{V_i^2(x) + \frac{\xi_i}{m_f}\frac{\rho_0 D^2 S_f}{(1-2k)(k+1)}\left[\frac{r_0^2 k}{k+1}\right]^k \left[(k_{2,i}L - k_{2,i}x + k_{r2,i}r_0)^{1-2k} - r_0^{1-2k}\right]} & (x > x_0) \end{cases} \tag{28}$$

For the above derivation process, each coefficient has its own physical meaning. For $k_{1,i}$ and $k_{2,i}$, i represents the influence of sparse waves on the fragments at the initiation and termination ends of the i-th layer explosion. For $kr_{1,i}$ and $kr_{2,i}$, i represents the influence of the explosives on the initiation and termination fragments of the i-th layer explosion, and ξ_i represents the utilization of the detonation energy by the i-th layer fragments.

4.2.4. Direction Angle of Double-Layer Prefabricated Fragments

The direction angle model of the double-layer spherical prefabricated fragment is shown in Figure 11. The initial velocity obtained by the prefabricated fragments in the first stage is $V_i(x)$, and the scattering direction angle of the fragments is $\Phi_i(x)$. The initial velocity obtained by the fragments after three acceleration processes is $v_i(x)$.

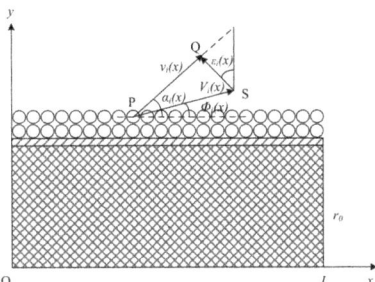

Figure 11. Model of dispersion angle of prefabricated fragments.

In the model, i is the direction angle of the prefabricated fragment in layer i, v_i is located along the PQ direction, and ε_i is the angle between the SQ direction and the axis direction. An equation can be established by using the velocity triangle:

$$\frac{v_i(x)}{\sin\left[\frac{\pi}{2} - \varepsilon_i(x) + \varphi_i(x)\right]} = \frac{V_i(x)}{\sin\left[\frac{\pi}{2} - \alpha_i(x) + \varepsilon_i(x)\right]} \tag{29}$$

and the direction angle is:

$$\alpha_i(x) = \arccos\left\{\frac{V_i(x)}{v_i(x)}\cos[\varphi_i(x) - \varepsilon_i(x)]\right\} + \varepsilon_i(x) \tag{30}$$

Therefore, the axial distribution model of the direction angle of double-layer prefabricated fragments can be expressed as:

$$\alpha_i(x) = \begin{cases} \arccos\left\{\left[\frac{\xi_1 e_k(x)}{m_f V_1^2(x)} + 1\right]^{-0.5}\cos\left[\mathrm{atan}\left(\frac{\tan\theta_1(x)}{2}\right) - \varepsilon(x)\right]\right\} + \varepsilon(x) & (i=1) \\ \arccos\left\{\left[\frac{\xi_2 e_k(x)}{m_f V_2^2(x)} + 1\right]^{-0.5}\sin\varepsilon(x)\right\} + \varepsilon(x) & (i=2) \end{cases} \tag{31}$$

where ε_i is described by using the inverse proportional function:

$$\varepsilon_i(x) = \frac{1}{k_{3,i}\frac{x}{L} + k_{4,i}} - k_{5,i} \tag{32}$$

where ε_i is the radian system, and $k_{3,i}$, $k_{4,i}$ and $k_{5,i}$ are the correction coefficients.

4.3. Simulation Results of the Three-Stage Detonation Drive Model

The finite element simulation results were used to obtain the correction coefficient in Equations (28) and (31). The quarter finite element model of the double-layer prefabricated fragments warhead was established. The mechanical properties of the lining materials are described by the Johnson–Cook constitutive model [25], and the mechanical properties of the prefabricated fragment material are described by the Elastic model. The loading radius of the warhead is 38 mm; the loading length is 100 mm; the diameter of the prefabricated fragment is 2.5 mm, and the thickness of the steel lining is 1.3 mm. The parameters of the explosives are shown in Table 2, and the parameters of the metallic materials are shown in Table 3.

Table 2. Parameters of the JWL equation of state for explosives.

Explosive	$\rho_0/(\text{g·cm}^{-3})$	$D/(\text{m·s}^{-1})$	P_{CJ}/GPa	A/GPa	B/GPa	R_1	R_2	Ω	E_0/GPa
8701	1.695	8450	29.66	854.5	20.49	4.6	1.35	0.25	9.5

Table 3. The parameters of the metal materials.

Parameters	45 Carbon Steel	Tungsten Alloy
density/(g·cm^{-3})	7.83	17.6
shear modulus/GPa	77.0	136
Young's modulus/GPa	200	350
Poisson ratio	0.32	0.286
A	792	/
B	510	/
N	0.26	/
C	0.014	/
M	1.03	/
$a_m/(\text{m/s})$	3574	/
b_m	1.92	/
$a_2/(\text{m/s})$	/	4029
b_2	/	1.237

The dispersion of the double-layer prefabricated fragments obtained by finite element simulation is shown in Figure 12. The explosive detonates, and the detonation wave propagates from bottom to top. The fragments undergo detonation waves, interactions between fragments, and acceleration of detonation products to obtain velocity. At 15 μs, the detonation wave is transmitted to the upper end of the charge, and the interaction between the fragments behind the wavefront is complete. The detonation products have performed work on the prefabricated fragments at the lower end of the charge, and the sparse wave has also been transmitted to the detonation products. Therefore, the initial velocity of the upper fragment is low, while the initial velocity of the lower fragment is high. The fragments close to the initiation point are greatly affected by the sparse wave, so the velocity of these fragments is low. At 30 μs, the acceleration process of the detonation wave on the prefabricated fragments and the interaction between the fragments are complete. The effect of the detonation product on the upper fragment has just begun, so the initial velocity of the upper fragment is smaller than that of the lower one. At 45 μs, the acceleration effect of the detonation product on the fragments is almost complete, and at this time, the influence of sparse waves on the velocity of the fragments at both ends is very obvious.

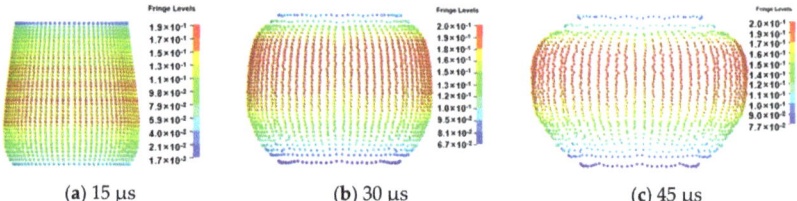

(a) 15 μs   (b) 30 μs   (c) 45 μs

Figure 12. Dispersion of prefabricated fragments.

The values of the coefficients in the model (27) and model (30) fitted by the average value of the simulation velocity of each column of prefabricated fragments are shown in Table 4. The coefficients in the model (27) and model (30), whose values were obtained via fitting, are listed in the table.

Table 4. Correction coefficient in three-stage detonation driving model.

Model	Parameters	Values
The axial distribution model of the initial velocity of double-layer prefabricated fragments	$k_{1,1}$	0.0805
	$k_{2,1}$	0.1450
	$k_{r1,1}$	1.0251
	$k_{r2,1}$	1.0354
	ξ_1	1.3887
	$k_{1,2}$	0.1135
	$k_{2,2}$	0.2510
	$k_{r1,2}$	1.0245
	$k_{r2,2}$	1.0424
	ξ_2	1.1207
The axial distribution model of the direction angle of double-layer prefabricated fragments	$k_{3,1}$	5.9252
	$k_{4,1}$	0.5143
	$k_{5,1}$	0.2878
	$k_{3,2}$	5.7807
	$k_{4,2}$	0.6312
	$k_{5,2}$	0.2717

It is found that the energy utilization rate of detonation products acting on the fragments from the inner layer of fragments is 69%, while the rate for the fragments from the outer layer of fragments is 56%. Therefore, $k_{1,1}$ is smaller than $k_{1,2}$, and $k_{2,1}$ is smaller than $k_{2,2}$. The deceleration effect of sparse waves on the outer layer of fragments is smaller than that on the inner layer.

As shown in Table 4, the incoming angle at the initiation end of each layer is smaller than that at the termination end. The integral radius r_{max} of the initiation end of each layer is less than the termination end, indicating that the workability of explosives on prefabricated fragments is proportional to the distance between the fragments and the initiation end. The energy utilization rate of the inner layer is higher than that of the outer layer after considering the mass of the explosive and the leakage of the detonation products. Negative numbers exist for the angle ε_i, which indicates that as the distance between the prefabricated fragments and the initiation point increases, ε_i gradually decreases to 0 and increases afterwards.

The initial velocity of the double-layer prefabricated fragments obtained from Equation (28) is shown in Figure 13.

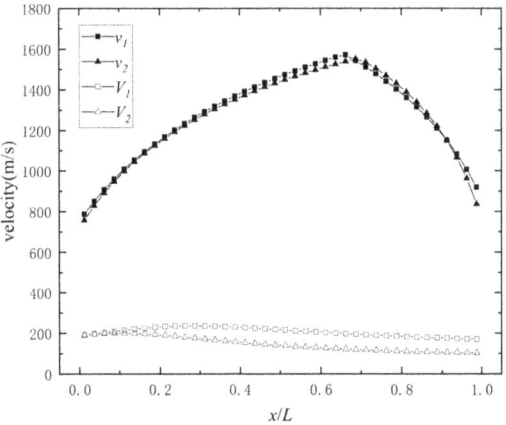

Figure 13. The axial distribution of the initial velocity of prefabricated fragments.

In Figure 13, the final initial velocities v_1 and v_2 are similar after the "detonation products acceleration stage", and the maximum value of v_i is around $x_0/L = 0.67$. As x increases, v_i shows increasing regularity with decreasing slope when $x < x_0$, and v_i shows decreasing regularity when $x > x_0$. The V_i of the prefabricated fragments before the "detonation product acceleration stage" shows a first increasing and then decreasing pattern. The ratio of v_i to V_i is between 4 and 13; therefore, the energy obtained by the prefabricated fragment in the "detonation products acceleration stage" and the energy obtained by the prefabricated fragment in the "metal–medium interaction stage" is between 15 and 168, which indicates that the "detonation product acceleration stage" is the main stage at which the initial velocity of the prefabricated fragments is obtained. This shows that the acceleration ability of the detonation products to the prefabricated fragments is much greater than that of the detonation wave and the shock wave [26].

The correction coefficient is brought into Equation (31) and the initial direction angle of each column is obtained as shown in Figure 14.

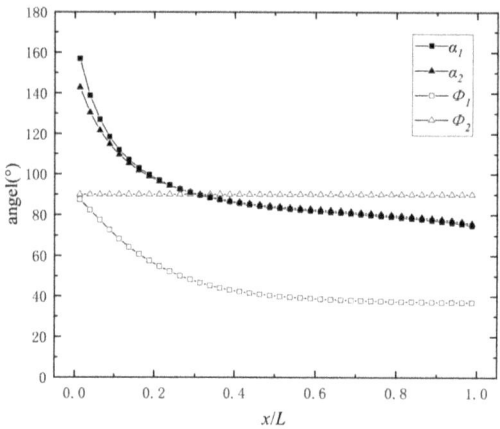

Figure 14. The axial distribution of the divergence angle of prefabricated fragments.

In Figure 14, the values α_1 and α_2 of the direction angle of the two layers after the "detonation products acceleration stage" almost coincide. As x increases, α_i is quickly reduced to $x/L = 0.2$ and then slowly to $x/L = 1$. The value of α_i is greater than 90 at the initiation end, indicating that some of the prefabricated fragments near the initiation end disperse in the opposite direction of the detonation wave. The scatter direction Φ_1 of the inner layer is similar to the direction angle α_1, and $\alpha_1 > \Phi_1$, but the difference between Φ_2 and α_2 decreases with increasing x, and increases inversely after decreasing to zero.

4.4. Comparison between Experimental Results and the Three-Stage Detonation Drive Model Calculation

The penetration depth of the prefabricated fragments on the steel target was measured, and the results of the initial velocity distribution along the axis of the prefabricated fragments are shown in Figure 15. The experiment cannot distinguish which layer of the prefabricated fragments hits the target, so the experimental results show the average velocity of the prefabricated fragments.

With the increase in the distance between the prefabricated fragment and the initiation end, the initial velocity of the prefabricated fragment increases first and then decreases, and the maximum velocity of 1557 m/s appears at around $x/L = 0.66$. The calculation results of Equations (3), (4) and (28) all show a trend of initial increase followed by a decrease with increasing distance, and the maximum value appears at around $x/L = 0.67$, which is consistent with the test results. Equation (2) cannot reflect the pattern of velocity change with the axial direction. Equation (3) has a maximum error greater than 900 m/s, and

Equation (4) is only similar to the test results near the maximum velocity. In conclusion, Equations (2)–(4) show large errors when used for the calculation of the initial velocity of the double-layer prefabricated fragments. The calculation results of the three-stage detonation driving model almost coincide with the test results.

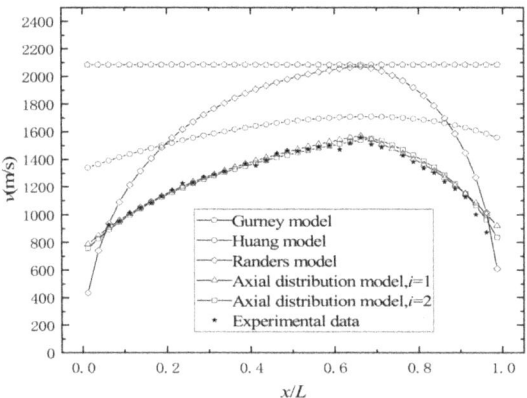

Figure 15. Comparison diagram of the initial velocity distribution of prefabricated fragments along the axial direction.

The results of the initial direction angle of the prefabricated fragment distribution along the axis of the prefabricated fragments are shown in Figure 16.

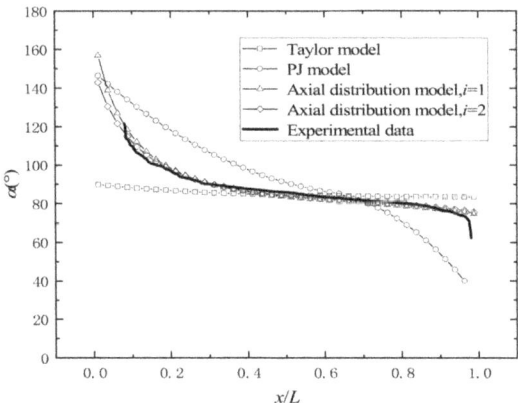

Figure 16. Comparison diagram of the axial distribution of the dispersion direction angle of prefabricated fragments.

With the increase in the distance between the prefabricated fragment and the initiation end, the initial direction angle of the prefabricated fragment decreases to around 90°, then decreases steadily. Afterwards, it shows a trend of rapid decrease again near the termination end of the explosion. The calculation results of Equation (6) show a trend of initial decrease and then increase, which is inconsistent with the test results. The calculation results of Equations (7) and (31) are similar to the test results. The calculated results of Equation (7) differ from the test results at the explosion termination end of the warhead. In conclusion, Equations (6) and (7) cannot be used to calculate the angle of the fragments. The calculation results of the three-stage detonation driving model have the smallest error, and the calculation results almost coincide with the test results.

4.5. Remarks about the Three-Stage Detonation Drive Model and Future Work

For the initial velocity model, the Gurney model did not consider the axial distribution of velocity or the phenomenon of reduced energy utilization due to the effect of detonation products on multi-layer fragments. Therefore, the Gurney model results in a higher velocity and only one value. The Huang model and Randers model are models obtained from experiments, where the values of each coefficient are fitted based on the experimental results of a specific warhead (which are all single-layer fragments), and are not applicable to the double-layer fragment structure of the warhead studied in this research.

Considering the scattering direction angle, the Taylor model did not consider the effect of lateral coefficient waves on the scattering direction angle of fragments, and the calculated results were all around 90°, which is inconsistent with the actual situation. The coefficients in the PJ model were obtained based on specific warhead results, and the results of the coefficients are not applicable to the double-layer fragment structure of the warhead studied here.

The three-stage driving model proposed in this article calculates the initial velocity and dispersion direction angle of fragments. It is suitable for the dual-layer fragment structure of the warhead studied in this article, and the calculation results fit well with the experimental results.

In the derivation process, it is shown that when deriving the "metal–medium interaction stage" of fragments in various arrangement modes, due to the effect of the inner-lining medium on the inner-layer fragments, the medium state of the inner-layer fragments is different from that of the outer-layer fragments. However, in order to simplify the calculation, this study assumes that the medium state of the inner- and outer-layer fragments is the same at this time, which causes errors in the calculation results. In order to simplify the calculation model, the fragment initial velocity theory calculation model introduces coefficients for modifying the shape of the warhead, correcting the sparse wave input angle and energy utilization rate. Although these coefficients have their own physical meanings, their specific physical calculation methods need to be further improved.

5. Conclusions

A three-stage detonation driving model was built by considering the metal–medium interaction stage between the layers of fragments, and a static explosion test was performed. The following conclusions can be obtained:

(1) The energy of detonation products is enormous; therefore, the "detonation product acceleration stage" is the main stage in the three-stage detonation driving model of double-layer prefabricated fragments.

(2) The De Marre formula can accurately describe the relationship between the initial velocity of prefabricated fragments and the penetration depth on the target.

(3) With the increase in the distance between the prefabricated fragments and the initiation end, the initial velocity of the prefabricated fragment increases first and then decreases, while the maximum initial velocity of the prefabricated fragment appears at around $x/L = 0.66$.

(4) The obstruction of the outer-layer fragment during the flow of detonation products reduces detonation products leaking from the inner-layer fragment area. The outer-layer fragments are surrounded by air, and there is no obvious obstruction to the outward flow of detonation products from the outer-layer fragments. Therefore, the energy utilization rate of the inner-layer fragment is 69%, which is higher than the value for the outer-layer fragments (56%).

(5) The initial parameters of double-layer prefabricated fragments calculated from the three-stage detonation driving model are in good agreement with the test results. The model can provide theoretical support and a design scheme for the initial parameter design of double-layer prefabricated fragments warheads.

Author Contributions: Conceptualization, Y.H. (Yuan He) and C.W.; Data curation, Y.H. (Yuan He), L.G. and J.D.; Formal analysis, L.G. and C.W.; Investigation, Y.H. (Yuan He), L.G. and C.W.; Methodology, Y.H. (Yuan He), L.G. and H.W.; Project administration, Y.H. (Yong He); Resources, Y.H. (Yong He) and H.W.; Supervision, C.W. and Y.H. (Yong He); Validation, J.D. and H.W.; Writing—original draft, Y.H. (Yuan He); Writing—review and editing, Y.H. (Yuan He), Y.H. (Yuan He), L.G., C.W., J.D. and H.W. All authors have read and agreed to the published version of the manuscript.

Funding: This work was supported by the National Natural Science Foundation of China [NO.U2241285 and NO. JCJQ-2019-00-011].

Institutional Review Board Statement: Not applicable.

Informed Consent Statement: Not applicable.

Data Availability Statement: Not applicable.

Conflicts of Interest: The authors declare no conflict of interest.

References

1. Liu, H.; Huang, G.Y.; Guo, Z.W.; Feng, S.S. Fragments velocity distribution and estimating method of thin-walled cy-lindrical improvised explosive devices with different length-to-diameter ratios. *Thin Wall Struct.* **2022**, *175*, 109212. [CrossRef]
2. Gao, Y.; Feng, S.S.; Xiao, X.; Feng, Y.; Huang, Q. Fragment characteristics from a cylindrical casing constrained at one end. *Int. J. Mech. Sci.* **2023**, *248*, 108186. [CrossRef]
3. Zou, S.; Gu, W.; Ren, W.; Shen, C.; Chen, Z.; Hao, L. Functional damage assessment method for preformed fragment warheads to evaluate the effect on the phased-array antenna. *Electronics* **2023**, *12*, 1907. [CrossRef]
4. Cullis, I.G.; Dunsmore, P.; Harrison, A.; Lewtas, I.; Townsley, R. Numerical simulation of the natural fragmentation of explosively loaded thick walled cylinders. *Def. Technol.* **2014**, *10*, 198–210. [CrossRef]
5. Dhote, K.D.; Murthy, K.; Rajan, K.M. Dynamics of multi layered fragment separation by explosion. *Int. J. Impact Eng.* **2015**, *75*, 194–202. [CrossRef]
6. Danel, J.F.; Kazandjian, L. A few remarks about the Gurney energy of condensed explosives. *Prop. Explos. Pyrotech.* **2004**, *29*, 314–316. [CrossRef]
7. Snyman, I.M.; Mostert, F.J. Computation of fragment velocities and projection angles of an anti-aircraft round. In Proceedings of the South African Ballistics Organisation Conference, Pretoria, South Africa, 29 September–1 October 2014.
8. Gurney, R.W. *The Initial Velocities of Fragments from Bombs, Shell and Grenades*; BRL Report No. 405; Aberdeen Proving Ground: Aberdeen, MD, USA, 1943.
9. Felix, D.; Colwill, I.; Stipidis, E. Real-time calculation of fragment velocity for cylindrical warheads. *Def. Technol.* **2019**, *15*, 264–271. [CrossRef]
10. Qian, L.X.; Liu, T.; Zhang, S.Q. Fragment shot-line model for air-defense warhead. *Prop. Explos. Pyrotech.* **2000**, *25*, 92–98.
11. Huang, G.; Li, W.; Feng, S. Axial distribution of fragment velocities from cylindrical casing under explosive loading. *Int. J. Impact Eng.* **2015**, *76*, 20–27. [CrossRef]
12. Pehrson, G.R. An Improved Equation for Calculating Fragment Projection Angle. In Proceedings of the 7th International Symposium on Ballistics, Hague, The Netherlands, 19–21 March 1976.
13. Taylor, G.I. *Analysis of the Explosion of a Long Cylindrical Bomb Detonated at One End*; Cambridge University Press: Cambridge, UK, 1963; pp. 277–286.
14. König, P.J. A correction for ejection angles of fragments from cylindrical wareheads. *Prop. Explos. Pyrotech.* **1987**, *12*, 154–157. [CrossRef]
15. Li, R.; Li, W.B.; Wang, X.M. Effects of control parameters of three-point initiation on the formation of an explosively formed projectile with fins. *Shock Waves* **2018**, *28*, 191–204. [CrossRef]
16. Wang, X.; Kong, X.; Zheng, C.; Wu, W. Effect of initiation manners on the scattering characteristics of semi-preformed fragment warhead. *Def. Technol.* **2018**, *14*, 578–584. [CrossRef]
17. Hirsch, E. Improved gurney formulas for exploding cylinders and spheres using "hard core" approximation. *Prop. Explos. Pyrotech.* **1986**, *11*, 81–84. [CrossRef]
18. Lindsay, C.M.; Butler, G.C.; Rumchik, C.G.; Schulze, B.; Gustafson, R. Increasing the utility of the copper cylinder expansion test. *Prop. Explos. Pyrotech.* **2010**, *35*, 433–439. [CrossRef]
19. Backofen, J.E.; Weickert, C.A. Obtaining the Gurney energy constant for a two-step propulsion model. *AIP Conf. Proc.* **2002**, *620*, 958–961.
20. Wang, W.; Zeng, L. Research status of numerical simulation of underwater explosion. *Chin. J. Nav. Acad. Aero. Eng.* **2006**, *2*, 209–216.
21. Li, X.; Wang, W.; Liang, Z.; Ruan, X. Research progress on acceleration ability of explosive detonation to metal shells. *Chin. J. Missile Arrow Guid.* **2022**, *42*, 7–15.

22. Wang, F.M.; Lu, F.Y.; Li, X.Y. Research on the projection characteristics of fragments under the loading of the oblique shock wave. *Chin. J. Natl. Univ. Def. Technol.* **2013**, *35*, 60–64.
23. Chapman, D.L.; Oxon, B.A., VI. On the Rate of Explosion in Gases. *Philos. Mag.* **1899**, *47*, 90–104. [CrossRef]
24. Charron, Y.J. Estimation of Velocity Distribution of Fragmenting Warheads Using a Modified Gurney Method. Ph.D. Thesis, Air Force Inst of Tech Wright-Pattersonafb Oh School of Engineering, Wright-Patterson Air Force Base, OH, USA, 1979.
25. Grisaro, H.; Dancygier, A.N. Numerical study of velocity distribution of fragments caused by explosion of a cylin-drical cased charge. *Int. J. Impact Eng.* **2015**, *86*, 1–12. [CrossRef]
26. Backofen, J.E. Modeling a material's instantaneous velocity during acceleration driven by a detonation's gas-push. *AIP Conf. Proc.* **2006**, *845*, 936–939.

Article

The Effect of Surface Electroplating on Fragment Deformation Behavior When Subjected to Contact Blasts

Yuanpei Meng, Yuan He *, Chuanting Wang , Yue Ma, Lei Guo, Junjie Jiao and Yong He *

School of Mechanical Engineering, Nanjing University of Science and Technology, Nanjing 210094, China; ctwang@njust.edu.cn (C.W.)
* Correspondence: heyuan@njust.edu.cn (Y.H.); yonghe1964@163.com (Y.H.)

Abstract: Preformed fragments can deform or even fracture when subjected to contact blasts, which might lead to a reduction of the terminal effect. Therefore, to solve this problem, the effect of surface electroplating on the fragment deformation behavior under contact blasts was analyzed. Firstly, blast recovery tests were carried out on uncoated and coated fragments. After the contact blast, the two samples produced different deformation behaviors: the uncoated fragments were fractured, while the coated fragments maintained integrity. The tests were simulated by finite element simulation, and the deformation behavior of the different samples matched well with the test results, which can explain the protective effect of the coating after quantification. In order to further reveal the dynamic behavior involved, detonation wave theory and shock wave transmission theory in solids were used to calculate the pressure amplitude variation at the far-exploding surface of the fragments. The theoretical results showed that the pressure amplitude of the uncoated samples instantly dropped to zero after the shock wave passed through the far-exploding surface, which resulted in the formation of a tensile zone. But the pressure amplitude of the coated samples increased, transforming the tensile zone into the compression zone, thereby preventing the fracture of the fragment near the far-exploding surface, which was consistent with the test and simulated results. The test results, finite element simulations, and theories show that the coating can change the deformation behavior of the fragment and prevent the fracture phenomenon of the fragment. It also prevents the material from missing and a molten state of the fragment in the radial direction by microscopic observation and weight statistics.

Keywords: contact blast; coatings; fragment deformation behavior; dynamic response; protection performance

Citation: Meng, Y.; He, Y.; Wang, C.; Ma, Y.; Guo, L.; Jiao, J.; He, Y. The Effect of Surface Electroplating on Fragment Deformation Behavior When Subjected to Contact Blasts. *Materials* **2023**, *16*, 5464. https:// doi.org/10.3390/ma16155464

Academic Editors: Pawel Pawlus and Young Gun Ko

Received: 22 June 2023
Revised: 1 August 2023
Accepted: 2 August 2023
Published: 4 August 2023

1. Introduction

The fragments had different degrees of deformation and even fracturing behavior after blast loading, which would affect the integrity [1,2], the terminal effect [3], and the initial velocity [4,5] of the preformed fragments. However, preformed fragments were often considered rigid when their dispersion characteristics and terminal effect were studied under different modes of detonation (concave [6], unsymmetrical [7], and conventional [8]). This has led to poor agreement between many studies and test results.

The deformation and fragmentation behavior of preformed fragments under contact blasts actually belonged to the problem of the dynamic response of the material under a high strain rate [1]. Many scholars have studied the fracture behavior of different metallic materials under high strain rates. The spalling phenomenon of common metals has been studied by scholars in the last century [9]. The fracture characteristics of metals at high strain rates could also be predicted by using simulation software [10,11]. In addition to the macroscopic study and prediction of fracture behavior, the microstructure changes of materials during high strain rate loading have also been studied [12,13]. Electron microscopic analysis of the recovered preformed fragments revealed that the aggregation of microporosity caused the failure [1,2]. The fragments fractured when the tensile stress generated inside the fragment was greater than the fracture stress [14].

Therefore, two methods have been used to prevent the deformation, fracture, and even fragmentation behavior of preformed fragments under the blast: adding linings and improving the mechanical properties of the fragments. To ensure the completeness of the fragments during the blast loading, the lining was provided between the fragments and the explosive to attenuate the shock wave and the load of the primary explosive [3,15]. The presence of the lining could increase the pulse width and reduce the deformation speed of the shock wave, but this would affect the initial velocity and terminal effect [16]. For example, the mechanical properties of metallic materials could be enhanced by adding different particles [17,18]. The dynamic failure of Fiber Reinforced Metal Tubes (FRMTs) under inner blast load was experimentally investigated [19]. The mechanical properties of metals were also improved by microstructure optimization [20]. These methods on the dynamic mechanical properties of the material could avoid fracture failure under a high strain rate. But these studies were hardly universal and needed to be more cost-effective.

Protective coatings, such as polyurea [21] and polymers [22], were also suggested to reduce the deformation and fracture of metals under blast. However, this approach was still influenced by the fracture strength of the metal material itself due to the low impedance of the polyurea and polymers coatings themselves, which cannot change the direction of the shock wave at the back-blast surface [23,24]. The material would still produce deformation and even fracture at a sufficiently high strain rate. Therefore, there is an urgent need to investigate a generalized method that can change the deformation behavior and prevent it from fracturing at high strain rates without affecting its initial velocity and terminal effect.

This study was initiated to protect the fragments under contact blasts via an alternative approach: surface electroplating of high-impedance coatings. The effect of surface electroplating and impedance matching on fragment deformation behavior subjected to contact blasts was thus investigated.

2. Experimental and Simulation Methods

2.1. Preparation Methods

The uncoated test samples (UC-10L) were made with commercial purity zirconium in the shape of a cylinder ($\varphi 11$ mm $\times$ $h13$ mm). The coated sample (C-10L) was prepared by coating a thick layer of nickel (commercial purity) on the uncoated sample. Since the matrix in the coated sample is the uncoated sample, it is referred to as a zirconium fragment when comparing the discussion of the matrix (in C-10L) and the uncoated sample (in UC-10L). The procedure of coating preparation is shown in Figure 1. Nickel metal with a thickness of ~1 mm was coated outside the uncoated test samples: Firstly, the surface of the preformed fragments was degreased and activated. Then, the preformed fragments were coated with coating solutions A and B (Nanjing WANQING chemical Glass ware & Instrument Co., Ltd. Nanjing, China), in turn, repeated ~40 times until the thickness of the coating reached ~1 mm. The formulations of coating solutions A and B are shown in Table 1, with PH values of 3.4 and 4.5, respectively. The microstructure was observed using the JSM-IT500HR (Tokyo, Japan) scanning electron microscope with a working voltage of 20 KV.

Table 1. Coating solution formulations (g/L).

Type	$NiSO_4$	$NiCl_2$	$NaCl_2$	H_3BO_3	$C_7H_5NO_3S$	$C_{12}H_{25}SO_4Na$	$C_6H_5SO_2Na$	$C_4H_6O_2$
Coating solution A	300	40	/	40	0.8	0.05	/	/
Coating solution B	300	/	10	35	0.8	0.05	0.5	0.4

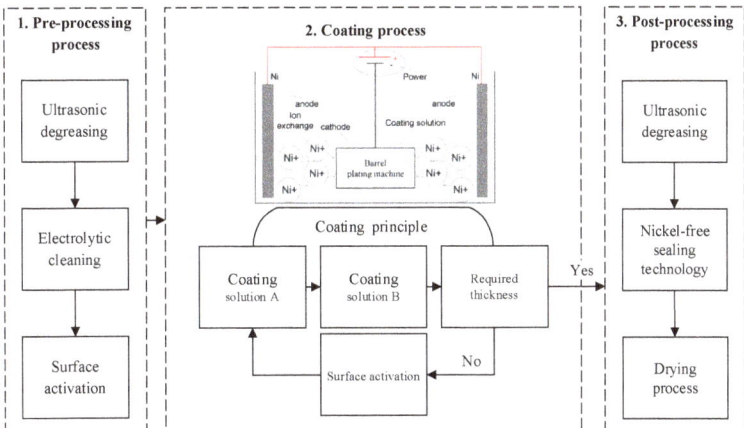

Figure 1. Flow chart of coating preparation procedure.

Figure 2a shows the samples before and after coating (UC-10L and C-10L). Figure 2b shows the cross-section microstructure of the C-10L. The interface between the zirconium fragment and the Ni coating is clearly visible, and the two parts are marked in Figure 2c. The hardness of the coating was tested by HV-1000A microhardness tester. The hardness of the coating is 3960 ± 60 MPa, while the hardness of the zirconium fragment was 1270 ± 40 MPa.

| (a) | (b) | (c) |

Figure 2. Characterization of samples (UC-10L and C-10L). (**a**) uncoated and coated samples, (**b**) cross-section of the C-10L, (**c**) electron microscope images.

2.2. Experimental Method

Uncoated samples (UC-10L) and coated samples (C-10L) were used in contact explosion tests. A previous study found that the Kevlar/epoxy lining material was beneficial in reducing the fragment deformation degree, preserving the initial velocity of fragments, and maintaining the fragments' quality integrity [15,25]. The schematic diagram and physical diagram of the test layout are shown in Figure 3, in which the warhead axis is parallel to the horizontal plane, and the multilayer wood boards are placed on the right to recover the test samples after the explosive drive. An appropriate distance was selected to ensure that more fragments were recovered, which was set to 40 cm. The warhead structure is shown in Figure 3b. The simulated warhead (φ72 mm × h50 mm) was a condensed charge (8701). The tests were carried out using single-point detonation in the center of the end face. The placement of fragments is shown in Figure 3c. It was sealed with adhesive tape to ensure the tight arrangement of fragments. Because of the weak strength of the tape, the binding force on the fragments scattering during the contact explosion drive could be ignored. The specific method was as follows: A column booster (φ15 mm × h5 mm) was built in the center of the left end of the explosive, and an electric detonator was connected to the column booster. The charge, lining (Kevlar/epoxy composite), and the two test samples fit snugly with each other.

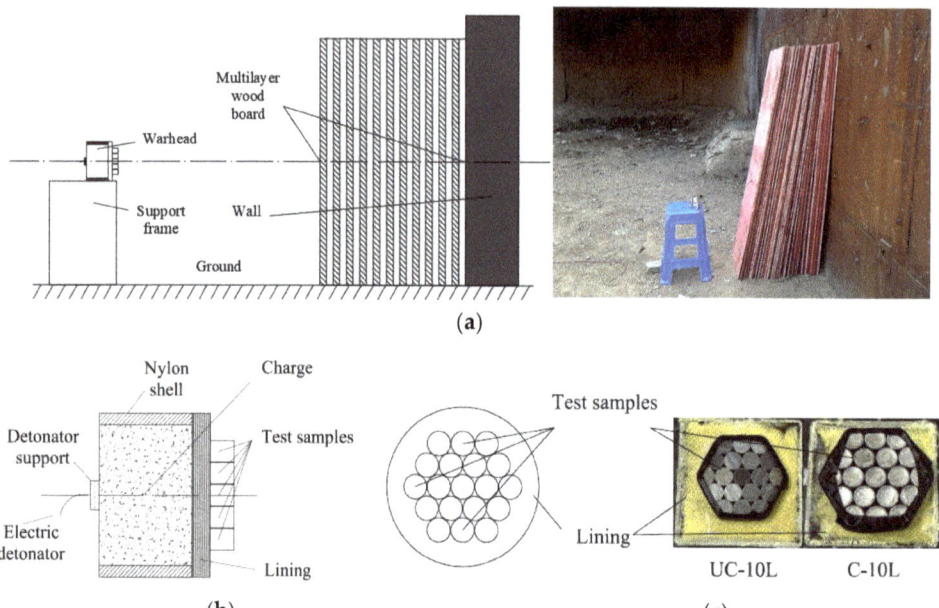

Figure 3. Schematic and layout diagram of the blast recovery test. (**a**) Experimental design, (**b**) Warhead assembly drawing, (**c**) The arrangement of the tests.

The lining was made as follows: First, Kevlar was cut into a square with a side length of 72 mm and put into the box one by one. Afterward, the resin glue was applied to each layer of Kevlar, and a heavy object was used to press it to ensure a tight fit between Kevlar layers [25,26]. The test arrangement is listed in Table 2. The recovered fragments were then microscopically observed by the FEI Quanta 250F.

Table 2. Test arrangement.

Test Method	Fragmentation Type	Lining Thickness
UC-10L	uncoated samples	10 mm
C-10L	coated samples	10 mm

2.3. Simulation Model

The simplified warhead simulation model is shown in Figure 4, which consists of the charge, lining, shell, and test samples. Charge (φ72 mm× h50 mm) used 8701, and lining (φ72 mm× h10 mm) was above the charge. The shell used nylon, wrapped in the cylindrical surface of the explosive and the bottom; its thickness was 2 mm. This paper used the ALE algorithm to numerically simulate the blast driving process by finite element simulation(FEM). The calculation process ignored the impact of the column booster on the detonation. Charge, detonation products, air, lining, and samples were used in the multi-matter Euler grid, and Kevlar/epoxy composite material was selected as the lining. To analyze the change of wave pressure at different locations of Kevlar/epoxy composite media, eight reference points are set equidistantly at the intersection of the lining and charge, as shown in Figure 4. The material parameters used in the simulation are shown in Tables 3–8.

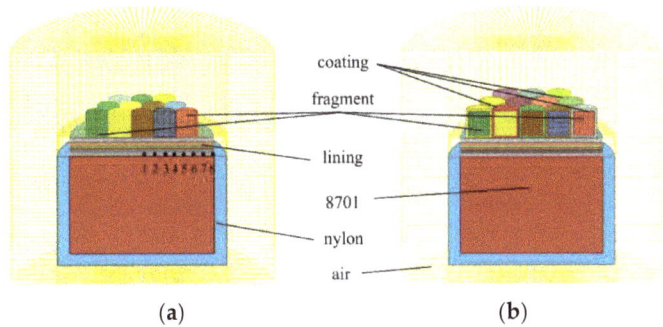

Figure 4. Simulation modeling of the blast recovery tests. (**a**) UC-10L, (**b**) C-10L.

Table 3. Material parameters of nylon.

$\rho/g\cdot cm^{-3}$	E/GPa	v	σ_0/MPa	E_{tan}/MPa	F_s
1.1	4.5	0.375	98	4.5	1.0

Table 4. Material parameters of explosive.

$\rho/g\cdot cm^{-3}$	$D/m\cdot s^{-1}$	P_{CJ}/GPa	$a/m\cdot s^{-2}$	b	R_1	R_2	*OMEG*
1.68	8800	29.75	4818	0.213	4.602	1.653	0.5

Table 5. Material parameters of air.

$\rho/g\cdot cm^{-3}$	C_0/GPa	C_1/GPa	C_2/GPa	C_3	C_4	C_5	C_6
1.1845	0	0	0	0	0.4	0.4	0

Table 6. Material parameters of lining.

$\rho/g\cdot cm^{-3}$	E_1/GPa	E_2/GPa	E_3/GPa	V_{12}	V_{13}	V_{23}
1.44	18.5	18.5	6	0.25	0.33	0.33

Table 7. Material parameters of the zirconium fragment.

$\rho/g\cdot cm^{-3}$	$C_v/J\cdot kg^{-1}\cdot K^{-1}$	T_{emit}/K	A/MPa	B/MPa	n	C	m
6.484	270	1473	303.8	549.12	0.65	0.027	0.827

Table 8. Material parameters of the Ni coating.

$\rho/g\cdot cm^{-3}$	$C_v/J\cdot kg^{-1}\cdot K^{-1}$	T_{emit}/K	A/MPa	B/MPa	n	C	m
8.9	446	1726	163	648	0.33	0.006	1.44

The macroscopic homogeneous model was used to model the Kevlar/epoxy composite in this paper [27], and 054/055 material in the finite element material model was used to reproduce the macroscopic orthotropic anisotropic mechanical properties, including failure criteria for the fiber and the epoxy matrix [28,29].

As shown in Figure 4, the lining is modeled with a single-layer mesh of 40 layers, and the number of meshes per layer in the axial direction is 1. The adjacent Kevlar layers needed to apply bonding forces due to the presence of the epoxy medium. Similarly, in the plated samples, there is a bond between nickel and zirconium, so the binding contact element was introduced in the finite (*CONTACT_AUTOMATIC_SURFACE_TO_SURFACE_TIEBREAK). This keyword allowed the Kevlar of two adjacent layers to remain bound at the beginning of the FEA [29,30]. But the binding keyword automatically degraded to face-to-face contact when the detonation wave pressure reached the destructive forces of tension and compression defined above.

3. Experimental and Simulation Results

3.1. Axial Deformation and Fracture

After the contact blast, the UC-10L and C-10L samples were recovered, as shown in Figure 5a. The UC-10L samples fractured and were divided into two main pieces (long and short) in axial length. Their fracture surfaces were not flat, and the average value was taken when measuring the length, while the recovered C-10L samples were complete, and no fractures were found. Comparison of simulation and test results on the morphology of fragments are shown in Figure 5b,c. After the fracture failure of the UC-10L samples occurred, the fracture location was selected as the benchmark for comparing the two since the mesh in the simulation would be deleted where the blank position in the simulation is deleted by the FEM after the mesh failure. The consistency between the test results and the simulation of both samples was acceptable, indicating that the simulation can predict the results of the test to some extent.

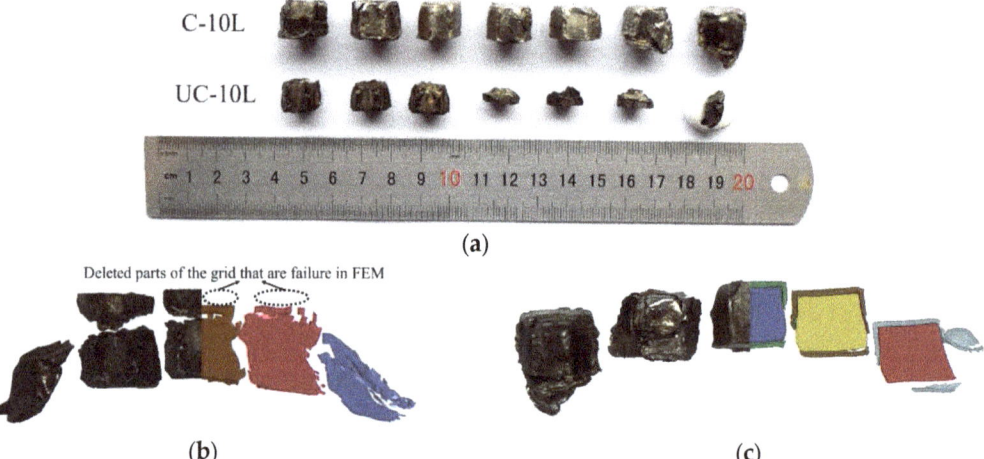

Figure 5. Comparison of recovered samples and simulations. (**a**) Typical samples recovered by the two methods, (**b**) Comparison of simulation and test results of UC-10L, (**c**) Comparison of simulation and test results of C-10.

The axial lengths of recovered typical samples are shown in Figure 6a, and the height of the sample is lower than its original height (13 mm for the sample before the UC-10L samples and 15 mm for the sample before the C-10L samples). For the UC-10L samples, the complete samples were composed of long and short samples [5]. The sum of the average value of long samples (9.17 mm) and the average value of short samples (2.82 mm) was 11.99 mm. The sum was lower than that of the original sample (13 mm).

The average axial length of the samples recovered from the test method of C-10L was 11.88 mm. The value was not only lower than 15 mm (before the C-10L samples) but even lower than the sum of the axial length of the fracture samples and the short samples after the C-10L samples. The compression ratio of the C-10L samples was around 20.8%, higher than the compression rate received by the UC-10L samples (around 7.8%).

Combining the comparison results in Figure 5, it was speculated that the coating changed the deformation behavior.

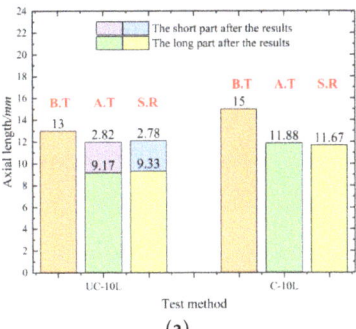 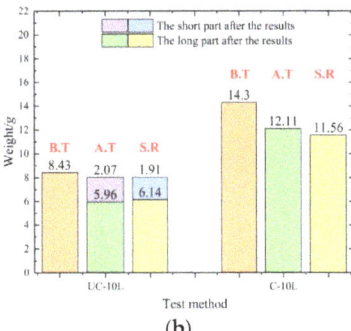

(a)   (b)

Figure 6. Axial length and weight of samples after blast. (a) Axial length of samples after blast, (b) Weight of samples after blast. ("B.T" means "before the test", "A.T" means "after the test", and "S.R" means "simulation results").

The weights of recovered samples are shown in Figure 6b. The sum of the average value of long samples (5.96 g) and the average value of short samples (2.07 g) was 8.03 g. The sum was less than 8.43 ± 0.05 g (before the test), similar to the statistical results of axial length. This was because the fragment also occurred in the radial direction. The specific analysis will be discussed in Section 3.2.

In summary, after comparing the axial length and weight of the two sets of experiments, it could be preliminarily inferred that due to the presence of the coating, the deformation behavior of the fragment under the contact explosion was changed.

3.2. Radial Local Fragmentation and Melting

In order to compare the protective effect of the coating on the zirconium fragment radially, the surface nickel coating was manually removed from the recovered samples of C-10L. The radial local fragmentation and melting situation are shown in Figures 7a and 8a, and there are six straight "ridges" in the circumferential axis of all recovered samples, which correspond to the hexagonal shape of their near-explosive surface. This is because although the samples are placed as close to each other as possible, there are still small gaps between the fragments in the radial direction. Therefore, the adjacent fragments in the radial direction collide to form a "ridge" when the shock wave passes through the samples [5].

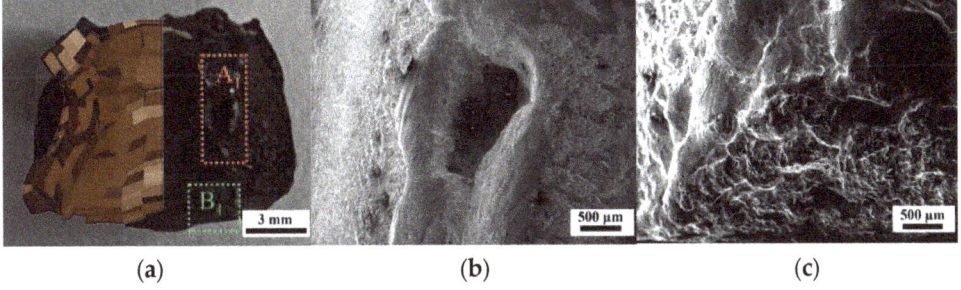

(a)   (b)   (c)

Figure 7. Optical and electron micrograph of the sample recovered from the test method of UC-10L. (a) Comparison of simulation and test results, (b) Electron micrograph of position A_1, (c) Electron micrograph of position B_1.

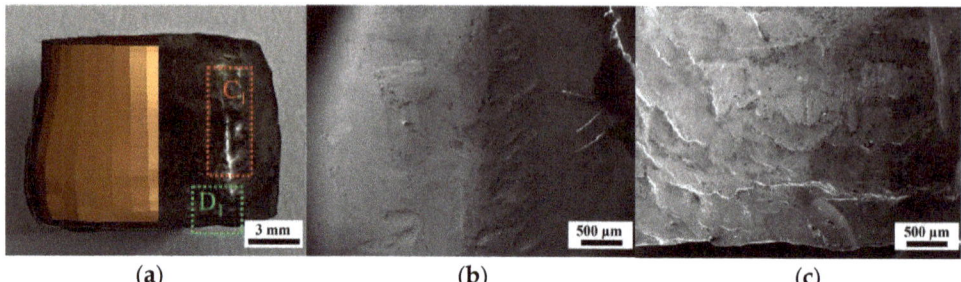

Figure 8. Optical and electron micrograph of the sample recovered from the test method of C-10L. (**a**) Comparison of simulation and test results, (**b**) Electron micrograph of position C_1, (**c**) Electron micrograph of position D_1.

As shown in Figure 7b, the cylindrical surfaces of the recovered uncoated samples have a lot of material missing and a molten state. This is due to the "welding effect" caused by the collision between adjacent fragments in the test method of UC-10L, so there is a molten state on the "ridge" [5], and due to the fragmentation caused by the detonation action of the fragment, resulting in the separation of the fragments welded together and the material is thus partially missing [1,5]. As shown in Figure 8b, because the samples from the test method of C-10L have the coatings, even if the coatings come off in this condition, the nickel replaces the impact "welding effect" of the fragments in the radial direction. Thus, no molten state and material is missing on the surface in the radial direction, which protects the fragments from fracture in the radial direction.

As shown in Figure 7c, the surfaces of the bottom of the "ridge" of the samples recovered from the UC-10L samples have traces of upward flow in addition to the molten state [1,25]. But as shown in Figure 8c, the surfaces of the bottom of the "ridge" of the C-10L samples only have a transverse texture produced by compression. The reason is that the surface of the UC-10L samples was not protected by the coating, and the high-temperature gas flow (explosive detonation product) generated by the explosion caused it to be prone to upward plastic flow [25]. The surface of the C-10L samples was coated to replace this plastic flow.

Thus, combined with the weight statistics, it shows that the coating not only causes a change in the deformation pattern of the zirconium fragments in the axial direction but also prevents it from local fragmentation and melting in the radial direction. To further illustrate the protective effect of the coating on the zirconium fragments, Figure 9 shows the kinetic energy–time variation curves of the zirconium fragments in UC-10L and C-10L. As shown in points A and B of Figure 9, the kinetic energy increase in C-10L lags behind that of UC-10L. This is because Figure 9 shows the kinetic energy change curve of zirconium fragments, and the shock wave reaches the coating first in the C-10L. The kinetic energy change curve of zirconium fragments for UC-10L decreases sharply at point C, which is due to the fracture of zirconium at this time. This is due to the fact that the part of the fracture is deleted directly in the simulation, which leads to this situation. This also proves that the presence of the coating ensures the integrity of the zirconium fragments and retains more of their kinetic energy.

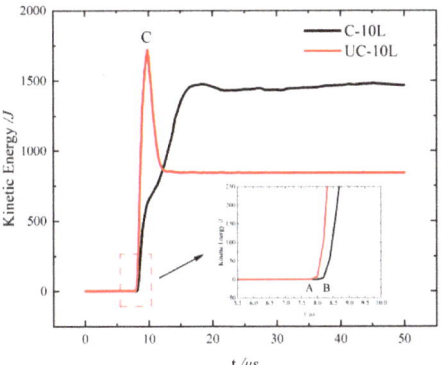

Figure 9. Kinetic energy–time variation curves of the zirconium fragments.

4. Analysis and Discussion

4.1. Detonation Wave Transmitted to the Lining

The impact effect of high-speed detonation products on solids was different from that of general static loads. Thus, it must be studied from a dynamic perspective and wave concept [14]. The experiment in this study was the case of an axial drive fragment, so the radial detonation wave was ignored. The wavefront of the detonation wave was spherical, so the detonation wave was considered oblique incidence in the lining. According to the angle of incidence and the magnitude of the wave impedance, the transmission reflection generated by oblique incidence at the interface was divided into normal oblique incidence, informal oblique incidence, and Prandtl–Meyer (P-M) expansion.

The wave impedance of Kevlar/epoxy lining is less than that of explosives, so it belongs to P-M expansion at the interface between the explosive and lining. The flowing image is shown in Figure 10, where OI is an oblique detonation wave front, the angle between OI and the interface of the contact medium is φ_0; OT is the oblique transmission shock wave front in the incoming medium, and the angle between OT and the initial interface of the medium is φ_3; the lining medium is deformed under the action of detonation, and the angle between the interface after the medium moves and the initial interface of the medium is δ. In this way, the oblique detonation wave, oblique reflection expansion wave, oblique transmission shock wave, and interface divide the entire flow into six regions: (0) area is unexploded, (1) area is the area of detonation product after oblique detonation wave, (2) area is the expansion area of detonation product, (3) area is the area of detonation product after expansion, (m_0) region is the initial medium, and (m) region is the area of medium disturbance after the oblique transmission shock wave [14,25].

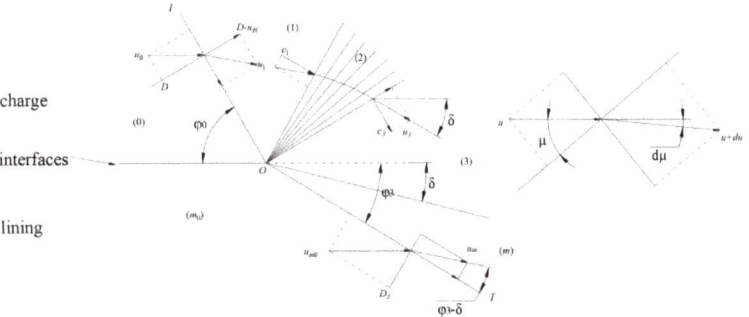

Figure 10. Schematic of the blast wave flowing into the lining.

This study assumed that the detonation wave was stable and self-sustaining detonation in the explosive. The state parameters of the detonation wave generated are taken from the parameters of the C-J point. The wavefront is a circular arc in the two-dimensional case. The effect of circumferential blast wave transverse reflection on the axial direction is ignored. The relationship between the parameters in the (3) region and the known parameters can be obtained according to the conservation and flow law of the detonation wave front, as shown in the following two equations:

$$M_3^2 = \frac{M_1^2 + \frac{2}{k-1}}{\left[\frac{(k+1)\rho_{m0}\sin\varphi_3}{b\rho_0\sin\varphi_0}\left(\frac{\sin\varphi_3}{\sin\varphi_0} - \frac{a}{D}\right)\right]^{\frac{k-1}{k}} - \frac{2}{k-1}} \tag{1}$$

$$\left[\sqrt{\frac{k+1}{k-1}}arctg\sqrt{\frac{k-1}{k+1}\left(M_3^2 - 1\right)}arctg\sqrt{M_3^2 - 1}\right] -$$

$$\left[\sqrt{\frac{k+1}{k-1}}arctg\sqrt{\frac{k-1}{k+1}\left(M_3^2 - 1\right)}arctg\sqrt{M_3^2 - 1} + \theta\right] \tag{2}$$

$$= arctg\left[\frac{\left(1 - \frac{a\sin\varphi_0}{D\sin\varphi_3}\right)tg\varphi_3}{b + \left(b - 1 + \frac{a\sin\varphi_0}{D\sin\varphi_3}\right)tg^2\varphi_3}\right]$$

where M_3 is the Mach number of the (3) zone, and M_1 is the Mach number of the (1) zone; ρ_0 and ρ_{m0} are the initial densities of the explosive and lining, respectively; k is the thermal insulation index of the explosive detonation product; a and b are the empirical constants of impact compression of the lining medium; D is the explosive detonation rate; φ_0 is the angle between the incident and the initial interface of the medium; φ_3 is the angle between the transmitted wave and the initial interface of the medium; and θ is the flow folding angle.

The calculation for the Mach number M_1 is shown in the following equation:

$$M_1 = \sqrt{1 + \left(\frac{k+1}{k}\right)^2 ctg^2\varphi_0} \tag{3}$$

The calculation for the flow bending angle θ is shown in the following equation:

$$tg\theta = \frac{tg\varphi_0}{1 + k(1 + tg^2\varphi_0)} \tag{4}$$

The calculation for the adiabatic index k of the explosive detonation product is shown in the following equation:

$$\begin{cases} k = 1.25 + k_0\left(1 - e^{-0.546\rho_0}\right) \\ k_0 = \frac{\sum\limits_{L}^{i=1}\frac{\mu_i}{M_i}}{\sum\limits_{L}^{i=1}k_{0i}M_i} \end{cases} \tag{5}$$

where k_0 is the total adiabatic index of the mixed explosive detonation product; ρ_0 is the charge density (g/cm^3) of the mixed explosive; μ is the mass percentage of component i of the explosive mixture; and M_i is the molar mass of component i of the explosive mixture.

The composition of the explosives used in the test is shown in Table 9. The adiabatic index (k) of 8701 explosives can be calculated as 2.85 [25]. The values of the parameters required for the above calculation are shown in Table 10.

Table 9. Explosive composition and its mass fraction.

μ (RDX)%	μ (Nitrotoluene)%	μ (Vinyl Acetate) %	Stearic Acid
95	3	2	trace

Table 10. The material parameters used in calculations.

ρ_0/g·cm^{-3}	ρ_{m0}/g·cm^{-3}	a/m·s^{-2}	b	D/m·s^{-2}	k_0/RDX	k_0/Vinyl Acetate	k_0/Nitrotoluene
1.68	1.273	2610	1.42	8800	2.65	2.78	2.78

Note: ρ_0 is the density of the explosive; ρ_{m0} is the density of Kevlar/epoxy composite lining; a and b are the empirical constants of the lining's shock compression relation; D is the detonation wave velocity; k_0 is a part of k, which is related to density [31,32].

Combining Equations (1)–(5), the parameters M_3 and φ_3 in the (3) region can be obtained.

The (m) zone parameter can be obtained from Equations (6)–(8):

$$\frac{\rho_{m0}}{\rho_m} = 1 - \frac{1}{b}\left(1 - \frac{a\sin\varphi_0}{D\sin\varphi_3}\right) \tag{6}$$

$$p_m = \frac{\rho_{m0}D^2\sin\varphi_3}{b\sin\varphi_0}\left(\frac{\sin\varphi_3}{\sin\varphi_0} - \frac{a}{D}\right) \tag{7}$$

$$u_m = \frac{D\sin\varphi_3}{\sin\varphi_0}\left[\left(1 - \frac{1}{b} + \frac{a\sin\varphi_0}{bD\sin\varphi_3}\right)^2 + tg^2\varphi_3\right]^{\frac{1}{2}} \tag{8}$$

where ρ_m, p_m, and u_m are the state parameters of the post-wave medium. The transmitted wave velocity is: $D_m = \frac{D\sin\varphi_3}{\sin\varphi_0}$.

Eight reference points were chosen based on the angle between the detonation wave and the lining. As shown in Figure 11a, the variation of pressure amplitude for eight references on the lining plane, the time to reach the pressure peak increases as the angle between the shock wave and the axis becomes larger. The pressure peaks at these eight reference points are compared with the theoretically calculated curves, as shown in Figure 11b. Finite element simulation using macroscopic Kevlar/epoxy lining modeling method with an error of 6% or less. The anastomosis is good and can reflect the pressure change in the lining to a certain extent.

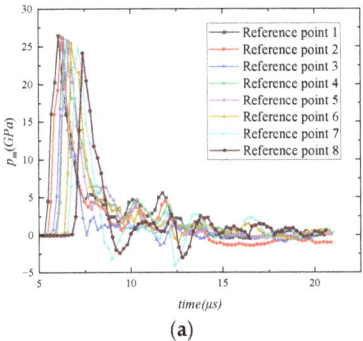
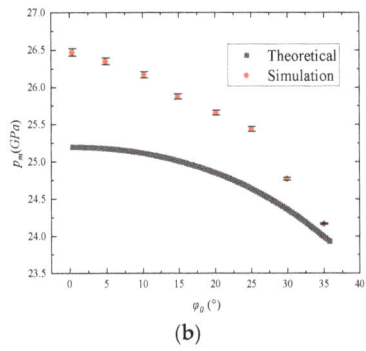

(a) (b)

Figure 11. Pressure amplitude changes in the lining. (**a**) Pressure variation curve of reference points, (**b**) Comparison of theoretical and simulated pressure value.

Therefore, the macroscopic model established in this paper can be employed to calculate the pressure value at various positions of the lining. This pressure value could be regarded as the input pressure transmitted to the samples.

4.2. Detonation Wave Transmitted to the Samples

When the shock wave travels through the liner and reaches the samples, the shock wave is assumed to be a plane wave when they propagate, which is based on the following: 1. The uncoated fragments and the coated fragments are in contact with the lining. 2. The explosive core is along the axis of the samples. 3. A one-dimensional wave is used to simplify the analysis when near the symmetry axis of complex space. 4. As shown in Figure 11a,b, the detonation wave at different angles is introduced into the lining with little effect at different locations.

Thus, the one-dimensional plane strain wave correlation theory is applied to analyze the propagation of shock waves in a multilayer medium to analyze the deformation behavior of fragments in two different test samples.

Figure 12 is a schematic diagram of the different interfaces of the UC-10L and C-10L samples. The arrows in Figure 12 indicate the propagation of the shock wave. Figure 12a is the UC-10L samples, where the purple interface is the interface between the lining and the zirconium fragment, and the red interface is the interface between the zirconium fragment and air. Figure 12b shows the C-10L samples, where the yellow interface is the interface between the lining and the nickel coating, the blue interface is the interface between the nickel coating and the zirconium fragment, the green interface is the interface between the zirconium fragment and nickel coating, and the orange interface is the interface between nickel coating and air.

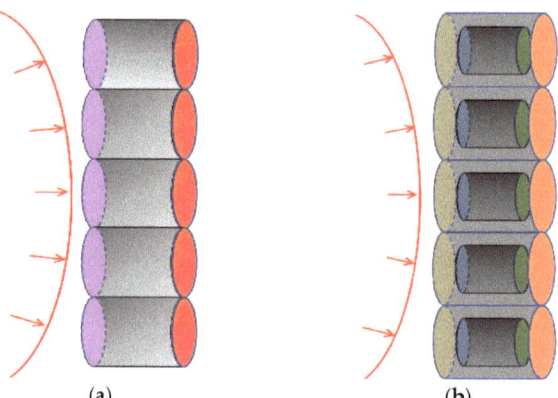

(a)   (b)

Figure 12. Schematic diagram of different interfaces of the test samples. (**a**) UC-10L, (**b**) C-10L.

As the impact impedance of the right medium of purple (yellow) is greater than the impact impedance of the lining on its left side, when the shock wave propagates to the interface of purple (yellow) in Figure 12, the pressure amplitude of the right medium of the purple (yellow) interface will be higher than the initial medium pressure amplitude of the lining.

Then, the propagation theory of shock waves between the two mediums is used to solve the following questions [33]. For the left wave D_L in the lining, taking the left wave as the observation point, the following relationship can be obtained from the fundamental equation of shock wave:

$$u_{a0} - u_{a1} = \sqrt{(p_{a1} - p_{a0})(v_{a0} - v_{a1})} \tag{9}$$

$$D_L + u_{a0} = v_{a0}\sqrt{\frac{p_{a1} - p_{a0}}{v_{a0} - v_{a1}}} \tag{10}$$

Similarly, for the right wave D_R in the right medium, the following relationship can be obtained from the fundamental equation of shock wave:

$$u_{b1} - u_{b0} = \sqrt{(p_{b1} - p_{b0})(v_{b0} - v_{b1})} \tag{11}$$

$$D_R - u_{b0} = v_{b0}\sqrt{\frac{p_{b1} - p_{b0}}{v_{b0} - v_{b1}}} \tag{12}$$

At the interface of the two mediums, it can be obtained by the continuity condition:

$$u_{a1} = u_{b1}$$
$$p_{a1} = p_{b1} \tag{13}$$

In the above equations, subscripts a and b represent the medium material on the left and right sides of the purple (yellow) interface, respectively, and subscripts 0 and 1 represent the parameters before and after the wave, respectively. The initial parameters ($u_{a0}, \rho_{a0}, P_{a0}$) of the dielectric material on the left side of the purple (yellow) interface are calculated by the previous Equations (1)–(8). The initial velocity (u_{b0}) in the right medium of purple (yellow) is 0, and the initial density (ρ_{b0}) is a known parameter. The following relationship can be obtained from the Hugoniot relationship between the shock wave velocity and the post-wave particle velocity in the condensed medium:

$$D_L = a_1 + b_1 u_{a1}$$
$$D_R = a_2 + b_2 u_{b1} \tag{14}$$

where a_1 and b_1 are the Hugoniot parameters of the left dielectric material of the purple (yellow) interface, and a_2 and b_2 are the Hugoniot parameters of the right dielectric material of the purple (yellow) interface.

The unknown parameters of the shock wave and particle in the two mediums can be obtained by solving Equations (6)–(14), where p_{b1} is the initial pressure amplitude obtained by the material under the action of the shock wave. Because this study considered the influence of trans-reflection on the particle parameters of the medium when the shock wave propagated in a different medium, the attenuation of the shock wave in the condensed medium was ignored.

The solution of the parameters on both sides of the blue interface is similar to the solution on both sides of the purple (yellow) interface. When the shock wave propagates further to the blue interface, the right medium pressure amplitude of the blue interface is lower than the initial medium pressure amplitude on its left side. This is because the shock impedance of the medium on the left side of the blue interface is greater than that on the right side.

According to the above analysis methods, it is shown that the shock propagation expressions in the shock wave are consistent with the above expressions. The values of the parameters represented by a_1 and b_1 need to be changed to the parameters of the medium on the left side of the blue interface, and a_2 and b_2 need to be changed to the parameters of the material on the right side of the blue interface. Similarly, this method is still used in the green interface. The parameters [14,34] of the different mediums used in the equations are shown in Table 11.

Table 11. Parameters for different mediums.

Medium Type	$\rho_{m0}/\text{g·cm}^{-3}$	$a/\text{m·s}^{-2}$	b
Nickel coating	8.9	4590	1.44
Zirconium fragment	6.5	4240	1.015

Thus, combined with the results obtained in Section 4.1 (Figure 11), the pressure amplitude in the zirconium fragments can be calculated based on the above equations and parameters.

4.3. Changes of Deformation Behavior Caused by the Ni Coating

The air is considered an incompressible medium, and the red interface (UC-10L in Figure 12a) is regarded as a free surface [33]. Thus, the particle state should be solved by the interaction between the shock wave and the free surface. When the shock wave propagating along the medium reaches the free surface, the pressure of the wavefront immediately drops to zero. Then, the medium begins to expand and move forward, and a tensile wave is reflected in the medium compressed by the shock wave. Then, the medium obtains another velocity increment in the original direction of motion. At this time, the particle velocity is doubled; that is, the shock wave is twice the speed criterion of the free surface [35]. Currently, the velocity of the left side of the red (orange) interface is twice as large, and the mass pressure amplitude is zero.

According to the above theoretical analysis, the state parameters of the left and right medium can be obtained when the shock wave travels through several interfaces. Figure 13a shows the variation curves of pressure amplitude with the incident angle before and after passing through the purple interface in the uncoated sample. Figure 13b shows the variation curve of pressure amplitude with incident angle before and after passing through the green interface in the coated sample. The shaded parts in both figures are the variation values of pressure amplitude. The shaded parts in both figures are the value change of pressure amplitude.

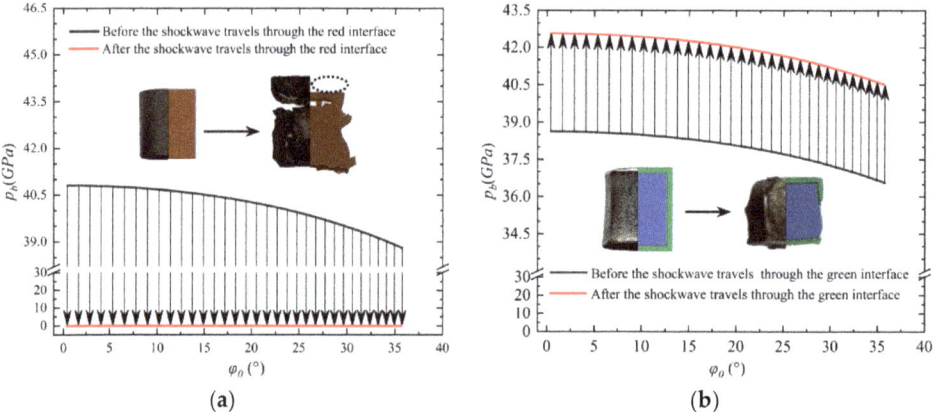

Figure 13. Change of pressure parameters at the different sample surfaces. (**a**) Change of pressure parameters at the red interface, (**b**) Change of pressure parameters at the green interface.

Before the shock wave inside the fragments reaches the green interface (coated samples) and red interface (uncoated samples), the pressure amplitude of the mass inside the fragments is greater than 0, as shown in the curves in Figure 13a,b. Because the sample as a whole is subjected to the compression effect generated by the shock wave, its length will be smaller than the length before the detonation in a one-dimensional plane perspective. The length of the samples recovered from the test is shown in Figure 6. It is shown that for all the tests and simulations, the fragment length is less than its original length, which can prove the correctness of the theory.

For the UC-10L samples, when the shock wave is transmitted to the red interface, the pressure amplitude instantly drops to 0. The shaded part of the arrow direction of Figure 13a indicates that it is subjected to stretching when the tensile effect is stronger than its dynamic elastic limit. It produced the fracture phenomenon, so two samples with different fracture lengths were recovered in the UC-10L samples.

For the C-10L samples, when the shock wave to the green interface, as shown in Figure 13b, the pressure amplitude of the mass point on the left side of the interface instantaneously increases. This results that the far-exploding surface will be subject to greater compression than before and the UC-10L samples. The upward arrow in the shaded part of Figure 13b indicates the compression effect on the green interface relative to the rest of the fragments. Therefore, the compression rate of the recovered sample in the C-10L samples (20.8%) was higher than that in the UC-10L samples (7.8%), which proved the consistency between the theoretical analysis and the C-10L samples.

It is reasonable that the deformation behavior will not change when the wave impedance of the coating is lower than that of the fragments but will only reduce the amount of deformation [21,23]. Then, the far-exploding surface of the fragments will have a stretching effect. But the effect of the shock wave unloading caused by stretching is much lower than that of the fragment in the free surface. The effect of stretching gradually increases as the impedance of the coating decreases, and the stretch area decreases.

Therefore, impedance is the key factor that leads to the change in the deformation behavior combined with the above discussion. When there is a layer of medium with an impedance greater than that of the fragment outside the surface of the fragment, it can change the tensile deformation into compressive deformation.

4.4. Fracture Mechanism and Calculation of the Fracture Position

The previous section discussed qualitatively the mechanism of fracture of UC-10L samples in terms of the compression and tension zones generated by the shock wave on the far-exploding surface of the fragments. But in fact, the shock wave that causes the force direction to change is a triangular pulse (pressure–time curve) with a wavelength of λ. It takes some time for the shock wave to affect the various parameters of the mass inside the fragment, and the time is related to wavelength. Thus, there is a process during the mass pressure amplitude to become 0. This section explains the fracture mechanism from the wavelength perspective and calculates the fracture position.

The process principle is shown in Figure 14. When the detonation wave acts on the near-exploding surface, a triangular stress wave will propagate in the fragment, keeping it in a compressed state. According to the principle of a one-way strain plane wave, as the stress wave propagates to the right, the amplitude increases and the wavelength decreases, but the wavefront is still triangular. When the wavefront surface of the stress wave reaches the free surface, a stretching wave comparable to the incident compressional wave will be reflected. The direction of this stretching wave is opposite to the direction of the incident wave. At this time, the incident wave interferes with the reflected wave, and the pressure amplitude gradually drops to 0. Since the tail of the incident wave is still within the free surface, the material within the free surface is kept in compression. After that, the incident wave continues to move outward while the reflected stretch wave moves continuously into the fragment, and the two waves constantly interfere with each other. The material is transferred from the original compression state to the tensile state within the free surface, where the reflected wavefront goes. And as the distance of the tensile wave to the free surface increases, the tensile stress also gradually increases. Fracture begins when the value of tensile stress reaches the critical fracturing stress of the material.

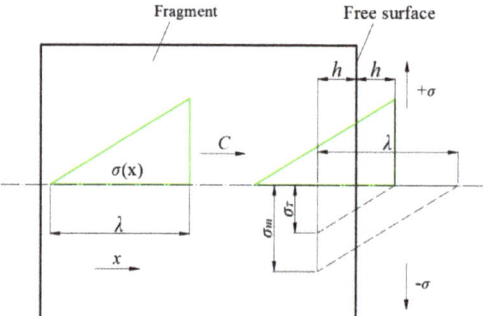

Figure 14. Schematic diagram of the fracture process of the C-10L.

So, in order to facilitate the calculation of the location of the fracture, it is assumed that the shock wave of the fragment is an elastic wave, the wave speed of the fragment of material does not change with the compression, and the wavelength of the compression wave does not change with the distance. That is, the wave speed and wavelength of the shock wave in the fragment are invariant. Therefore, the triangular compressional wave within the fragment propagates without attenuation at the elastic longitudinal wave speed [36]. The peak stress is σ_m, and the dynamic failure stress is σ_T. The incident wave interferes with the reflected wave after the triangular incident stress wave reaches the free surface. Suppose fracture occurs at a distance h (mm) from the far-exploding surface, and the geometric relationship can obtain according to Figure 14:

$$\frac{\sigma_T}{\sigma_m} = \frac{2h}{\lambda} \Rightarrow h = \frac{\sigma_T}{2\sigma_m}\lambda \tag{15}$$

where λ is the wavelength of the compression wave, and h is the length of the fracture.

The value of λ in the formula is difficult to be found directly, and the empirical fitting is mainly carried out through experiments at this stage [36]. Thus, the approximation method is used in this paper. When the detonation wave travels along the charge to the contact surface, the pressure (P_m) on the fragmentation surface suddenly increases to the maximum value and then drops rapidly. Time elapsed for the pressure amplitude to drop to zero is expressed as the diameter of explosives(d) divided by the shock wave velocity (D_k). Thus, the equation for the wavelength (λ) is obtained:

$$\lambda = \frac{d}{D_k}C \tag{16}$$

where C is the longitudinal wave velocity of the shock wave in the material.

$$C = \sqrt{\frac{\frac{E}{3(1-2v)} + \frac{2E}{3(1+v)}}{\rho}} \tag{17}$$

where v is the Poisson's ratio, E is the modulus of elasticity, and ρ is the density.

The fracture will occur when the tensile action on pure zirconium under high pressure exceeds the fracture strength. Therefore, the dynamic fracture stress (σ_T) in the equation is used as the fracture strength of the material, and fracture occurs when the fracture strength at this strain rate range is exceeded. However, most of the current methods to obtain the material σ_T use the shock wave physics experimental technique by using a flat plate impact test with a lightweight air cannon. The pressure range covered by the chemical explosion contact blast is crossed with the flat plate impact test, so the required parameters are referred to the flat plate impact test [37]. Therefore, the above formula takes parameters as shown in Table 12.

Table 12. Table of parameters used for calculation.

$\rho/g\cdot cm^{-3}$	E/GPa	v	σ_T/GPa	d/mm
6.5	101	0.31	1.69	72

The value of C is 4520 m/s by Equation (17). In the case of neglecting the attenuation of the shock wave in the rupture, the variation curve of the fracture length (h) from the rupture far-exploding surface (red) with the angle of incidence is shown in Figure 15 according to Equation (15) and the calculation results of the pressure amplitude in the previous section.

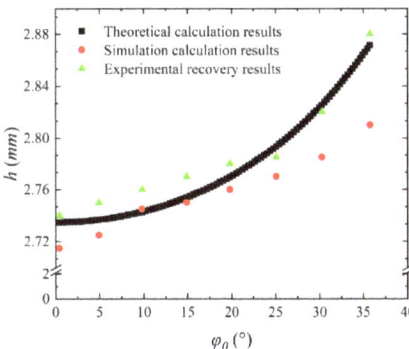

Figure 15. Variation of fracture length with incident angle.

The results of the simulation and the experiments are represented by the green and orange ranges, as shown in Figure 15. The theoretical results of h (2.73–2.86 mm) are consistent with the results of experimental recovery measurements (2.74–2.88 mm) and the results of simulated (2.71–2.78 mm), which confirms the reliability of the analysis. According to the recovered fracture samples, it is found that the fracture surface is not a relatively smooth surface in the height direction, and its fracture surface is uneven. This is because the fracture of the material is a process of damage accumulation, which can be expressed by the combination of macro micro simulation [38,39].

Therefore, it is difficult to accurately calculate the position of different fractures for materials during the accumulation process. Then, the approximate theoretical calculation is used here, and the measurement method takes the average height of the recovered fracture samples. Thus, the results of the theoretical calculations can account for a certain degree of agreement with the simulation and test results.

5. Summary

The effect of surface electroplating on the fragment deformation behavior under contact explosion was analyzed by the combination of theory, experiment, and simulation in this study. The following conclusions can be drawn:

1. The coating prevents the molten state and plastic flow of the fragments in the radial direction. In contrast to Kevlar/epoxy lining, the coating not only prevents the molten state of fragments due to collisions between neighboring fragments but also prevents the plastic flow of the fragments by preventing their contact with high-temperature gas;

2. The coating changes the deformation behavior. Because the wave impedance of the nickel coating is greater than that of the fragment, it changes from the stretch zone of the original interface with the air to the compression zone at the interface between the fragment and the coating, which solves the problem of tensile fracture of the fragments on the far-exploding surface (free surface);

3. By simplifying the shock wave as one-dimensional stress propagation, the axial fracture length of the uncoated samples is between 2.73–2.86 mm. The error is less than 6%

compared with the recovered samples from the test method of UC-10L. This can prove the applicability of the theory considering the error caused by the cumulative fracture of the material microdamage.

Author Contributions: Conceptualization, Y.M. (Yuanpei Meng), C.W., Y.M. (Yue Ma), J.J. and Y.H. (Yong He); Methodology, Y.M. (Yuanpei Meng), C.W. and Y.M. (Yue Ma); Validation, Y.M. (Yuanpei Meng), C.W., Y.M. (Yue Ma) and L.G.; Investigation, Y.M. (Yuanpei Meng); Resources, Y.H. (Yuan He), L.G. and Y.H. (Yong He); Data curation, Y.M. (Yuanpei Meng); Writing—original draft, Y.M. (Yuanpei Meng); Writing—review & editing, Y.H. (Yuan He), C.W., Y.M. (Yue Ma), L.G., J.J. and Y.H. (Yong He); Supervision, Y.H. (Yuan He) and J.J.; Project administration, Y.H. (Yuan He) and Y.H. (Yong He); Funding acquisition, Y.H. (Yuan He), C.W. and Y.H. (Yong He). All authors have read and agreed to the published version of the manuscript.

Funding: This work was supported by the National Natural Science Foundation of China [NO.U2241285 and NO.JCJQ-2019-00-011].

Institutional Review Board Statement: Not applicable.

Informed Consent Statement: Not applicable.

Data Availability Statement: Not applicable.

Conflicts of Interest: The authors declare no conflict of interest.

References

1. Charles, E.; Anderson, J.R. Shock propagation and damage in tungsten cubes subjected to explosive loading. *Int. J. Impact Eng.* **1989**, *8*, 69–81. [CrossRef]
2. Huang, B.; Chen, L.Q.; Qiu, W.B.; Yang, X.L.; Shi, K.; Lian, Y.Y.; Liu, X.; Tang, J. Correlation between the microstructure, mechanical/thermal properties, and thermal shock resistance of K-doped tungsten alloys. *J. Nucl. Mater.* **2019**, *520*, 6–18. [CrossRef]
3. Ma, Y.; He, Y.; Wang, C.T.; He, Y.; Guo, L. Response behavior of double layer tungsten fragments under detonation loading. *J. Northwest Polytech. Univ.* **2022**, *40*, 819–828. [CrossRef]
4. Cheng, X.W.; Wang, J.X.; Tang, K.; Ling, S.J.; Li, Y.L. Analysis on the initial velocity field of a multilayer spherical fragment driven by explosion. *J. Vib. Shock.* **2020**, *39*, 129–134. [CrossRef]
5. O'Donoghue, P.E.; Predebon, W.W.; Anderson, C.E. Dynamic launch process of performed fragments. *J. App Phys.* **1998**, *63*, 337–348. [CrossRef]
6. Shi, Y.P.; Zhou, T.; Guo, Z.W.; Huang, G.Y.; Feng, S.S. Velocity distribution of preformed fragments from concave quadrangular charge structure. *Int. J. Impact Eng.* **2023**, *176*, 104551. [CrossRef]
7. Li, Y.; Li, Y.H.; Wen, Y.Q. Radial distribution of fragment velocity of asymmetrically initiated warhea. *Int. J. Impact Eng.* **2017**, *99*, 39–47. [CrossRef]
8. Gurney, R.W. *The Initial Velocities of Fragments From Bombs, Shell and Grenade*; Ballistic Research Laboratory: Aberdeen, MD, USA, 1943.
9. Yellup, J.M. The computer simulation of an explosive test rig to determine the spall strength of metals. *Int. J. Impact Eng.* **1984**, *2*, 151–167. [CrossRef]
10. Hiroyoshi, I.; Yoichi, K.; Masuhiro, B. Fracture characteristics and damage prediction in flat steel beams under contact explosions. *Int. J. Impact Eng.* **2023**, *175*, 104540. [CrossRef]
11. Povarnitsyn, M.E.; Khishchenko, K.V.; Levashov, P.R. Simulation of shock-induced fragmentation and vaporization in metal. *Int. J. Impact Eng.* **2008**, *35*, 1723–1727. [CrossRef]
12. Tan, Y.; Wang, Y.W.; Cheng, H.W.; Cheng, X.W. Dynamic fracture behavior of $Zr_{63}Cu_{12}Ni_{12}Al_{10}Nb_3$ metallic glass under high strain-rate loading. *J. Alloys Compd.* **2021**, *853*, 157110. [CrossRef]
13. Li, C.; Yang, K.; Tang, X.C. Spall strength of a mild carbon steel: Effects of tensile stress history and shock-induced microstructure. *Mater. Sci. Eng. A* **2019**, *754*, 461–469. [CrossRef]
14. Zhang, S.Z. *Explosion and Shock Dynamics*; The Publishing House of Ordnance Industry: Beijing, China, 1993; ISBN 7-80038-480-2.
15. Yu, Q.B.; Wang, H.F.; Jin, X.K.; Yu, W.M. Influence of Buffer Material on Explosive Driven of Reactive Fragment Warhea. *Tran. Beijing Inst. Technol.* **2013**, *33*, 124–126. [CrossRef]
16. Chen, C.; Hao, Y.P.; Yang, L.; Wang, X.M.; Li, W.B.; Li, W.B. Research on Double Layer Medium Gap Test and Analysis of Shock Initiation Characteristics of Acceptor Explosive. *Acta Armamentarii* **2017**, *38*, 1957–1964. [CrossRef]
17. Li, J.C.; Chen, X.W.; Huang, F.L. On the mechanical properties of particle reinforced metallic glass matrix composite. *J. Alloys Compd.* **2018**, *737*, 271–294. [CrossRef]
18. Tang, X.C.; Jian, W.R.; Huang, J.Y.; Zhao, F.; Li, C.; Xiao, X.H.; Yao, X.H.; Luo, S.N. Spall damage of a Ta particles-reinforced metallic glass matrix composite under high strain rate loading. *Mater. Sci. Eng. A* **2018**, *711*, 284–292. [CrossRef]

19. Li, X.; Xu, R.; Zhang, X.; Zhang, H.; Yang, J.L. Inner blast response of fiber reinforced aluminum tubes. *Int. J. Impact Eng.* **2023**, *172*, 104416. [CrossRef]

20. Vasilev, E.; Miroslav, Z.; McCabe, R.J.; Knezevic, M. Experimental verification of a crystal plasticity-based simulation framework for predicting microstructure and geometric shape changes: Application to bending and Taylor impact testing of Zr. *Int. J. Impact Eng.* **2020**, *144*, 103655. [CrossRef]

21. Wang, X.; Ji, C.; Wu, G.; Wang, Y.T.; Zhu, H.J. Damage response of high elastic polyurea coated liquid-filled tank subjected to close-in blast induced by charge with preformed fragments. *Int. J. Impact Eng.* **2022**, *167*, 104260. [CrossRef]

22. Mohotti, D.; Fernando, P.L.N.; Weerasinghe, D.; Remennikov, A. Evaluation of effectiveness of polymer coatings in reducing blast-induced deformation of steel plates. *Def. Technol.* **2021**, *17*, 1895–1904. [CrossRef]

23. Chu, D.Y.; Wang, Y.G.; Yang, S.L.; Li, Z.J.; Zhuang, Z.; Liu, Z.L. Analysis and design for the comprehensive ballistic and blast resistance of polyurea-coated steel plate. *Def. Technol.* **2023**, *19*, 35–51. [CrossRef]

24. Chu, D.Y.; Li, Z.J.; Yao, K.L.; Wang, Y.G.; Tian, R.; Zhuang, Z.; Liu, Z.L. Studying the strengthening mechanism and thickness effect of elastomer coating on the ballistic-resistance of the polyurea-coated steel plat. *Int. J. Impact Eng.* **2022**, *163*, 104181. [CrossRef]

25. Ma, Y.; He, Y.; Wang, C.T.; He, Y.; Guo, L.; Chen, P. Influence of lining materials on the detonation driving of fragments. *J. Mech. Sci. Technol.* **2022**, *36*, 1337–1350. [CrossRef]

26. Bresciani, L.M.; Manes, A.; Ruggiero, A.; Iannitti, G.; Giglio, M. Experimental tests and numerical modelling of ballistic impacts against Kevlar 29 plain-woven fabrics with an epoxy matrix: Macro-homogeneous and Meso-heterogeneous approaches. *Compos. B Eng.* **2016**, *88*, 114–130. [CrossRef]

27. Gower, H.L.; Cronin, D.S.; Plumtree, A. Ballistic impact response of laminated composite panels. *Int. J. Impact. Eng.* **2008**, *35*, 1000–1008. [CrossRef]

28. Jack, V.H. Modelling of Impact Induced Delamination in Composite Materials. Ph.D. Thesis, Carleton University, Ottawa, ON, Canada, 1999. [CrossRef]

29. Nayak, N.; Banerjee, A.; Panda, T.R. Numerical study on the ballistic impact response of aramid fabric- epoxy laminated composites by armor piercing projectile. *Procedia Eng.* **2017**, *173*, 230–237. [CrossRef]

30. Feli, S.; Asgari, M.R. Finite element simulation of ceramic/composite armor under ballistic impact. *Compos. B Eng.* **2011**, *42*, 771–780. [CrossRef]

31. Guo, H.F.; Wu, Y.Q.; Li, J.D.; Qu, K.P.; Xiao, W.; Chen, P. Meso Numerical Simulation of Composite Explosives Under High Strain Rate. *Trans. Beijing Inst. Technol.* **2019**, *39*, 1311–1314. [CrossRef]

32. Mochalova, V.; Utkin, A.; Savinykh, A.; Garkushin, G. Pulse compression and tension of Kevlar/epoxy composite under shock wave action. *Compos. Struct.* **2021**, *273*, 114309. [CrossRef]

33. Wang, L.L. *Basis of Stress Waves*; Defense Industry Publishing: Beijing, China, 2005.

34. Greeff, C.W. Phase Changes and the Equation of State of Zr. *Mater. Sci. Eng.* **2005**, *13*, 1015. [CrossRef]

35. Li, W.X. *One-Dimensional Indeterminate Flow and Shock Waves*; Defense Industry Publishing: Beijing, China, 2003.

36. Yang, C.Q. *Dynamic Fracture Study of Two Structural Steels*; Shanxi North Central University: Taiyuan, China, 2020.

37. Li, Y.H.; Cai, L.C.; Zhang, L.; Li, Y.L. Dynamic Behavior of α-Zirconium at Low Pressure. *Chin. J. Phys.* **2007**, *21*, 188–192. [CrossRef]

38. Ikkurthi, V.R.; Chaturvedi, S. Use of different damage models for simulating impact-driven spallation inmetal plates. *Int. J. Impact Eng.* **2004**, *30*, 275–301. [CrossRef]

39. Meng, Y.P.; Guo, Z.P.; Wang, C.T.; He, Y.; He, Y.; Hu, X.B. Research on Taylor impact fracture behavior of ZrCuAlNiNb amorphous alloy. *J. Northwest Polytech. Univ.* **2021**, *39*, 1296–1303. [CrossRef]

Article

Research on the Formation Characteristics of the Shaped Charge Jet from the Shaped Charge with a Trapezoid Cross-Section

Bin Ma [1,*], Zhengxiang Huang [1,*], Yongzhong Wu [2], Yuting Wang [1], Xin Jia [1] and Guangyue Gao [3]

1   School of Mechanical Engineering, Nanjing University of Science and Technology, Nanjing 210094, China
2   China Ordnance Industry Navigation and Control Technology Research Institute, Beijing 100089, China
3   Shanxi Jiangyang Chemical Co., Ltd., Taiyuan 030041, China
*   Correspondence: mabin@njust.edu.cn (B.M.); huangyu@njust.edu.cn (Z.H.)

Abstract: The formation characteristics of the shaped charge jet (SCJ) from the shaped charge with a trapezoid cross-section is analyzed in this work. A theoretical model was developed to analyze the collapsing mechanism of the liner driven by the charge with a trapezoid cross-section. Based on the theoretical model, the axial and radial velocities of the SCJ from different trapezoid cross-section charges. The pressure model was employed to calculate the velocity for the subcaliber shaped charge, which was verified through numerical simulation. The results show that the influence of the angle of the trapezoidal charge (acute angle) on the axial velocity of the SCJ is not distinct, whereas the variation of the radial velocity of the shaped charge jet is obvious as the change in the angle of the trapezoidal charge. In addition, the related X-ray experiments were conducted to verify the theory. The theoretical results correlate with the experimental results reasonably well.

Keywords: shaped charge; trapezoid cross-section; formation; X-ray

Citation: Ma, B.; Huang, Z.; Wu, Y.; Wang, Y.; Jia, X.; Gao, G. Research on the Formation Characteristics of the Shaped Charge Jet from the Shaped Charge with a Trapezoid Cross-Section. *Materials* **2022**, *15*, 8663. https://doi.org/10.3390/ma15238663

Academic Editor: Aniello Riccio

Received: 28 October 2022
Accepted: 30 November 2022
Published: 5 December 2022

Publisher's Note: MDPI stays neutral with regard to jurisdictional claims in published maps and institutional affiliations.

1. Introduction

As technology advances, many precision-guided weapons are emerging such as cruise projectiles, unmanned aerial vehicles, etc., [1] which are the basis of precision striking in the battlefield of the future. The structural characteristics of these precision-guided weapons present largely an asymmetrical cross-section in consideration of the aerodynamics [2]. Due to these structural constraints, the kinetic warhead and fragmentation warhead are captured extensively for these precision-guided weapons [3–5].

Shaped charge is one effective weapon against armored and concrete targets, and it has been given great attention around the world. It is well known that the shaped charge jet (SCJ) tip velocity can reach 6000–8000 m/s, even up to 10,000 m/s, and the tail element can fly a velocity of approximately 2000 m/s [6]. For the traditional axisymmetric shaped charge, the SCJ can be elongated considerably due to the existence of the strain rate ranging from 10^4 to 10^5 s^{-1} [7]. However, shaped charge is very sensitive to structural characteristics, and the asymmetry of the asymmetric section weapons is a challenge to the design of the shaped charge. Barry Stewart et al. [8] studied the feasibility of emerging a smear-compensated over-fly top attack SCJ. In their work, a novel side-mounted initiation train was designed to optimize space and maintain an antistructures emplacement capability. Experiments validated the initiation and two design variants and obtained a smear-compensated SCJ. Li Y-D et al. [9] studied the influence of an axially asymmetric shaped charge on the SCJ through a numerical simulation. Their conclusions showed that the detonation radius over the longitudinal axis—restricted by the charge radius over the same axis—as well as the detonation wave in the charge and the force acting on the liner, were affected, eventually influencing the jet velocity and shape. Wang Y et al. [10] obtained the morphology of the SCJ from shaped charges with square and circular cross-sections based on X-ray experiments. The results showed that the SCJ from the square cross-section shaped charge presented partially discrete phenomenon; they thought it could adjust the compaction rate of the SCJ from the square cross-section shaped charge through changing the inscribed circle diameter of the charge cross-section. In addition, Żochowski, Paweł et al. [11,12] researched and characterized the main parameters of the shaped charge jet formed and the penetration capability of the more contemporary high-hardness (500 HB) ARMSTAL 30PM steel armor based on a simulation and experiments. They showed that their numerical model of the warhead was defined more accurately than in previously published studies, since it was based on the real grenade dimensions and technical documentation.

While the influence of the asymmetry factors such as asymmetrical initiation, assembly error, etc., on the SCJ performance has been investigated extensively and some experiments related to the shaped charge with noncircular cross-section have been carried out, there is limited research exploring the formation characteristics of the SCJ from shaped charge with a trapezoid cross-section. Therefore, this paper aims to improve upon this work. Here, the X-ray experiments were carried out for the SCJ formation process of the shaped charge with two trapezoid cross-sections. A theoretical model for calculating the axial and radial velocities is then developed to describe the velocity characteristics of the SCJ.

2. Formation Characteristics of the SCJ from the Shaped Charge with a Trapezoid Cross-Section

2.1. Liner Collapsed Theory for the Shaped Charge with a Trapezoid Cross-Section

The SCJ formation for the shaped charge with a trapezoid cross-section is not an axisymmetric 2-dimensional problem, but a complex 3-dimensional problem. To explore the SCJ formation characteristics of the shaped charge with a trapezoid cross-section, the coordinate system as shown in Figure 1 was established, in which the rectangular coordinate system and the cylindrical coordinate system are all involved. In Figure 1, the highlighting represents the infinitesimal of the liner taken as the research object.

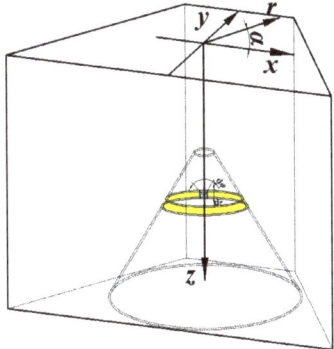

Figure 1. Three-dimensional system of coordinates in the shaped charge with a trapezoid cross-section.

The infinitesimal of the liner for the study is shown individually in Figure 2. In this model, $d\varphi$ and dz are the degree and the height along z axis. For the infinitesimal element, the mass can be expressed as:

$$dm = \rho_m \pi \left(r_e^2 - r_i^2\right) dz \frac{d\varphi}{2\pi} \tag{1}$$

where ρ_m is the density of liner material, and r_e and r_i are the external radius and the inner radius, respectively.

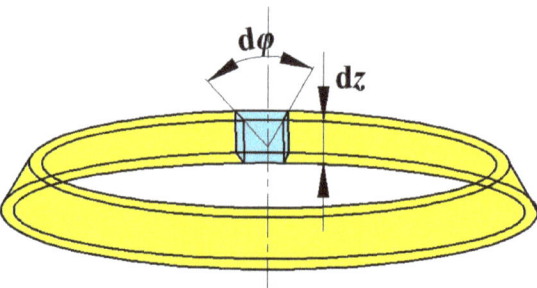

Figure 2. The infinitesimal element of the liner.

The process of the explosive driving the liner can be divided into two phases based on the detonation products action process. In the first stage, the shaped charge is detonated, the detonation wave propagates in the explosive with a spherical wave front, and the detonation wave arrives at the surface of the liner at time t_0. The state of the detonation product on the surface of the liner starts static to kinetic and is accompanied by the rarefaction wave. Furthermore, the infinitesimal element of the liner begins moving under the combined action of the detonation product and the rarefaction wave. In the second stage, the detonation wave propagates to the side of the charge, and a series of lateral rarefaction waves appear and diffuse to the detonation product. At time t_b, the lateral rarefaction waves spread to the surface of the liner, in which the reflected waves are also formed on the surface of the liner. The liner is collapsed to form the SCJ under the comprehensive function of many factors.

During the research, the initial coordinate of the liner infinitesimal element was set as (x_0, y_0, z_0), and t_0 and can be written:

$$t_0 = \frac{\sqrt{x_0{}^2 + y_0{}^2 + z_0{}^2}}{D_J} \tag{2}$$

where D_J is the detonation velocity of the explosive.

When the detonation wave arrives at the surface of the liner, the initial pressure on the surface of the liner is the C-J pressure of the explosive and can be obtained as follows [13,14]:

$$p_0 = \frac{1}{k+1}\rho_0 D_J{}^2 \tag{3}$$

where k and ρ_0 are the adiabatic exponent and the initial density of the explosive, respectively.

The initial state of the detonation products on the surface of the liner can be described as [15]:

$$\begin{cases} u_0 = 0 \\ c_0 = \frac{1}{2}D_J \end{cases} \tag{4}$$

where u_0 is the initial velocity of the detonation product particle and c_0 is the sound velocity of the detonation product.

This assumes that the motion of the detonation product on the surface of the liner is an isentropic process [16] and γ is the isentropic exponent. The pressure of the detonation product on the surface of the liner can be derived as follows:

$$p(t) = p_0 \cdot \left(\frac{c}{c_0}\right)^{\frac{2\gamma}{\gamma-1}} \tag{5}$$

where c is the sound velocity of the detonation product on the surface of the liner at time t.

Furthermore, the motion equation of the liner infinitesimal element can be linked to the pressure of the detonation product on the basis of Newton's Second Law, in which it can be given as:

$$dm\frac{dV}{dt} = p(t)ds \tag{6}$$

where ds is the area of the liner infinitesimal element contacting with the detonation product and V is the velocity of the liner infinitesimal element.

2.2. Calculation Model for the State Parameters of the Detonation Product on the Liner Surface

The shaped charge with a trapezoid cross-section has a 3-dimensional structure, and the profile of the shaped charge was selected as the research object to carry out the theoretical analysis. In the process of the liner being collapsed, the rarefaction wave is a critical factor.

When the detonation product is not influenced by the rarefaction wave, the distribution of the detonation product can be expressed as in Figure 3.

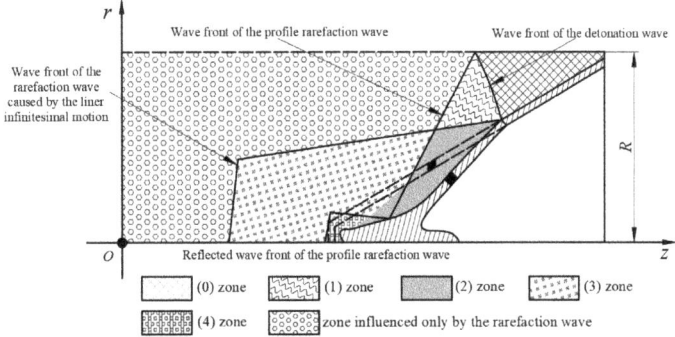

Figure 3. Distribution of the detonation product that is not influenced by the rarefaction wave.

The state of the detonation product on the surface of the liner can be rewritten as:

$$\begin{cases} u - \frac{2}{\gamma-1}c = -\frac{2}{\gamma-1}c_0 \\ x = (u+c)(t-t_0) + F_2(u,c) \end{cases} \tag{7}$$

where u is the velocity of the detonation product particle and $F_2(u, c)$ is a function related to the motion characteristic of the liner infinitesimal element.

Considering that the velocity of the detonation product particle on the surface of the liner is equal to the motion velocity of the corresponding liner infinitesimal element, it can be derived as follows:

$$\frac{dc}{dt} = \frac{\gamma-1}{2}\frac{ds}{dm}p_0c_0^{\frac{2\gamma}{\gamma-1}}c^{\frac{2\gamma}{\gamma-1}} \tag{8}$$

Therefore, the sound velocity of the detonation product is obtained from the time integration of Equation (8):

$$c = (\frac{\gamma+1}{2}\frac{ds}{dm}p_0c_0^{\frac{2\gamma}{\gamma-1}}(t-t_s) + c_0^{\frac{1+\gamma}{1-\gamma}})^{\frac{1-\gamma}{1+\gamma}} \tag{9}$$

where t_s is the time of the rarefaction wave entering the detonation product.

The detonation wave propagates to the profile of the charge, and the rarefaction wave is formed to start to spread into the detonation product. The appearance of the rarefaction wave influences the collapsing process of the liner. The diagram of the arising moment of the rarefaction wave is shown in Figure 4.

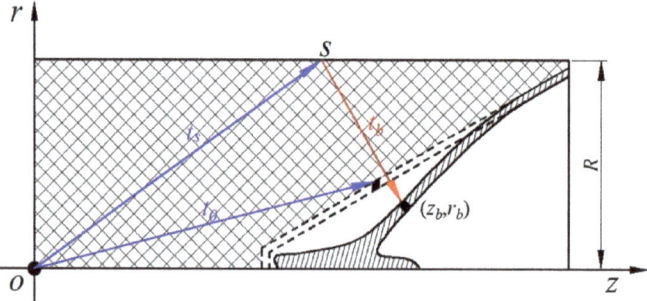

Figure 4. Calculation for the state-transforming moment of the detonation product.

The shortest path of the profile of the rarefaction wave arriving at the liner infinitesimal element is perpendicular to the generatrix of the liner (red line in Figure 4) and the profile rarefaction wave enters from s at time t_s. This rarefaction wave arrives to the surface of the liner at time t_b. Based on the geometrical relationship from Figure 4, the coordinates of the liner infinitesimal element at time t_b can be calculated as:

$$\begin{cases} z_b = z_0 + \int_{t_0}^{t_b} V(t) \sin(\alpha + \delta(t)) dt \\ r_b = r_0 - \int_{t_0}^{t_b} V(t) \cos(\alpha + \delta(t)) dt \end{cases} \tag{10}$$

where 2α is the cone angle of the liner, (z_0, r_0) is the initial coordinate of the liner infinitesimal element in *roz* coordinate system, and $V(t)$ and $\delta(t)$ are the velocity and direction angle of the liner infinitesimal element at time t, respectively.

After derivation, t_b can be expressed as:

$$t_b = \frac{\sqrt{[z_b - (\sqrt{R - r_b}) \tan \alpha]^2 + R^2}}{D_J} + \frac{R - r_b}{c_s \cos \alpha} \tag{11}$$

where R is the charge radius in section and c_s is the velocity of the rarefaction wave. This can be also written as:

$$c_s = \frac{k}{k+1} D_J \tag{12}$$

Due to the emergence of the rarefaction wave, the collapsing process of the liner is influenced. The distribution of the detonation product with the influence of the rarefaction wave is shown in Figure 5.

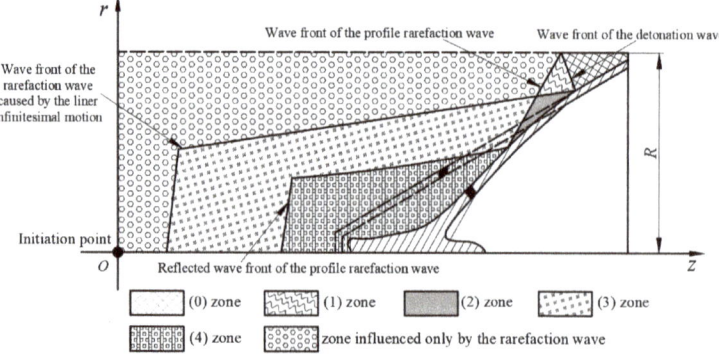

Figure 5. Distribution of the detonation product influenced by the rarefaction wave.

The state of the detonation product under the action of the rarefaction wave can be described as:

$$\begin{cases} x = (u - c)(t - t_s) + x_s \\ x = (u + c)(t - t_0) + F_4(u, c) \end{cases} \tag{13}$$

where x_s is the distance between the incoming point of the rarefaction wave and the initial position of the liner infinitesimal element, and $F_4(u, c)$ is a function related to the motion characteristic of the liner infinitesimal element.

Combining $\frac{dx}{dt} = u = V$, the sound velocity of the detonation product on the liner surface can be derived as follows:

$$c = \left(E(t - t_s)^{\frac{\gamma+1}{\gamma-1}} - \frac{\gamma+1}{2} \frac{ds}{dm} p_0 c_0^{\frac{2\gamma}{\gamma-1}} (t - t_s) \right)^{\frac{1-\gamma}{1+\gamma}} \tag{14}$$

where E is the integration constant, which depends on the parameters of the detonation product from II zone.

2.3. Velocity Analysis of the SCJ from the Shaped Charge with a Trapezoid Cross-Section

Chou [17] indicated that the collapsing velocity of the liner along its generatrix was variational, as is shown in Figure 6a. In this figure, V is the velocity value of the liner infinitesimal element, δ is the angle between the motion direction and the normal of the liner generatrix, and θ is the angle between the initial tangent and the current tangent on the liner surface. In time dt, the velocity vector of the liner infinitesimal element, $\overrightarrow{V}(t)$, has an increment $d\overrightarrow{V}$, and the velocity of the liner infinitesimal element is marked as $\overrightarrow{V}(t + dt)$ at time $t + dt$.

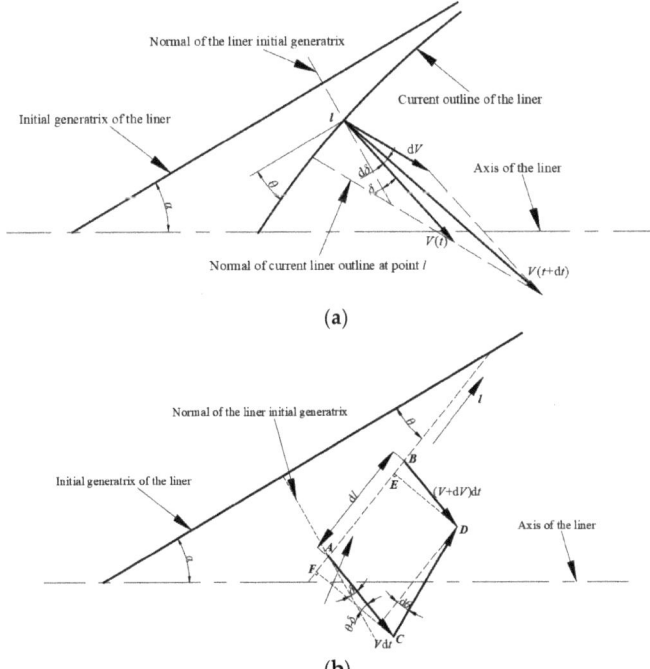

(a)

(b)

Figure 6. Velocity vector of the liner infinitesimal element. (**a**) Velocity vector of the liner infinitesimal element; (**b**) Relationship between deflection angle and velocity for the liner infinitesimal element.

This assumes that the detonation product always acts vertically on the surface of the liner infinitesimal element, so the increment value of the velocity can be obtained as:

$$dV = \left|\overrightarrow{dV}\right| \cos(\theta - \delta) \tag{15}$$

The relation between θ and the velocity of the liner infinitesimal element is shown in Figure 6b.

According to Figure 6, this can be obtained as:

$$d\delta = \frac{\left|\overrightarrow{dV}\right|}{V} \sin(\theta - \delta) \tag{16}$$

In the process of the research, $\theta(l)$ was defined as the angle between the tangent line at A and the initial generatrix of the liner. According to the geometrical relationship shown in Figure 6b, this can be derived from:

$$\frac{d\theta}{dt} = -\frac{dV}{dl} \cos(\theta - \delta) \tag{17}$$

The collapsing characteristics of the liner can be described through collapsing parameters of the liner infinitesimal element, such as collapsing velocity V and ejection angle δ.

Considering the pressure model, the expression can be derived as follows:

$$\frac{\left|\overrightarrow{dV}\right|}{dt} = \frac{ds}{dm}p \tag{18}$$

This equation integrates Equation (15), and the velocity of the liner infinitesimal element at time t can be rewritten as:

$$V = \int_0^t \frac{ds}{dm} p \cos(\theta - \delta) dt \tag{19}$$

Combining Equations (16) and (18), we can obtain the expression:

$$\frac{d\delta}{dt} = \frac{ds}{dm} p \frac{\sin(\theta - \delta)}{V} \tag{20}$$

Based on the above analysis, we can calculate the parameters V, δ, and θ, through which the collapsing process of the liner can be further defined.

For the shaped charge with the trapezoid cross-section, the collapsing velocities of the liner infinitesimal elements at the same height of the liner are inconsistent. During the analysis, the *roz* plane is a symmetry plane in a coordinate system. As is shown in Figure 7, elements A and B are symmetrically-distributed liner infinitesimal elements about the *roz* plane.

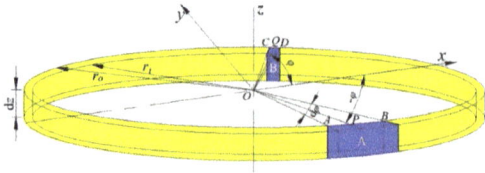

Figure 7. Symmetrically-distributed liner infinitesimal elements about the *roz* plane.

This takes the annular element as the research object, and dM_j and dM_s represent the mass of the SCJ and the slug from the element, respectively. Thus, the expression can be given as:

$$dM_j = \sum_{i=1}^{N} dm_{ji}$$
$$dM_s = \sum_{i=1}^{N} dm_{si}$$

(21)

where $N = \pi/d\varphi$ and $d\varphi$ are the angles of the liner infinitesimal element, as shown in Figure 7.

Considering the law of the conservation of momentum [18], the axial and the radial velocities of the SCJ and the slug from the annular liner element can be derived as follows:

$$V_{jz} = \frac{\sum_{i=1}^{N} dm_{ji} v_{jzi}}{\sum_{i=1}^{N} dm_{ji}}, \; V_{sz} = \frac{\sum_{i=1}^{N} dm_{si} v_{jsi}}{\sum_{i=1}^{N} dm_{si}}$$
$$V_{jx} = \frac{\sum_{i=1}^{N} 2dm v_{xi}}{dM_J + \frac{V_{sz}}{V_{jz}} dM_s}, \; V_{sx} = \frac{\sum_{i=1}^{N} 2dm v_{xi}}{dM_J + \frac{V_{jz}}{V_{sz}} dM_s}$$

(22)

2.4. Numerical Simulation

In this work, the shaped charge with the trapezoid cross-section was different from the cylindrical-shaped charge, so the diameter of the liner smaller than the diameter of the charge (subcaliber) appears. During the simulation, the Φ112 mm shaped charge was carried out to verify the correctness of the pressure model for the subcaliber shaped charge. The structure of the Φ112 mm shaped charge is shown in Figure 8.

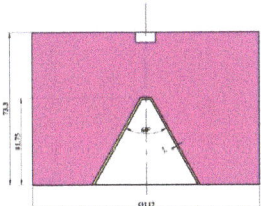

Figure 8. Structure of the Φ112 mm shaped charge.

According to the structure of the shaped charge from Figure 8, the simulation model was built through AUTODYN. In the model, the material of the liner is the oxygen-free high-conductivity copper (OFHC) and the charge is the CMOP B explosive. For obtaining the collapsing velocity in different liner infinitesimal elements, the Gaussian points (a total of 42) were set, as is shown in Figure 9.

Figure 9. Simulation model of the Φ112 mm shaped charge.

Based on the model built, the collapsing process of the liner at a typical moment can be acquired, as shown in Figure 10. These are the collapsing states of the liner at the initial moment, 10 μs, 20 μs and 30 μs after charge detonation. In the deformation process of the liner, the Gaussian points move as the material flows, and they can record the parameter variation of the liner infinitesimal elements.

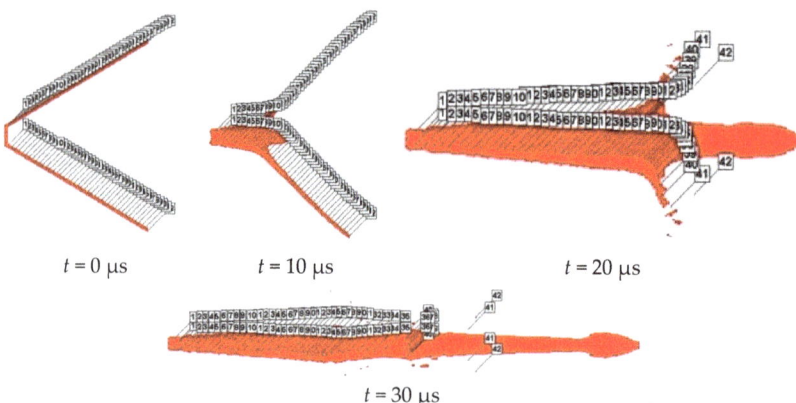

$t = 0$ μs $t = 10$ μs $t = 20$ μs

$t = 30$ μs

Figure 10. Liner collapsed process for the Φ112 mm shaped charge.

According to the simulation results, the Gaussian points numbered as 5, 15, 25, and 35 were selected to capture the collapsing velocity of the corresponding infinitesimal element, in which the collapsing velocities from the simulation were compared with calculated results employed by the pressure model. These the results are shown in Figure 11, which indicates that the theoretical results correlated with the simulation results reasonably well. Therefore, the pressure model employed to calculate the velocity for the subcaliber shaped charge is feasible.

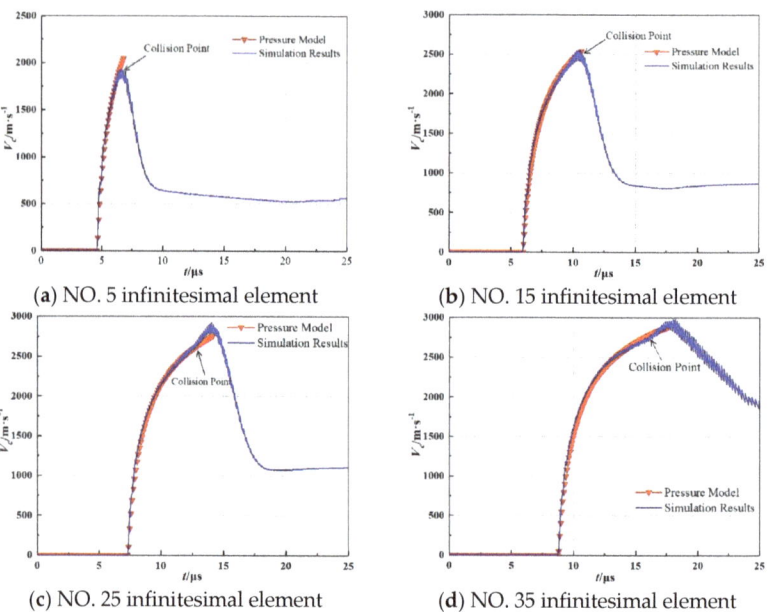

(**a**) NO. 5 infinitesimal element   (**b**) NO. 15 infinitesimal element

(**c**) NO. 25 infinitesimal element   (**d**) NO. 35 infinitesimal element

Figure 11. Acceleration process comparison between the calculation and the simulation for different liner infinitesimal elements.

During the analysis, the liner infinitesimal elements numbered 5, 15, 25, and 35 were selected to compare the acceleration process from the theoretical calculation and the simulation. The infinitesimal elements selected reflect the overall collapsing process of the liner. As shown in Figure 11, the theoretical calculation adopting the pressure model was consistent with the simulation results for the collision velocity of the liner, which indicates that the pressure model for calculating the collapsing velocity of the liner is feasible.

The ultimately collapsing velocities from the calculation and the simulation are shown in Figure 12, which indicates a reasonably good correlation, especially for the first 32 infinitesimal elements. In addition, some variation between the theoretical and the simulation results of the collapsing velocity appear after the NO. 32 infinitesimal element. The reasons for the discrepancies are that the influence of the rarefaction waves is not considered in the theoretical model, and the influence is more obvious near the liner bottom. In total, the ultimately collapsing velocities from the calculation are consistent with those from the simulation; the pressure model was employed to calculate the ultimately collapsing velocities of the liner, which proved to be reasonable.

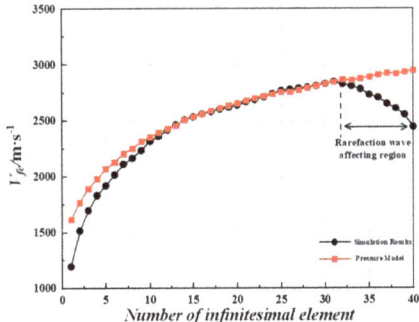

Figure 12. Ultimately collapsing velocity of liner infinitesimal elements from the pressure model and the simulation.

2.5. Results and Discussion

During the research, the shaped charge with the isosceles trapezoid cross-section was analyzed and used to explore the influence of the non-axial symmetry of the charge on the formation characteristics of the SCJ. In the shaped charge with the isosceles trapezoid cross-section, the explosive employed was Comp. B and the material of the liner was oxygen-free high-conductivity copper (OFHC). The structure of the liner is shown in Figure 13.

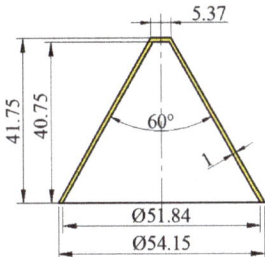

Figure 13. Structure of the liner used in this study.

For the shaped charges used in the calculation, the cross-sections are shown in Figure 14. To research the influence of the variation of the acute angle (base angle of the trapezoid) on the SCJ formation, the structure of the shaped charges with the different base angles of the trapezoid was calculated.

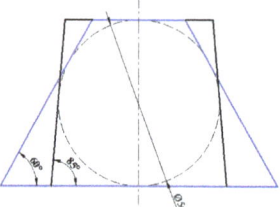

Figure 14. Planform of the shaped charge with different acute angles.

For the shaped charge, the inscribed circle diameter of the trapezoid cross-section is 56 mm, and the height of the charge is 73.3 mm.

Based on the theoretical model established in this work, the axial velocity of the SCJ with different base angles of the trapezoid was calculated, and the result is shown in Figure 15. According to the axial velocity of the shaped charge jet formed by different liner infinitesimal elements, the distance between the original position on the liner and cone axis is not large enough for the elements to accelerate to their final collapse velocity in the region near the cone apex of the liner, and the inverse velocity gradient of the SCJ elements from the apex end of the liner leads to the accumulation in SCJ tip. During the calculation, the acute angle of the trapezoid increased in intervals of 5° from 60° to 85°, and the SCJ axial velocity changed from 6636 to 6533 m/s, which is not a significant change.

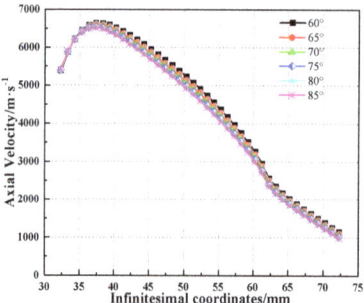

Figure 15. Variation of the SCJ axial velocity with infinitesimal coordinates under different base angles of the trapezoid.

The SCJ radial velocity is an important factor affecting the SCJ stability, so the SCJ radial velocities were calculated based on the theoretical model for different charge structures. The SCJ radial velocities were also obtained, as shown in Figure 16.

According to Figure 16, the variation of the SCJ radial velocity due to the effect of the variation of the base angle of the trapezoid is obvious. The maximum of the SCJ radial velocity was 461.6 m/s for the acute angle of the trapezoid being 60°, which decreased to 408.2, 347.3, 279.9, 208.3 and 105.0 m/s when the base angle increased by 5° from 65° to 85°, respectively. Figure 14 shows that as the base angle changed from 65° to 85°, the asymmetry of the charge gradually weakened, and the symmetry of the detonation waves produced by the detonation of the explosive enhanced gradually.

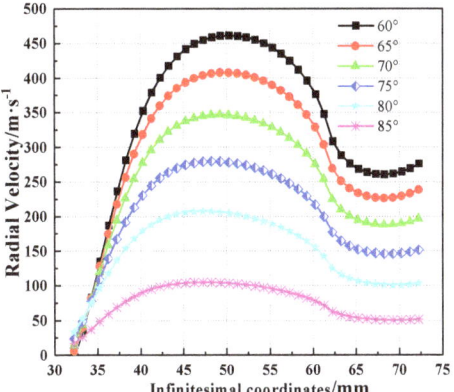

Figure 16. Variation of the SCJ radial velocity with infinitesimal coordinates under different base angles of the trapezoid.

3. Experiments and Results

3.1. The Shaped Charge with a Trapezoid Cross-Section Used in the Experiments

The cross-section of the shaped charge used in the experiments was a trapezoid with a length of 73 mm. In the experiments, the SCJ formations of the shaped charge with the acute angles of 60° and 75° were filmed, in which the thickness of the conical copper liner was 0.8 mm and the cone angle was 60°. In addition, the explosive used was a Comp. B high explosive, and a fusion cast process was adopted. The structures and the physicals are shown in Figure 17.

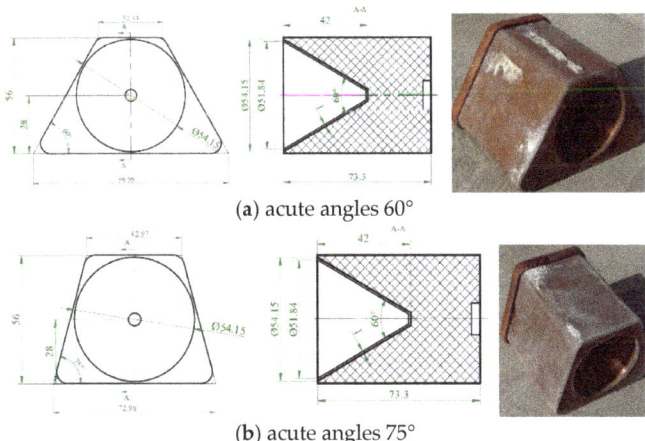

(a) acute angles 60°

(b) acute angles 75°

Figure 17. Shaped charge with a trapezoid cross-section.

Two 450 kV flash X-ray exposures were used to visualize the shape of the SCJ produced by the shaped charge with a trapezoid cross-section. The four films protected by a protective cassette were used to capture images of the SCJ. To calculate the SCJ velocity based on the data from the film, a mark was made in the film protected cassette, which can give a benchmark for the SCJ tip and tail at different times. The diagram of the experimental setup is shown in Figure 18. The exposure time of the flash X-ray in experiments was set as 30 and 40 μs after the jet tip arrived at the bottom of the liner for the shaped charge with the acute angles of 60°, and 40 and 50 µs for the shaped charge with the acute angle of 75°, respectively.

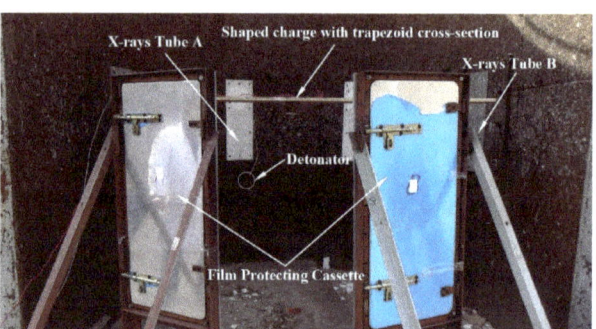

Figure 18. Experimental setup for the X-ray test.

3.2. Results and Discussion

To obtain the characteristics of the SCJ produced by the shaped charge with a trapezoid cross-section, the X-ray experiments were carried out. The flash X-ray radiographs of the SCJ are shown in Figure 19. In the process of the experiments, the bottom of the shaped charge was used as a reference point, and the moment of the detonation wave reaching the bottom of the shaped charge was time zero. For catching the morphological characteristics and kinetic parameters, the X-ray radiographs at 27.7 and 38.2 µs were finally captured for the SCJ produced by the shaped charge with cross-section acute angle of 60°, and were 36.9 and 52.4 µs for the cross-section acute angle of 75°.

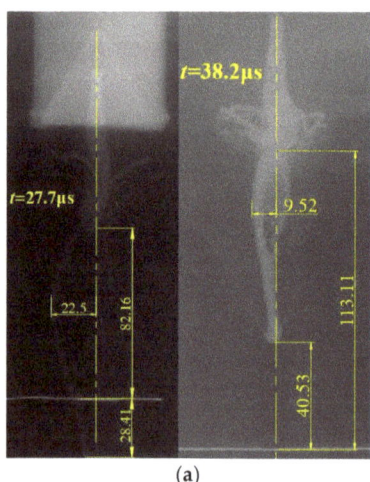 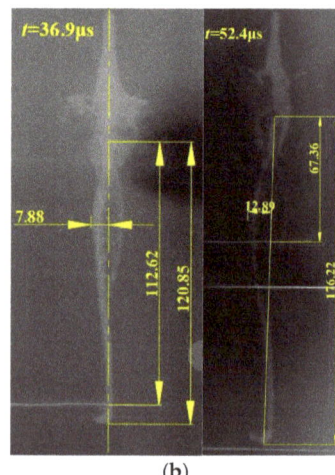

$$(a) \qquad (b)$$

Figure 19. X-ray image of the SCJ from the shaped charge with a trapezoid cross-section. (**a**) SCJ produced by the shaped charge with an acute angle of 60°, (**b**) SCJ produced by the shaped charge with an acute angle of 75°.

It should be noted that the variation between the setting exposure time and the real exposure time exists due to control precision of the X-ray system, in which the variations of the exposure time present no effect on the analysis of the experimental results. During the calculation of the related parameters, the real exposure time was used. The related measurement parameters based on the flash X-ray radiographs are shown in Table 1.

Table 1. Experimental results.

No.	t_e/μs	Z_{tip}/mm	Z_{tail}/mm	R/mm	v_{tip}/m·s^{-1}	v_{tail}/m·s^{-1}	v_R/m·s^{-1}	
1	27.7	40.53	113.11	9.52	6570	2948	343.68	466.35
2	38.2	−28.41	82.16	22.5			589.00	
1	36.9	−8.23	112.62	7.39	6490	2920	213.55	229.77
2	52.4	−108.91	67.36	12.89			245.99	

According to the X-ray radiographs, the axial and radial velocities were calculated. The axial velocities of the SCJs from the shaped charge with the acute angles of 60° and 75° are 6.57 and 6.49 m/s, and the theoretical values are 6636 and 6558 m/s, respectively. The errors between these values were 1% and 1.05%, respectively. In addition, the average radial velocities were 466.35 and 279.9 m/s for these two angles, and the theoretical values were 461.6 and 229.77 m/s, respectively. The errors between these values are 1.02% and 17.9%, respectively. The measurement error may contribute to the variation observed. Based on the outcome analysis, the theoretical results correlate with the experimental results reasonably well.

4. Conclusions

The formation characteristics of the SCJ from the shaped charge with a trapezoid cross-section was researched on the basis of the theory and X-ray experiments. Based on the research output presented above, the following conclusions may be drawn:

1. A theoretical model was employed to describe the formation characteristics of the SCJ from the shaped charge with a trapezoid cross-section. Based on the model, we analyzed the axial and radial velocities of the SCJs produced by the trapezoid cross-section shaped charge with different acute angles.
2. The acute angle of the trapezoid increased in intervals of 5° from 65° to 85°, and the SCJ axial velocity changed from 6636 to 6533 m/s, which is not a significant change.
3. The maximum of the SCJ radial velocity was 461.6 m/s for the acute angle of the trapezoid being 60°, which decreased to 408.2, 347.3, 279.9, 208.3 and 105.0 m/s when the base angle increased in intervals of 5° from 65° to 85°, respectively. This indicated that the asymmetry of the charge is gradually weakened as the acute angle increases, and the symmetry of the detonation waves produced by the detonation of the explosive is gradually enhanced.
4. The X-ray experiments were conducted to verify their theoretical validity. The results showed that the axial velocities of the SCJs from the shaped charge with the acute angles of 60° and 75° were 6570 and 6490 m/s, and the errors with the theoretical values were 1% and 1.05%, respectively. The average radial velocities were 466.35 and 279.9 m/s for those two cases, and the errors were 1.02% and 17.9%, respectively. This provides reasonable support for the current theory.

Author Contributions: Supervision, B.M. and Z.H.; project administration, B.M. and Z.H.; funding acquisition investigation, B.M. and Z.H.; validation, B.M. and Z.H.; writing—review B.M. and Z.H.; editing of the final version, B.M. and Z.H.; methodology, B.M., Y.W.(Yuting Wang) and X.J.; software, B.M., Y.W.(Yuting Wang) and X.J.; resources and formal analysis, Y.W.(Yongzhong Wu) and G.G. All authors have read and agreed to the published version of the manuscript.

Funding: This research was funded by [National Natural Science Foundation of China] grant number [11972196] and [Youth Fund of Jiangsu Natural Science Foundation] grant number [BK20190433].

Institutional Review Board Statement: Not applicable.

Informed Consent Statement: Not applicable.

Data Availability Statement: Not applicable.

Conflicts of Interest: The authors declare no conflict of interest.

References

1. Jing, D.; He, B.; Silin, L.; Yi, Z.; Cen, W.; Yao, W. Research on the Intelligent Combat of Cruise Missile. In Proceedings of the 2019 5th International Conference on Computing and Data Engineering, Association for Computing Machinery, Shanghai, China, 4–6 May 2019; pp. 11–14.
2. Feng, Y.; Tang, W.; Gui, Y. Aerodynamic configuration optimization by the integration of aerodynamics, aerothermodynamics and trajectory for hypersonic vehicles. *Chin. Sci. Bull.* **2014**, *59*, 4608–4615. [CrossRef]
3. Dong, H.; Liu, Z.; Wu, H.; Gao, X.; Pi, A.; Huang, F. Study on penetration characteristics of high-speed elliptical cross-sectional projectiles into concrete. *Int. J. Impact Eng.* **2019**, *132*, 103311. [CrossRef]
4. Ding, L.; Li, Z.; Lu, F.; Li, X. Rapid assessment of the spatial distribution of fragments about the D-shaped structure. *Adv. Mech. Eng.* **2018**, *10*, 1687814018777594. [CrossRef]
5. Guo, Z.W.; Huang, G.Y.; Liu, H.; Feng, S.S. Fragment velocity distribution of the bottom part of d-shaped casings under eccentric initiation. *Int. J. Impact Eng.* **2020**, *144*, 103649. [CrossRef]
6. Ma, B.; Huang, Z.; Guan, Z.; Jia, X.; Xiao, Q.; Zu, X. Theoretical analysis of the acceleration effect of the magnetic field on the shaped charge jet. *Int. J. Mech. Sci.* **2017**, *133*, 283–287. [CrossRef]
7. Jia, X.; Huang, Z.X.; Zu, X.D.; Gu, X.H.; Xiao, Q.Q. Theoretical analysis of the disturbance of shaped charge jet penetrating a woven fabric rubber composite armor. *Int. J. Impact Eng.* **2014**, *65*, 69–78. [CrossRef]
8. Stewart, B.; Lewtas, I. Integration of an Over-Fly Top Attack Shaped Charge Jet within a Kinetic Energy Penetrator for Enhanced Multiple-Effects. In Proceedings of the 30th International Symposium on Ballistics, Long Beach, CA, USA, 11–15 September 2017; Chocron, S., Ed.; DEStech Publications, Inc.: Arlington, TX, USA, 2017; pp. 1656–1670.
9. Li, Y.D.; Dong, Y.S.; Feng, S. Numerical simulation of the influence of an axially asymmetric charge on the impact initiation capability of a rod-like jet. *Combust. Explos. Shock. Waves* **2012**, *48*, 713–717. [CrossRef]
10. Wang, Y.; Huang, Z.; Zu, X.; Ma, B.; Cai, Y.E. Experimental Study on Jet Formation and Penetration Characteristics of Square cross-section Shaped Charge. *Lat. Am. J. Solids Struct.* **2021**, *18*. [CrossRef]
11. Żochowski, P.; Warchoł, R.; Miszczak, M.; Nita, M.; Pankowski, Z.; Bajkowski, M. Experimental and Numerical Study on the PG-7VM Warhead Performance against High-Hardness Armor Steel. *Materials* **2021**, *14*, 3020. [CrossRef] [PubMed]
12. Żochowski, P.; Warchoł, R. Experimental and numerical study on the influence of shaped charge liner cavity filing on jet penetration characteristics in steel targets. *Def. Technol.* **2022**. [CrossRef]
13. Lorenz, K.T.; Lee, E.L.; Chambers, R. A simple and rapid evaluation of explosive performance–The disc acceleration experiment. *Propellants Explos. Pyrotech.* **2015**, *40*, 95–108. [CrossRef]
14. Higa, Y.; Iyama, H.; Shimojima, K.; Higa, O.; Itoh, S. Numerical Simulation for Soil Surface Explosion Problem-A study of fragments controlling effect using liner plate application. In Proceedings of the 2018 Third International Conference on Engineering Science and Innovative Technology (ESIT), North Bangkok, Thailand, 19–22 April 2018; pp. 1–4.
15. Urtiew, P.A.; Hayes, B. Parametric study of the dynamic JWL-EOS for detonation products. *Combust. Explos. Shock. Waves* **1991**, *27*, 505–514. [CrossRef]
16. Wojewodka, A.; Witkowski, T. Methodology for simulation of the jet formation process in an elongated shaped charge. *Combust. Explos. Shock. Waves* **2014**, *50*, 362–367. [CrossRef]
17. Chou, P.C.; Carleone, J.; Flis, W.J.; Ciccarelli, R.D.; Hirsch, E. Improved formulas for velocity, acceleration, and projection angle of explosively driven liners. *Propellants Explos. Pyrotech.* **1983**, *8*, 175–183. [CrossRef]
18. Brown, J.; Curtis, J.P.; Cook, D.D. The formation of jets from shaped charges in the presence of asymmetry. *J. Appl. Phys.* **1992**, *72*, 2136–2143. [CrossRef]

Article

Study on Penetration Mechanism of Shaped-Charge Jet under Dynamic Conditions

Yizhen Wang, Jianping Yin *, Xuepeng Zhang and Jianya Yi

College of Electromechanical Engineering, North University of China, Taiyuan 030051, China
* Correspondence: yjp123@nuc.edu.cn; Tel.: +86-13994208931

Abstract: Aiming at the dynamic penetration process of a shaped-charge jet, we proposed a mathematical model for the penetration of a jet under dynamical conditions based on the theory of virtual origin and the Bernoulli equation taking into account the jet and target intensities. The dynamic penetration process of the jet was divided according to the penetration channel of the jet into the static target. The dynamic penetration model of the jet based on the unperturbed section and perturbed section was established. The penetration depth variation in the shaped-charge jet vertically penetrating target plates with different moving speeds (150~400 m/s) was analyzed by finite element software. The dynamic penetration model shows that with the increase in the target moving speed, the disturbed time of the jet continuously advances, and the dynamic penetration depth continuously decreases; as the velocity of the target increases, the penetration length of the unperturbed jet decreases and then becomes stable, while the penetration length of the perturbed jet decreases. The results showed that the mathematical model is consistent with the finite element simulation, and that the mathematical model can effectively characterize the penetration depth of the unperturbed and disturbed jet portions, adequately explain the dynamic response behavior of the jet penetrating a moving target, and effectively predict the dynamic penetration depth of the jet under the influence of the target movement.

Keywords: jet; dynamic conditions; virtual origin; penetration mechanism

Citation: Wang, Y.; Yin, J.; Zhang, X.; Yi, J. Study on Penetration Mechanism of Shaped-Charge Jet under Dynamic Conditions. *Materials* **2022**, *15*, 7329. https://doi.org/10.3390/ma15207329

Academic Editors: Chuanting Wang, Yong He, Wenhui Tang, Shuhai Zhang, Yuanfeng Zheng and Xiaoguang Qiao

Received: 29 September 2022
Accepted: 18 October 2022
Published: 20 October 2022

Publisher's Note: MDPI stays neutral with regard to jurisdictional claims in published maps and institutional affiliations.

1. Introduction

With the development of military technology, a large number of destructive elements with strong penetration capabilities have emerged to destroy underground military targets reinforced by the enemy. Short- and medium-range air defense and antimissile defense are vital methods in protecting our own targets from attack [1]. How to effectively destroy incoming targets has been the main research direction of relevant scholars at home and abroad. Currently, antimissile countermeasures in various countries rely primarily on high-velocity fragmentation to intercept incoming targets, that is, using kinetic energy to destroy the damage element to interfere with the target, directly penetrate it, destroy its critical components or aerodynamic shape, and cause it to deviate from its established trajectory. Alternately, its internal charge is detonated by kinetic impact, pre-detonating the warhead and thus the antimissile. At present, it is commonly believed that directly detonating the incoming warheads is the best [2].

In recent years, the technology of deep-penetrating warheads has been continuously developed, and the thickness of their shells has been increasing. Meanwhile, alloyed steel with great strength and toughness has been used to increase penetration. The internal charge is also increasing, effectively damaging the target. Traditional fragmentation or kinetic bar warheads produce damage elements that do not penetrate the shell effectively due to their large velocity, thickness, and strength. Even if the shell is penetrated, its internal charge cannot be reliably detonated. This situation has posed a serious challenge to the protection of our strategic locations [3,4]. Therefore, it is necessary to use JET as a damage element to effectively damage the incoming target [5–7]. The shaped charge has superior penetration capability, as well as the ability to induce a warhead charge, which can be used for air defense and antimissile missions. The shaped jet uses the Monroe effect to generate elevated temperature and a high-speed shaped jet to damage armored targets [8,9]. As an anti-armor method that has been widely used for a long time, it is used in a variety of anti-armor applications, and there is a relatively well-established theoretical basis for the use of anti-armor in combat missions [10–13]. The existing research on jet penetration under dynamic conditions mainly focuses on the interaction between shaped-charge jets and explosive reactive armor (ERA). The thickness of the face and back plates of the explosive reactive armor is tiny, so the effects of jet velocity decay and jet break and deflection due to the motion of the target plates are normally ignored when the jet penetrates this thin target dynamically. Currently, the penetration mechanism of shaped-charge jets into a moving target with high-velocity thick walls is poorly investigated. Jia Xin et al. established theoretical models such as the lateral drift velocity of the jet when it is interfered with and the penetration depth of the disturbed jet based on the virtual source point theory and the differential element method [14]. Tian Lili and others found through finite element simulations that the higher the velocity of the moving steel target, the larger the velocity loss after jet penetration. In order to ensure the ability to strike high-speed moving targets, it is necessary to use high explosives as the shaped charge and reasonably design the explosive height [15]. According to the published literature at home and abroad, most of the researchers have conducted simulation research on a shaped-charge jet penetrating reactive armor [16–19], while the research on the penetration mechanism of a shaped-charge jet under dynamic conditions, especially on the penetration mechanism of thick-walled moving targets, has barely been published. Therefore, in this paper, based on the theory of virtual origin of shaped-charge jets and the analysis of finite element simulations of jet penetration in static conditions, we divide the jet penetration process into two phases: unperturbed and perturbed. A theoretical model of jet dynamical penetration was developed based on the Bernoulli equation taking into account the target and jet strengths. The calculation method of jet dynamic penetration depth is obtained, and the theoretical model is verified by simulation results.

2. Typical Penetration Theory

2.1. Static Armor-Piercing Theory Based on Virtual Origin

The PER theory shows that for a metal jet, there is always a velocity gradient in the length direction, the jet velocity decreases gradually from beginning to end, and the jet velocity distribution is typically linear [20]. Allison and Vitali [21] proposed that it can be assumed that there is a virtual origin, on which the velocity of each jet micro-element has a linear change in space.

According to the virtual origin theory, the jet micro-elements start from the virtual origin, which is the origin of all jets. The jet velocity is linearly distributed along the jet length, and the jet micro-element velocity does not alter during the movement [22]. Thus, the slope of the line formed by the point on the jet penetration curve and the imaginary origin as a function of time during the jet penetration is the velocity of the microscopic element at the jet head at that moment, as shown by the blue line in the Figure 1. The penetration curve is the red dotted line in the figure, and the slope of a point on this line is the penetration velocity of the jet to the target at the corresponding time, as shown in Figure 1 [23].

This theory has certain accuracy for a general single-cone-shaped liner. The formula for calculating the static penetration depth of a jet given by Allison and Vitali [21] is as follows.

$$P = v_j t_0 \left(\frac{v_0}{v_j} \right)^{(1+\gamma)/\gamma} - z_0 \tag{1}$$

or

$$P = v_0 t \left(\frac{t_0}{t}\right)^{\gamma/(1+\gamma)} - z_0 \tag{2}$$

According to Formulas (1) and (2)

$$v_{j0} = \frac{v_0 \cdot z_0{}^{\gamma}}{(P(t) + z_0)^{\gamma}} \tag{3}$$

where P is the penetration depth; v_{j0} is the velocity of the head of the jet at the intersection of the projectile and target; v_0 is the velocity of the head of the jet as it contacts the target plate; t_0 is the time at which the jet head moves from the virtual origin to the target surface; z_0 is the distance from the imaginary origin to the target surface; $\gamma = \sqrt{\frac{\rho_t}{\rho_j}}$, ρ_t is the target density and ρ_j is the jet density; and $P(t)$ is a function of the penetration depth P of time t to the jet.

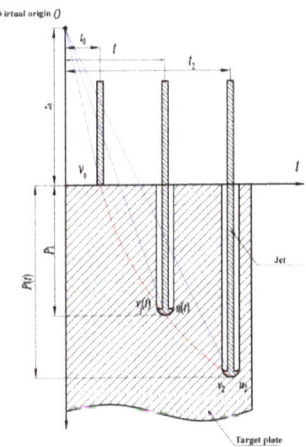

Figure 1. Jet penetration process diagram of virtual origin theory.

Regardless of the bullet/target strength, according to the Bernoulli equation:

$$\frac{1}{2}\rho_j(v_j - u)^2 = \frac{1}{2}\rho_t u^2 \tag{4}$$

where u is the penetration velocity. The penetration velocity can be expressed as:

$$u = \frac{v_0}{1+\gamma}\left(\frac{t_0}{t}\right)^{\frac{\gamma}{1+\gamma}} \tag{5}$$

Eichelberger thinks that the strength of the target plate and jet have an essential influence on the late process of penetration through experimental measurement, and adds an intensity term in Formula (4) [24]. The penetration velocity u of the jet is obtained as a function of the target impedance.

$$u = \frac{1}{1+\gamma}\left[v_j - \sqrt{\gamma^2 v_j^2 + (1-\gamma^2)\frac{2\sigma}{\rho_j}}\right] \tag{6}$$

where $\rho = \sigma_t - \sigma_j$, ρ is the difference between the plastic deformation impedance of the target and the jet.

2.2. *Analysis of Jet Aperture under Static Conditions*

In this part, the aperture of the jet penetrating the target in the quiescent state is analyzed by numerical simulations. In this paper, a truncated cone-shaped liner with a diameter of 100 mm and a charge length diameter ratio of two is selected, and its cone angle is 40°. The blast height is 100 mm, and its model is shown in Figure 2. For this finite element simulation analysis, we used the R11 LS-DYNA solver purchased by the school and single precision was used. The structural dimensions of the shaped charges and the partition of the model mesh are shown in Figure 3.

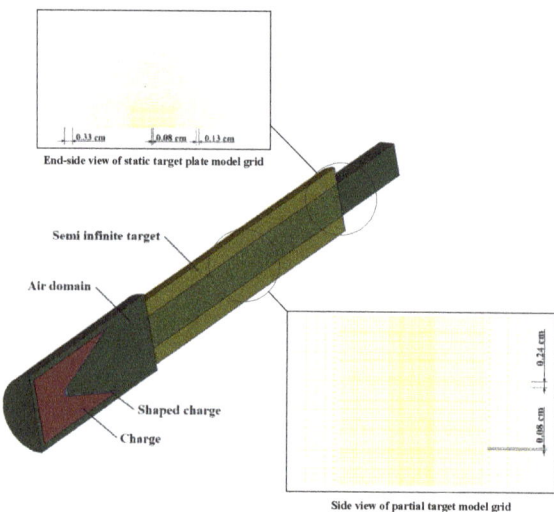

Figure 2. Schematic diagram of jet static penetration.

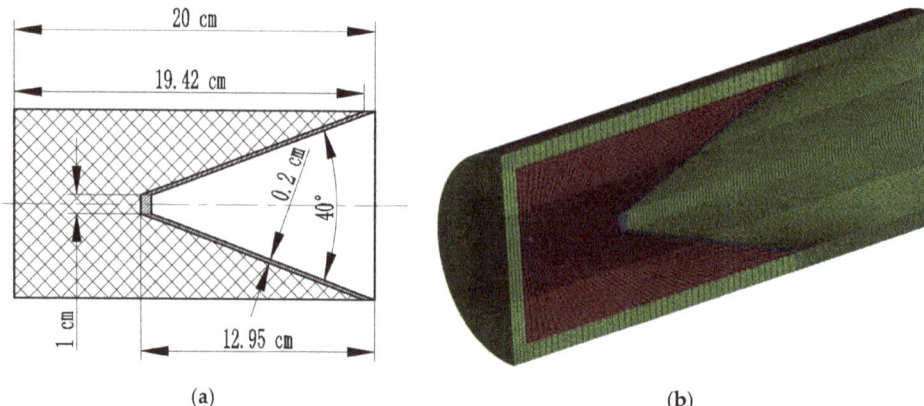

Figure 3. Dimension and grid diagrams of shaped jet. (**a**) Dimensional sketch of shaped jet; (**b**) grid generation of shaped jet.

The explosives are modeled by the HIGH_EXPLOSIVE_BURN constitutive model with a JWL equation of state. The material choice for the drug-type covers is copper and the target plate is 45 steel, and the JOHNSON_COOK model and the GRUNEISEN equation of state are used for the drug-type covers and target plates, with the parameters given in Tables 1 and 2.

Table 1. Explosive Material Parameters.

$\rho/(\text{g}\cdot\text{cm}^{-3})$	$D/(\text{m}\cdot\text{s}^{-1})$	A	B	R1	R2	ω
1.891	0.911	7.783	0.0707	4.2	1	0.3

Table 2. Material Parameters of Copper and 45 Steel.

Material	$\rho/(\text{g}\cdot\text{cm}^{-3})$	A/Mbar	B/Mbar	N	C
copper	8.96	0.0009	0.0292	0.31	0.025
45 steel	7.85	0.00507	0.0032	0.28	0.064

As shown in Figure 2, the static jet penetration adopts a half model, so in the simulation, symmetrical constraints are imposed on the symmetry plane, and the flow-out boundary is set for the air domain boundary.

In the finite element analysis of the static penetration of a jet into a semi-infinite target, to improve computational efficiency and avoid boundary effects affecting the accuracy of the analysis, a dense grid is partitioned for the part of the target that will be in direct contact with the jet. The mesh is a regular hexahedral cell with an edge length of 0.8 mm, and a larger mesh is used for the edges of the target. The two are connected by a transitional mesh.

When the jet first touches the target plate, the velocity of the micro-element at the jet head decreases, and the particle velocity of the target plate at the contact position remains the same. At this point, the percolating jet element has not yet consumed all the energy and mass, and the jet channel diameter is enlarged by the subsequent jet element boost.

During the pit-opening phase, the target plate near the impact point generates high-speed plastic deformation with a large strain rate. Therefore, in the "three high zones", the target plate near the jet channel has local hardening, especially in the pit-opening stage. Later, as the jet head velocity decreases, the hardening of the target plate gradually decreases, and the hardening of the target plate in the quasi-steady region is significantly lower than in the pit-opening phase. For semi-infinite targets, jets accumulate at the bottom of the hole during the termination phase. Due to the high temperature of the jet, the target plate in contact with it is tempered, which further reduces the intensity of the target plate in this part of the jet, and its hardness is lower than in the quasi-steady region.

At the beginning of the pit-opening phase, the diameter of the jet flowing through the surface of the ejecta target is about 0.66 cm. The jet produces a hole in the target surface with a diameter of about 2–2.5 times the jet diameter, which is about 1.47 cm. During the penetration process in this phase, the energy of the jet micro-element is large and the impedance of the target plate to the jet is relatively small. The diameter of the counterbore produced by the jet gradually increases, as shown in Figure 4 (the scale units in the figure are centimeters and similarly hereinafter), because the subsequent jet micro-element continuously compresses the previous jet micro-element in the radial direction of the jet.

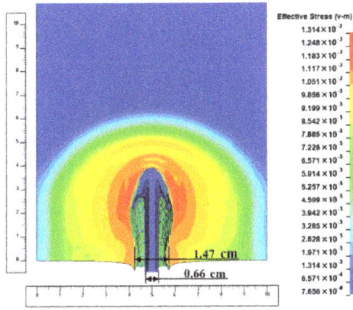

Figure 4. Schematic diagram of the ejecta aperture at the early stage of pit opening.

After that, the energy of each jet micro-element decreases continuously with the presence of the velocity gradient due to the decrease in the velocity of the subsequent jet micro-element and the relative decrease in the pressure at the contact position of the jet target plate. As a result, the remaining energy after the axial penetration of the jet is completed becomes smaller and smaller during the penetration of the target slab.

In addition, due to the severe plastic deformation of the target plate during the hole opening, a local hardening of the target plate is induced, which also increases the resistance of the target plate to the radial extension of the jet. Under the combination of the above two effects, the radial re-emission diameter of the jet decreases, as shown in Figure 5. Therefore, there is an extreme value of the jet reaming diameter during the pit-opening phase.

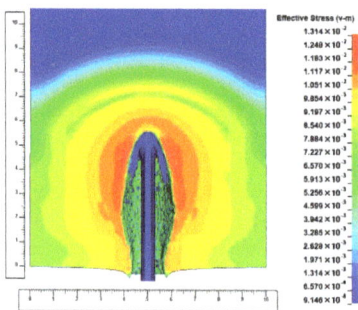

Figure 5. Schematic diagram of the jet aperture at the later stages of the pit-opening.

Since the hole digging process only accounts for a limited part of the total jet penetration process, the subsequent penetration enters the quasi-steady stage, and the energy gradient of the jet in this part is relatively slow and the residual energy of the jet micro-element in this part mainly remains stable after the completion of the axial penetration. Moreover, the amplitude of the material hardening in the quasi-steady region is inferior to that in the hole-opening phase, so that the aperture opened by the hole-expanding jet is mainly non-shifting with time and remains stable, as shown in Figure 6.

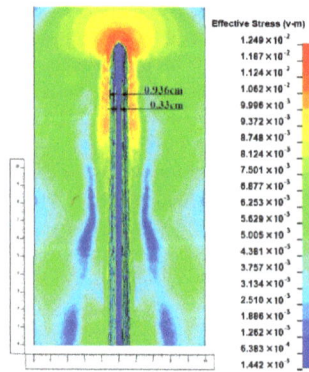

Figure 6. Schematic diagram of the jet aperture during the quasi-stationary phase of jet penetration.

In the quasi-steady phase, the diameter of the counterbore produced by the jet penetrating the target plate is about 0.936 cm, and the diameter of the jet at the corresponding position is about 0.33 mm, with an aperture about 2.5–3.0 times the diameter of the jet.

Thus, for a typical jet penetrating a homogeneous target, the reduction in the recoiling diameter is mainly present in the intermediate to quasi-steady phase of the pit-opening phase. Fundamentally, it first increases, then decreases and finally tends to stabilize. This is consistent with the equation of penetration and crater growth by a shaped-charge jet under the influence of a shock wave [25].

3. Dynamic Penetration Analysis of Jet

3.1. Determination of Interference Time

The target slab has a velocity v_t perpendicular to the penetration direction based on the theory of virtual origin and the jet penetration hole diameter law.

For the interference of the dynamic target plate to the jet, due to the strong role of jet reaming in the pit-opening stage, which provides favorable conditions for jet penetration, and due to the small opening of the jet on the surface of the target plate, and that in the early stage of the pit-opening stage the diameter of the reaming is constantly increasing, under dynamic conditions, the interference of target plate to jet appears near the surface of target plate. Then, the gap between the two, Δr, can be expressed as:

$$\Delta r = R_b - r_j \tag{7}$$

where r_b is the channel radius near the surface of the target plate, and r_j is the jet radius at the same position.

Then, the disturbed time t_h can be expressed as

$$t_h = \frac{\Delta r}{v_t} \tag{8}$$

where v_t is the velocity of the dynamic target plate.

Under the same operating conditions, when the parameters of the liner and the charge are in agreement, the jet formed by the liner is predominantly coherent, and the jet channel generated by its penetration into the static target is similar. Therefore, the interference time can be calculated based on the condition of the jet channel in static conditions.

3.2. Penetration Analysis of Disturbed Jet

Under the same operating conditions, when the parameters of the liner and the charge are in agreement, the jet formed by the liner is predominantly coherent, and the jet channel generated by its penetration into the static target is similar. Therefore, the interference time can be calculated based on the condition of the jet channel in static conditions. Assuming that the velocity distribution of the jet along the axis is linear, the velocity of the jet micro-element near the surface of the target slab can be obtained from Equations (2) and (3) as follows.

$$v_{jh} = \frac{z_0 \cdot v_{j0}\big|_{t=t_h}}{P\big|_{t=t_h} + z_0} \tag{9}$$

where v_{jh} is the velocity of the jet micro-element near the surface of the target slab; $v_{j0}\big|_{t=t_h}$ is the velocity of the jet head at the t_h moment; and $P\big|_{t=t_h}$ is the penetration depth at the t_h moment.

During the interference process, the fluid is assumed to be incompressible and inviscid. After the interferometric time, the pre-v_{jh} jet micro-elements have significant energy. After the completion of axial penetration, there is still sufficient energy to expand the target plate, so that Δr is greater than zero to overcome the influence of target movement on jet micro-element penetration; the micro-element of JET after v_{jh} will be firstly affected by the target plate movement due to its relatively little energy. The velocity of the jet itself in the direction of motion of the target slab is negligible. It is assumed that the target slab will collide with the jet micro-element where the v_{jh} is located out of the jet axis as a whole, and the post-v_{jh} jet micro-element can be regarded as re-penetrating the target slab.

Compared to the unperturbed part of the head, the velocity of the perturbed jet profile has less kinetic energy and the diameter of the jet profile is relatively larger. The position interfered by the movement of the target plate is mainly concentrated in the area from the near-surface of the target plate to the end of the pit-opening stage δ inside.

Then, the disturbed jet part is equivalent to the head velocity v_{jh}, and the thickness of the jet section pair with linear velocity gradient is δ. The interference part of the hole penetrates and then continues to flow to the bottom of the hole. After the penetration of the jamming part is completed, the remnant jet head velocity v_{ju} of the jet in this section is:

$$v_{ju} = \frac{v_{jh} \cdot z_h{}^{\gamma}}{(\delta + z_h)^{\gamma}} \tag{10}$$

where v_{ju} is the residual head velocity; v_{jh} is the jet velocity at the radial interference position of the target at t_h; z_h is the distance from the imaginary origin to the t_h moment of motion of the microscopic elements of the head of the jet; and δ is the distance from the near surface of the target plate to the end of the pit-opening stage.

For the jet in the perturbed part, it is necessary to find the part that still contributes to the tunneling bottom penetration. Therefore, the length of the jet section from the effective jet velocity v_{jk} to the residual head velocity v_{ju} of the disturbed jet Δl can be expressed as:

$$\Delta l = (1 - \frac{v_{jk}}{v_{ju}})[z_0 + P\big|_{t=t_h} + \delta] \tag{11}$$

where Δl is the length of the jet section from the effective jet velocity to the residual head velocity of the disturbed jet; v_{jk} is the effective jet velocity threshold:

$$v_{jk} = v_t{}^{\frac{1}{1+\gamma}} + \frac{\gamma}{1+\gamma} v_{ju} \tag{12}$$

Considering that the energy of the perturbed jet cross section is already predominantly consumed when it overcomes the perturbation and flows to the hole bottom, it is necessary to consider the effect of the jet and target strengths on the jet penetration. The penetration velocity u_r of the perturbed part is written according to Equation (6):

$$u_r = \frac{1}{1+\gamma}\left[v_{ju} - \sqrt{\gamma^2 v_{ju}{}^2 + (1-\gamma^2)\frac{2\sigma}{\rho_j}}\right] \tag{13}$$

where u_r is the penetration velocity of the perturbed part.

Since the velocity gradient of the jet in the disturbed section is relatively slow, v_{ju} can be considered as the velocity of the jet in this section, then the penetration depth of the disturbed section to the hole bottom ΔP_r is expressed as:

$$\Delta P_r = \frac{u_r \cdot \Delta l}{v_{ju} - u} \tag{14}$$

where ΔP_r is the penetration depth of the disturbed section.

3.3. Penetration Analysis of Undisturbed Jet

This part of the jet has already penetrated the target before the time of the perturbation. During the penetration of the jet into the target, it is the head part of the jet that touches the target. The energy carried by the jet is strong and the impact of the target slab on the jet is slight. The average velocity of the head and tail of the jet in this section is taken to be the velocity of the jet element in this section:

$$v_{jw} = \frac{v_{j0}\big|_{t=t_h} + v_{jh}}{2} \tag{15}$$

where v_{jw} is the velocity of the undisturbed jet; $v_{j0}|_{t=t_h}$ is the velocity of the jet head at the moment of t_h; and v_{jh} is the velocity of jet at disturbed location at the moment of t_h.

In order to better fit the actual situation, considering the influence of target plate and jet strength on the penetration process of the undisturbed jet, the penetration velocity u_w and penetration depth ΔP_w of undisturbed jet can be obtained according to Formulas (6) and (13)–(15):

$$u_w = \frac{1}{1+\gamma}\left[v_{jw} - \sqrt{\gamma^2 v_{jw}{}^2 + (1-\gamma^2)\frac{2\sigma}{\rho_j}} \right] \tag{16}$$

$$\Delta P_w = \frac{u_r \cdot P|_{t=t_h}}{v_{ju} - u_w} \tag{17}$$

where u_w is the penetration velocity of undisturbed jet; ΔP_w is the penetration depth of undisturbed jet.

Assuming that each section of the jet is ideally attached to the hole after the completion of radial penetration, and the accumulation of the jet at the hole bottom is negligible, the total penetration depth P of the jet to the moving target can be considered as the sum of the penetration depth of the disturbed section jet and the undisturbed section jet.

$$P = \Delta P_r + \Delta P_w \tag{18}$$

where P is the final dynamic penetration depth of the jet.

4. Finite Element Simulation Analysis and Verification

4.1. Model Establishment

The LS-DYNA finite element analysis software was used to establish the three-dimensional model of jet penetration into the dynamic target plate in order to obtain the parameters needed for theoretical calculation of jet head and tail velocity, velocity gradient, and jet diameter when the jet stretched to the surface of the target plate. The arbitrary Lagrangian Euler algorithm was used for explosive and shaped-charge liners, and the Lagrangian algorithm for target plates. Numerical simulations were performed between them via a fluid–solid coupling algorithm. The element grids of each section are hexahedral elements, and the explosive, shaped-charge liner, and air domain are common nodal Eulerian grids. The model diagram is shown in Figure 7.

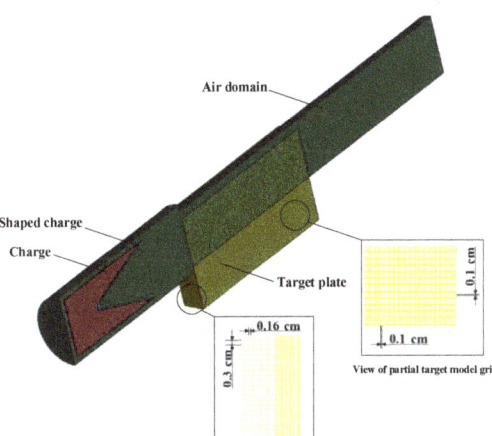

Figure 7. Schematic representation of the numerical simulation model.

The material model and equations of state used for the numerical simulations are the same as those used in the analysis of the jet penetration aperture in static conditions in

Section 1. In the dynamic penetration model, the symmetry constraint is also set on the symmetry plane of the half model, and the flow-out boundary is set on the air domain boundary, so that all substances can flow out freely. Consistent with the grid partition of the static penetration model, the dynamic target is also meshed with a transitional grid structure. The portion where the possible interaction with the jet in the target plate is divided into a regular hexahedron grid with a side length of 0.1 cm and distant portion is meshed in 0.3 cm × 0.3 cm × 0.16 cm.

4.2. Finite Element Simulation

In order to further verify, analyze, and study the penetration mechanism of the jet under dynamic conditions, under the condition of maintaining the blasting height and the same material parameters, the horizontal right velocity was applied to the target plate, which is 150 m/s, 200 m/s, 250 m/s, 300 m/s, 350 m/s, and 400 m/s. Combined with numerical simulations, the penetration of the jet into the target slab at different moving velocities was analyzed and the influence law of various factors was explored. The results of the numerical simulations are shown in Figure 8.

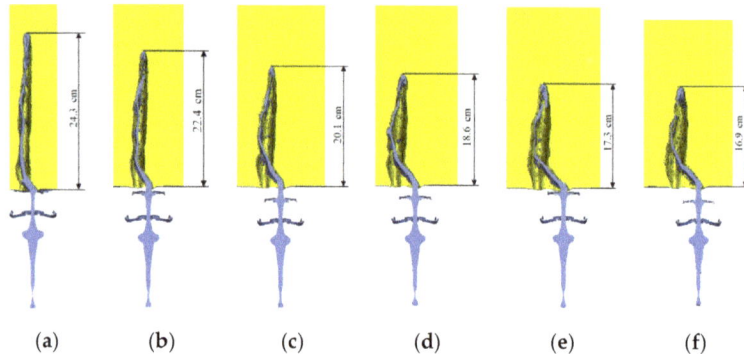

Figure 8. Penetration results at different target velocities: (**a**) v_t = 150 m/s; (**b**) v_t = 200 m/s; (**c**) v_t = 250 m/s; (**d**) v_t = 300 m/s; (**e**) v_t = 350 m/s; (**f**) v_t = 400 m/s.

In this simulation, the blast height of the shaped-charge warhead is the same as the diameter of charge, as the blast height is short. The jet does not break in the static armor-piercing process, and the jet penetrates the target continuously. In the process of dynamical penetration, the jet before the target interferes with the jet is also continuous, so it is also a continuous jet penetration process. It can be seen from Figure 7 that when the target plate speed is low, the jet will be destabilized later, and most of the jet that can penetrate has entered the duct, but it is interfered with by the lateral movement of the target plate, and squeezed and collided with the target plate duct, which affects the penetration ability of the jet. The time for the jet to intervene continues to advance as the velocity of the moving target plate increases. A growing number of armor-piercing jets are directly disrupted by the transverse motion of the target plate and fail to enter the penetration channel from the unperturbed jet to the target plate. Instead, the target plate needs to be re-penetrated, bending and stacking occur near the surface of the target plate, and then fracture and secondary collisions with the hole walls occur during the continued flow to the bottom of the pit. The penetration depth of the jet is clearly reduced.

4.3. Comparative Analysis of Theory and Simulation

Following the theoretical model developed in Section 2, the penetration depth of the jet under the corresponding dynamical conditions is obtained by calculating the relative errors between five sets of numerical simulation cases. See Table 3 and Figure 9 for a comparison between the penetration depth from numerical simulations of jets and the depth calculated theoretically.

Table 3. Comparison of jet penetration depth between theoretical calculations and numerical simulations.

Target Plate Speed v_t (m/s)	Theory Calculated Value P_j/cm	Numerical Value Simulation Value P_f/cm	Relative Error e_r (%)
150	24.5	24.3	0.8
200	22.9	22.4	2.2
250	20.3	20.1	1.0
300	18.5	18.6	0.5
350	17.0	17.3	1.7
400	15.6	16.9	7.7

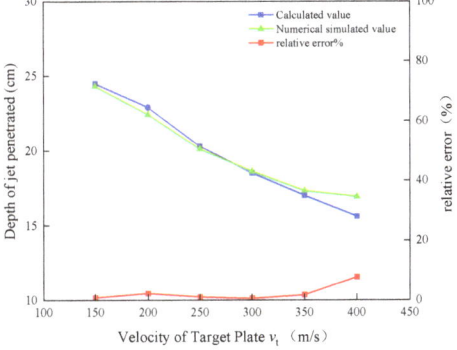

Figure 9. Comparison of jet penetration depth between theoretical calculations and numerical simulations.

As can be seen from Table 3, the error between the armor-breaking depth calculated by the theoretical model and the simulation results is minor. This method can be used to estimate the armor-breaking depth of a jet penetrating a moving target. As can be seen from Figure 9, the dynamical penetration depth of the jet into the target decreases with increasing velocity of the target.

In order to further study the dynamic penetration mechanism of the jet, according to the dynamic penetration model established previously, the penetration length P_w of the undisturbed section and the penetration length P_r of the disturbed section are extracted, and their proportions and changing rules in the dynamic penetration of the jet are analyzed.

As can be seen in Figure 10, the decay of the penetration length P_w of the unperturbed jet has a tendency to decrease significantly at first, and then to moderate as the velocity of the target is increased. Combined with the dynamic penetration model and numerical simulation analysis, the interference time keeps advancing as the velocity of the target increases, leading to fewer and fewer jets that can penetrate the target before the interference time. Thus, the penetration depth of the unperturbed jet profile decreases. Due to the elevated velocity of the jet head and the tremendous amount of energy it carries, the diameter of the target slab is relatively large during the opening phase. During the dynamic penetration, a part of the jet can avoid being disturbed by the motion of the target plate under the reaming of the head jet micro-element, so that it can follow the head jet micro-element to penetrate further into the target plate. As a result, this part of the jet is able to overcome the effect of the target motion to a certain extent and maintain the penetration capability of the target.

The penetration depth of the perturbed jet profile initially increases slightly and then decreases. This is because when the moving speed of the target plate is low ($v_t \leq 200$ m/s), the effective jet velocity threshold v_{jk} is also low, and the disturbed jet flowing to the bottom of the jet channel still has a certain penetration ability after the interference time, which makes the penetration ability of the disturbed jet increase on the whole, but because the penetration ability of this part is relatively weak, the penetration ability of the disturbed jet is not significantly improved. When the target plate speed is further increased ($v_t \geq 300$ m/s), because the effective jet velocity threshold v_{jk} is increased, the jet that can effectively penetrate the bottom of the jet crater is continuously reduced, and the penetration length of the disturbed jet is significantly reduced.

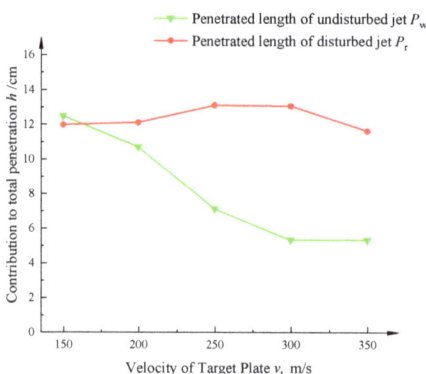

Figure 10. Curve of jet penetration length variation in the perturbed and unperturbed profiles.

Combining the regularity of the penetration lengths of the perturbed and unperturbed profiles, the ratio of the two in the total dynamical penetration depth of the jet is plotted in Figure 11.

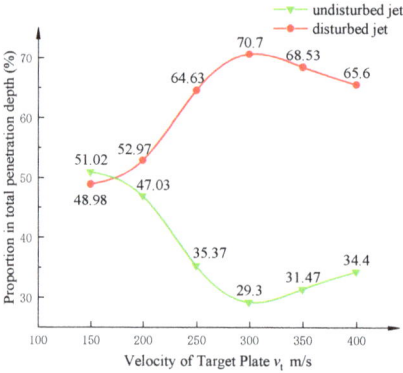

Figure 11. Curve of penetration length ratio of jet in perturbed and unperturbed sections.

Figure 11 shows that the ratio of the unperturbed jet to the overall penetration depth definitely decreases at first and then rises as the velocity of the target is increased. The penetration depth ratio of the perturbed jet is reversed. Based on the analysis of the dynamic penetration mechanism, it can be seen that when the moving speed of the target plate continues to increase, the interference time will advance, resulting in fewer and fewer jets in the undisturbed section, which directly affects the proportion of this part of the jet in the total penetration distance. At the same time, the penetration distance of the perturbed jet increases proportionally to the total dynamical penetration. However, as the target speed continues to increase, when the target speed is greater than 300 m/s, the effective length of the disturbed section jet continues to decrease due to the increase in the random moving speed of the effective jet velocity threshold, which makes the penetration ability of this part of the jet to the bottom of the jet channel begin to weaken. It can overcome the interference of the target slab and gradually reduce the effective part of the penetration that completes the bottom of the jet channel, such that the penetration length of the perturbed section of the jet decreases in proportion to the total penetration length.

However, the unperturbed profile of the jet has a higher energy due to its position at the jet head, which has the effect of expanding the jet channel. This is the reason why the subsequent part of the jet removes the effect of the motion of the target slab within a certain range, allowing the decay of the penetration length of the unperturbed part of the jet to begin to slow down. In addition, the penetration length of the perturbed jet decreases continuously, such that the fraction of the unperturbed jet in the overall fraction starts to rise for target velocities larger than 300 m/s.

The perturbed jet cross section does not fluctuate significantly at early times, and the unperturbed jet cross section continues to decrease in penetration length according to the overall penetration depth variation in two sections of jets. The penetration length of the perturbed jet profile starts to decrease, while the penetration length of the unperturbed jet profile tends to stabilize for target slab motion velocities larger than 300 m/s. Overall, the penetration depth of the jet into the target decreases with increasing velocity of the target.

5. Conclusions

In this paper, we established a mathematical model of the dynamic penetration mechanism of a shaped-charge jet based on the virtual origin theory and Bernoulli equation considering the strength of the jet and target. Combined with the formation of the jet channel in the process of jet static penetration, the theoretical model was verified with numerical simulation. We studied the variational law of the dynamic penetration of shaped-charged jets based on the mathematical model developed to characterize the dynamical penetration behavior of perturbed/undisturbed jets. The main conclusions are as follows:

1. When the velocity of the target is less than 300 m/s, the penetration ability of the disturbed jet has no obvious shift and the penetration ability of the undisturbed jet decreases linearly. When the speed of the target exceeds 300 m/s, the penetration length of the disturbed jet decreases linearly while the undisturbed jet tends to be stable.
2. The contribution of the undisturbed jet to the overall penetration of the jet takes 300 m/s target velocity as the inflection point, showing a parabola pattern of decreasing and then increasing; the contribution of the jet in the disturbed section takes the same target velocity as the inflection point, but the shift is opposite to that of the jet in the undisturbed section.
3. The relative error between the established mathematical model and the numerical simulations is less than 10 per cent, which is in agreement with the numerical simulations, compared to the simulations at six different groups of target velocities. We verified that the established mathematical model can reliably predict the penetration depth of the jet under dynamical conditions.

6. Discussion

Authors should discuss the results and how they can be interpreted from the perspective of previous studies and of the working hypotheses. The findings and their implications should be discussed in the broadest context possible. Future research directions may also be highlighted.

Author Contributions: Conceptualization, Y.W. and J.Y. (Jianping Yin); methodology, Y.W.; software, Y.W.; validation, Y.W.; formal analysis, Y.W., J.Y. (Jianping Yin), X.Z. and J.Y. (Jianya Yi); investigation, Y.W., J.Y. (Jianping Yin), X.Z. and J.Y. (Jianya Yi); data curation, Y.W. and J.Y. (Jianping Yin); writing—original draft preparation, Y.W.; writing—review and editing, J.Y. (Jianping Yin); supervision, J.Y. (Jianping Yin), X.Z. and J.Y. (Jianya Yi); project administration, Y.W. All authors have read and agreed to the published version of the manuscript.

Funding: The authors would like to acknowledge the financial support from the 2021 Basic Research Program of Shanxi Province (Free Exploration), grant number 20210302123207 and graduate Innovation Project of Shanxi Province, grant number 2022Y596.

Institutional Review Board Statement: Not applicable.

Informed Consent Statement: Not applicable.

Data Availability Statement: Not applicable.

Acknowledgments: The authors would like to thank the editor, associate editor, and the anonymous reviewers for their helpful comments and suggestions that have improved this paper.

Conflicts of Interest: The authors declare no conflict of interest.

References

1. Wang, Q.; Zhao, H.D.; Zhao, P.D.; Shao, X.F.; Zhang, X.D. Study on damage effectiveness of shaped warhead in air defense and anti-missile. *Ordnance Ind. Autom.* **2018**, *37*, 60–63+72.
2. Waggener, S. Relative performance of anti-air missile warheads. In Proceedings of the 19th International Symposium on Ballistics, Interlaken, Switzerland, 7–11 May 2001.
3. Liu, T.; Qian, L.X.; Zhang, S.Q. Study on fragment focusing mode of air-defense missile warhead. *Propellants Explos. Pyrotech.* **1998**, *23*, 240–243.
4. Held, M. Aimable fragmenting warhead. In Proceedings of the 13th International Symposium on Ballistics, International Ballistics Committee. Stockholm, Sweden, 1–3 June 1992; pp. 539–548.
5. Wang, L.X.; Zhou, T.; He, H.M.; Zhou, L. Numerical simulation and experimental investigation of initiation of shielded composition b impacted by shaped charge jet. *Explos. Mater.* **2015**, *44*, 56–60.
6. Zhang, X.P.; Ji, Q.; Yi, J.Y.; Yin, J.P. Numerical simulation of dynamic penetration damage effect of shaped charge warhead. *J. Ordnance Equip. Eng.* **2022**, *43*, 68–73.
7. Chen, Z.H. *Analysis of the Characteristics of the Jet from HEAT Projectile with the Error and Simulation Research on Penetration into the High-Speed Target in an Angle*; North University of China: Taiyuan, China, 2015.
8. Cao, P.; Zhang, G.W.; Qiao, T.T.; Kang, F.H. Numerical Simulation of Rod Jet Penetration Moving Target. *J. Ordnance Equip. Eng.* **2021**, *42*, 86–90.
9. Zhao, X.; Xu, Y.J.; Zheng, N.N.; He, H.F. Simulation and Optimization Research of Penetration Performance of Liner. *J. Ordnance Equip. Eng.* **2021**, *42*, 65–71.
10. Wang, C.; Yun, S.R.; Huang, F.L. An experimental study and numerical simulation on annular jet formation and penetration. *Acta Armamentarii.* **2003**, *24*, 451–454.
11. Kong, F.J.; Zhou, C.G.; Wang, Z.J.; Chen, Y.; Zhang, Y. Influence of top material on forming and penetration performance of double cone type charge. *Ordnance Mater. Sci. Eng.* **2021**, *44*, 30–35.
12. Yi, J.Y.; Wang, Z.J.; Yin, J.P.; Zhang, Z.M. Reaction characteristics of polymer expansive jet impact on explosive reactive armor. *e-Polymers* **2020**, *20*, 292–302. [CrossRef]
13. Zhao, X.; Xu, Y.J.; Dong, F.D.; Zheng, N.N.; He, H.F.; Wang, Z.J. Numerical simulation of efflux performance of double-layer and double-cone charge cover. *J. Ordnance Equip. Eng.* **2022**, *43*, 170–174.
14. Jia, X.; Huang, Z.X.; Xu, M.W.; Xiao, Q.Q. Theoretical Model and Numerical Study of Shaped Charge Jet Penetrating into Thick Moving Target. *Acta Armamentarii.* **2019**, *40*, 1553–1561.
15. Tian, L.L.; Zhang, T.; Qiao, L.G.; Ma, C.X. Numerical Simulation of Shaped Jet Penetrating Moving Steel Target. *Ordnance Ind. Autom.* **2021**, *40*, 6–9.
16. Eichelberger, R.J.; Pugh, E.M. Experimental verification of the theory of jet formation by charges with lined conical cavities. *J. Appl. Phys.* **1952**, *23*, 537–542. [CrossRef]
17. Ding, L.L.; Tang, W.H.; Ran, X.W. Simulation study on jet formability and damage characteristics of a low-density material liner. *Materials* **2018**, *11*, 72. [CrossRef] [PubMed]
18. Wang, L.X.; Sun, X.Y.; Li, G.; Zhang, H.L.; Wang, J.J. Technology status and development trend of anti-armor shaped warhead. *Aerodyn. Missile J.* **2016**, *8*, 91–96.
19. Li, Y.Q.; Gao, F.; Wang, C.X.; Gao, J.Y. Engineering algorithm for a new type of protective armor disturbing jet penetration. *Acta Armamentarii.* **2002**, *23*, 546–550.
20. Sui, J.Y.; Wang, S.S. *Terminal Effect*; National Defense Industry Publishing: Beijing, China, 2000.
21. Allison, F.E.; Vitali, R. *A New Method of Computing Penetration Variables for Shaped-Charge Jets*; Report No. 1184; Ballistic Research Laboratory Report: Aberdeen Proving Ground, MD, USA, 1963.
22. Wang, J.; Wang, C.; Ning, J.G. Theoretical model for shaped charge jets penetration and cavity radius calculation. *Eng. Mech.* **2009**, *26*, 21–26.
23. Wang, F.; Jiang, J.W.; Men, J.B. A penetration model for tunsgsten-copper shaped charge jet with non-constant jet. *Acta Armamentarii.* **2003**, *24*, 451–454.

24. Eichelberger, R.J. Experimental test of the theory of penetration by metallic jets. *J. Appl. Phys.* **1956**, *27*, 63–68. [CrossRef]
25. Xiao, Q.Q.; Huang, Z.X.; Gu, X.H. Equation of Penetration and Crater Growth by Shaped Charge Jet under the Influence of Shock Wave. *Chin. J. High Press. Phys.* **2011**, *25*, 333–338.

materials

Article

Formation of Shaped Charge Projectile in Air and Water

Zhifan Zhang [1,2], Hailong Li [1], Longkan Wang [3], Guiyong Zhang [1,4,]*** and Zhi Zong [1,4]

[1] State Key Laboratory of Structural Analysis for Industrial Equipment, School of Naval Architecture Engineering, Dalian University of Technology, Dalian 116024, China
[2] State Key Laboratory of Explosion Science and Technology, Beijing Institute of Technology, Beijing 100081, China
[3] China Ship Research and Development Academy, Beijing 100192, China
[4] Collaborative Innovation Center for Advanced Ship and Deep-Sea Exploration, Shanghai 200240, China
* Correspondence: gyzhang@dlut.edu.cn

Abstract: With the improvement of the antiknock performance of warships, shaped charge warheads have been focused on and widely used to design underwater weapons. In order to cause efficient damage to warships, it is of great significance to study the formation of shaped charge projectiles in air and water. This paper uses Euler governing equations to establish numerical models of shaped charges subjected to air and underwater explosions. The formation and the movement of Explosively Formed Projectiles (EFPs) in different media for three cases: air explosion and underwater explosions with and without air cavities are discussed. First, the velocity distributions of EFPs in the formation process are discussed. Then, the empirical coefficient of the maximum head velocity of EFPs in air is obtained by simulations of air explosions of shaped charges with different types of explosives. The obtained results agree well with the practical solution, which validates the numerical model. Further, this empirical coefficient in water is deduced. After that, the evolutions of the head velocity of EFPs in different media for the above three cases are further compared and analyzed. The fitting formulas of velocity attenuation of EFPs, which form and move in different media, are gained. The obtained results can provide a theoretical basis and numerical support for the design of underwater weapons.

Keywords: shaped charge projectile; velocity attenuation law; underwater explosion; trans-media

Citation: Zhang, Z.; Li, H.; Wang, L.; Zhang, G.; Zong, Z. Formation of Shaped Charge Projectile in Air and Water. *Materials* **2022**, *15*, 7848. https://doi.org/10.3390/ma15217848

Academic Editors: Chuanting Wang, Yong He, Wenhui Tang, Shuhai Zhang, Yuanfeng Zheng and Xiaoguang Qiao

Received: 23 September 2022
Accepted: 31 October 2022
Published: 7 November 2022

1. Introduction

With the widespread use of cabins near shipboard [1,2] and protection materials [3–6] for the design of warships, their explosion and shock resistance [7–9] is rapidly improved, which makes it very difficult for blast warheads to cause destructive attacks. However, due to the limitation of the dimensions of the warheads, the effect of increasing the charge weight on the improvement of the warhead power is minimal. Therefore, shaped charge warheads are gradually utilized to design the underwater weapon. In the traditional three types of shaped charges, explosively formed projectiles (EFP) [10,11] have the advantages of significant mass, small resistance, high velocity, and strong penetration ability, which are more suitable for underwater shaped charge warhead design. Therefore, it is significant to investigate the formation of EFP and its velocity attenuation in different media.

Many researchers studied the velocity attenuation law of shaped charge projectiles in air. Berner et al. [12] carried out a theoretical analysis of the flight characteristics of EFPs in air. Li et al. [13] made a theoretical analysis based on the EFP principle and flight dynamics principle and found that EFP aerodynamic resistance was significantly different when air density was different due to different temperatures. Liu et al. [14] designed a new two-wing EFP, which improved the penetration capability of the EFP. Olivera [15] proposed a numerical and analytical method for EFP maximum velocity performance estimation and verified the reliability of the analytical method through numerical simulation. Wu et al. [16] fitted the velocity attenuation equation of EFPs in air using numerical simulation. In addition, he experimentally verified the reliability of the fitting equation so that the flight distance and penetration capability of EFPs could be predicted. Du et al. [17] studied the attenuation law of the flight velocity of EFPs. However, little research about the velocity attenuation law of EFPs in water has been published.

The formation and velocity attenuation of shaped charge projectiles in water differ from those in air. Zhang et al. [18–25] systematically studied the underwater explosion and analyzed the damage of shaped charge projectiles to structures underwater. The results showed that the damage of shaped charge projectiles to structures underwater was more severe than that in the air. Cao et al. [26] studied the forming and velocity attenuation law of metallic jets in water but did not give the velocity attenuation law of metallic jets in water. According to Newton's Second Law, Lee et al. [27] introduced velocity attenuation and resistance coefficients and presented the classical theoretical formula for fragment entry into water. Tuo et al. [28] experimentally and numerically studied the evolution of the cavity and the velocity attenuation of a high-speed projectile entering water. The above studies mainly focus on velocity attenuation of the fragment without fracture, head deformation, and mass loss. However, its shape and mass are constantly changing as it moves in water. Therefore, it is of great significance to further study the velocity attenuation law of projectiles with complex shapes and high velocity in different media, especially in the water-entry process.

In order to make the projectile keep a better shape in its formation process in water, an air cavity is utilized at the bottom of the liner. The velocity of projectile information processes in air and water entry should be investigated. Sun et al. [29,30] analyzed the change of the missile's velocity across the medium from different incident angles and water entry speeds, but their studies focused on the low-speed interval. Wang et al. [31] analyzed the general law of EFP attenuation underwater. The effective velocity of EFP penetration in water was analyzed, but the attenuation formula was not given. Sun et al. [32] gave the optimal underwater torpedo air cavity length for underwater EFPs. However, they only considered the effect of the length of the air cavity on velocity, not whether the EFP breaks. Zhou et al. [33] studied the velocity attenuation of projectiles across the medium and established the physical model of the conical and spherical charge under the action of an underwater explosion. This model improved the residual velocity of EFPs in water but did not give the velocity attenuation equation across the medium. Mukhtar Ahmed et al. [34] recorded and calculated the velocity of the EFP by using Flash X-ray technology. The results show that the numerical simulation could reasonably predict the performance of the EFP to the underwater target. Most of the above scholars only analyzed the velocity attenuation of a projectile from air to water qualitatively, while they did not give the equation form of the velocity attenuation of projectiles.

In this paper, empirical formulas of EFPs in different media are modified or given. First, numerical models of shaped charges in air and water with/without air cavities are developed; and their formation processes are compared. After that, the effects of different charges on the maximum velocity of EFPs are discussed, with the empirical coefficient obtained in air. Based on that [35], the empirical coefficient of the maximum velocity of EFPs in water is given. Finally, the velocity attenuation law of EFPs in different media is studied. The velocity attenuation formulas of EFPs in water and from air to water are fitted by combining theoretical formulas with numerical simulation.

2. Basic Theory

2.1. Formation Velocity of Projectile in Air

An approximate analytical solution can estimate the maximum velocity of an EFP in its formation process. However, many assumptions are utilized for the analytical solution, leading to a deviation. In order to correct the analytical solution, many researchers combine experimental data with their empirical formula of the maximum velocity of EFPs in air, given by [35]:

$$u = 1 - \frac{1}{0.016\eta + 0.22\sqrt{\eta} + 1},$$ (1)

where $\eta = \frac{16}{27}\frac{\rho_e l_p}{\rho_m \xi}$; l_p and ξ are charge thickness at the midpoint of the liner and liner thickness, respectively; ρ_e and ρ_m are explosive and liner densities, respectively.

It is also pointed out that the EFP velocity calculated by the above approximate analytical solution is generally 30% higher than the measured value [35]. Therefore, the actual initial velocity of an EFP in the air can be obtained by:

$$V_{1-A} = 0.7 \times D \times u,$$ (2)

where u can be solved by Equation (1) and D is the detonation velocity of the explosive.

2.2. Attenuation Velocity of Projectile in Water

When the shaped charge projectile with high velocity moves in water, its surface shall be covered with supercavitation. Most of its surfaces do not directly contact water, and the friction resistance is negligible. The main factor affecting the projectile's movement is the differential pressure resistance which affects the shape of the projectile head. After moving in water for some time, its head develops into a "mushroom" shape. The head velocity of the shaped charge projectile rapidly declines. According to Newton's Second law [36]:

$$m\frac{d^2z}{dt^2} = m\frac{dv}{dt} = -\frac{1}{2}\rho_w A_0 C_d V_t^2,$$ (3)

where m is the mass of the shaped charge projectile; z is the distance that the shaped charge projectile advances; V_t is the projectile velocity at any time; A_0 is the projected area of the head of the shaped charge projectile in contact with water, and C_d is the resistance coefficient related to the cavitation number. Although the cavitation number and the resistance coefficient changes with the movement of the projectile, they are so small that they are set to a constant in this paper.

By integrating Equation (3), we get [36]:

$$V_t = \frac{V_0}{1 + V_0\beta t},$$ (4)

where V_0 is the initial velocity when it enters into the water, and constant β is the velocity attenuation coefficient and is defined as [36]:

$$\beta = \rho_w A C_d / 2m,$$ (5)

where β is related to the density of seawater; ρ_w the head area of the projectile; A the resistance coefficient, C_d and the mass m of the projectile. It is defined as a constant in this paper.

2.3. Attenuation Velocity of Projectile from Air to Water

Assume that a projectile with density ρ_m moves with velocity u_x in a Newtonian fluid with a viscosity coefficient of μ. The resistance on the projectile is F. It is assumed that the surface of the projectile is smooth and rotates asymmetrically, and its gravity, cavitation resistance, and temperature are not considered. According to the momentum equation and Newton's Second Law, the force and the mass can be given by [37]:

$$F dt = mu_x,$$ (6)

$$F = -Mdu_x/dt,$$ (7)

$$M = \rho_m SL, m = \rho S dx, \tag{8}$$

where M is the mass of the projectile, and m is the liquid mass acting on the front of the projectile head in time dt. Yang et al. [37] derived the velocity attenuation equation through the above formula and fitted it with polynomials. However, the physical meaning of the independent variables in the formula is ambiguous. On this basis, taking time as the independent variable and fitting with polynomials, this approach obtains good verification, and then the fitting formula of projectile velocity with time is expressed as [37]:

$$u_t = u_0 \exp\left(-At - Bt^2 + Ct^3\right), \tag{9}$$

where u_t is the projectile velocity at any time; u_0 is the initial velocity when it goes from air to water; t is the time, and A, B, and C are resistance constants.

2.4. Numerical Theory
2.4.1. Fluid Governing Equation

The Euler algorithm is used to simulate projectile formation in different media in this paper. The Euler grid is fixed, with materials transported in Figure 1. The primary calculation process is divided into three steps in AUTODYN. Conservation equations of mass, momentum, and energy are given by [38]:

$$\frac{\partial \rho}{\partial t} + \frac{\partial \rho u}{\partial x} + \frac{\partial \rho v}{\partial y} = 0, \tag{10}$$

$$\frac{\partial \rho v}{\partial t} + \frac{\partial \rho uv}{\partial x} + \frac{\partial \left(\rho v^2 + P\right)}{\partial y} = 0$$
$$\frac{\partial \rho u}{\partial t} + \frac{\partial \left(\rho u^2 + P\right)}{\partial x} + \frac{\partial \rho uv}{\partial y} = 0 \tag{11}$$

$$\frac{\partial E}{\partial t} + \frac{\partial u(E + P)}{\partial x} + \frac{\partial v(E + P)}{\partial y} = 0, \tag{12}$$

where x and y are coordinates, and ρ, v, u, E, and P are the density radial velocity, axial velocity, internal energy, and pressure of the fluid, respectively.

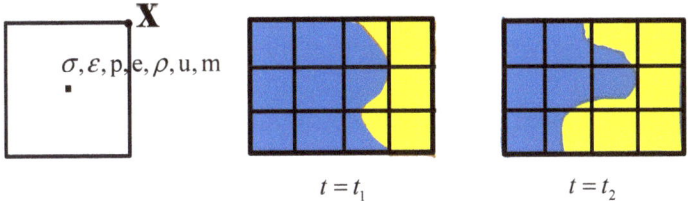

Figure 1. Materials transported in Euler.

2.4.2. Equation of State
(1) Equation of state for water

The shock equation of state is adopted for water, expressed as [38]:

$$U_S = C_0 + S_1 u + S_2 u^2, \tag{13}$$

where U_s is the shock wave velocity: u is the particle velocity; C_0, S_1 and S_2 are constants, and the specific values are set as $C_0 = 1647 (\text{m/s})$, $S_1 = 1.921, S_2 = 0$.

(2) Equation of state for air

The ideal gas equation is adopted for air, given by [38]:

$$p_{air} = (\gamma - 1)\rho_{air}\, e_{air}, \tag{14}$$

where p_{air} is air pressure; ρ_{air} is air density, adiabatic constant $\gamma = 1.4$, specific energy $e_{air} = 2.068 \times 10^5 \text{J/Kg}$.

(3) Equation of state for metal liner

Copper is used as the material of metal lining. The linear equation is used for the equation of state, and the Johnson-Cook equation is used for the strength model, expressed as [38]:

$$Y = \left[A + B\varepsilon_p^n\right]\left[1 + C\ln\varepsilon_p^*\right]\left[1 - T_H^m\right], \tag{15}$$

where Y is equivalent stress; A is initial yield stress; B is the hardening constant; ε_p^* is the plastic strain rate; n is the hardening index; C is the strain rate constant; m is the thermal softening index, and T_H is dimensionless temperature. The detailed parameters of the Johnson-Cook equation for copper are listed in Table 1.

Table 1. Parameters of the constitutive model of the liner [38].

A MPa	B MPa	N	C	M	*Bulk Modulus*/KPa	*Reference Temperature*/K	*Specific Heat* (J·kg^{-1}K^{-1})
90	292	0.31	0.025	1.09	1.29×10^8	300	383

(4) Equation of state for explosives

The JWL equation is adopted for explosives, given by [38]:

$$p_e = A\left(1 - \frac{\omega}{R_1\overline{V}}\right)e^{-R_1\overline{V}} + B\left(1 - \frac{\omega}{R_2\overline{V}}\right)e^{-R_2\overline{V}} + \frac{\omega E}{\overline{V}}, \tag{16}$$

where $\overline{V} = \rho_{b0}/\rho_b$, ρ_b, and ρ_{b0} are the density of detonation products and their initial density; E is the internal energy of explosive per unit volume; A, B, R_1, R_2, and ω are constants which are obtained by a specific experiment, and p_e is explosive detonation pressure. The detailed parameters of the JWL equation with different types of explosives are listed in Table 2 [38].

Table 2. Parameters of the JWL equation of different types of explosives [38].

Type	A GPa	B GPa	R_1	R_2	ω	ρ kg·m^{-3}	D_{CJ} m·s^{-1}	E GJ·m^{-3}	p_{CJ} GPa
TNT	373.77	3.75	4.15	0.90	0.35	1630	6930	6.0	21.0
COMPB	524.23	7.68	4.20	1.10	0.34	1717	7980	8.5	29.5
C4	609.77	12.95	4.50	1.40	0.25	1601	8193	9.00	28.0
HMX	778.28	7.07	4.20	1.00	0.30	1891	9110	10.5	42.0

3. Formation Process of Shaped Charge Projectiles in Different Media

3.1. Numerical Model

In order to study the formation law of shaped charge projectiles in different media, two-dimensional axisymmetric models of air and underwater explosions of shaped charges with spherical-segment liners were established. Denote three cases—air explosion and underwater explosions with and without air cavity—as Cases 1, 2, and 3, respectively. Four types of explosives were chosen: TNT, comp B, C4, and HMX. A numerical model of the air explosion of the shaped charge for Case 1 is shown in Figure 2. The charge had a height L of 40 mm and a diameter D of 20 mm. The liner was made of copper with variable thickness. Its inner and outer diameters were r = 13.99 mm and R = 12.20 mm. The dimension of the air cavity was variational. In order to avoid the reflection of shockwaves after reaching the boundary, the flow-out boundary was applied as a fluid boundary. The sub-option and preferred material for the flow-out boundary condition were flow-out (Euler) and all equal, respectively. The mesh size was determined after a convergence analysis.

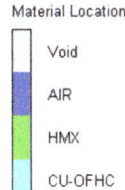
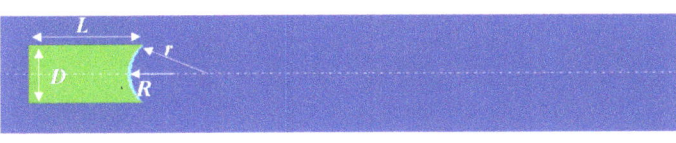

Figure 2. Numerical model of the air explosion of the shaped charge.

The numerical model of the underwater explosion of the shaped charge is similar to that of the air explosion, as shown in Figure 3.

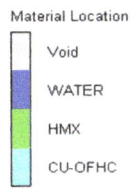

Figure 3. Numerical model of the underwater explosion of a shaped charge without an air cavity.

A light torpedo has an air cavity in its actual design. Therefore, a numerical simulation model of the underwater explosion of a shaped charge with an air cavity is developed in Figure 4. The length of this air cavity d dramatically affected the formation and velocity of the projectile. Sun et al. [32] found that when its length was three times the charge radius, the shape of the EFP in the formation process was better. In order to find out a better length of air cavity in this paper, three cases with d from two to four times the charge radius were chosen in Table 3.

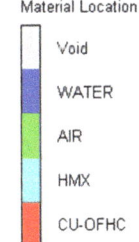
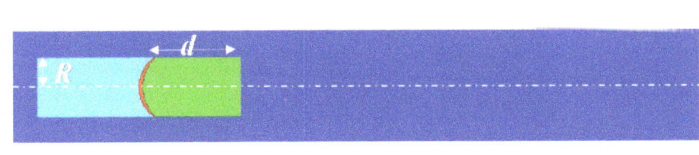

Figure 4. Numerical model of the underwater explosion of a shaped charge with air cavity.

Table 3. Cases for different numerical models.

Cases	Media	Air Cavity Length
1	Air	-
2	Water without air cavity	-
3	Water with air cavity	Twice the charge radius
4	Water with air cavity	Three times the charge radius
5	Water with air cavity	Four times the charge radius

3.2. Convergence Analysis

To ensure the reliability of Euler's algorithm, the velocity and morphology of the shaped charge projectile were simulated in this section. The experimental [39] and simulated values both agree well, as shown in Table 4 and Figure 5.

Table 4. Comparison of parameters between experimental [39] and simulated values.

Results	Time1 (µs)	Time2 (µs)	Head Velocity (km/s)	Tail Velocity (km/s)	Length (mm)
Experiment [39]	41.25	50.65	5.23	1.13	134.20
Simulation	40.00	50.00	4.71	1.09	145.00
Error/%	−3.00%	−1.28%	−9.94%	−3.54%	8.05%

In order to obtain a reasonable mesh size, a convergence analysis was carried out. Head X-velocities of shaped charge projectiles with different grid sizes and numbers were illustrated in Table 5 and Figure 6, respectively. The obtained results show that the head velocity with a grid size of 0.2 mm × 0.2 mm was similar to those of 0.1 mm × 0.1 mm and 0.12 mm × 0.12 mm. Taking calculation accuracy and efficiency fully into consideration, the grid size of 0.2 mm × 0.2 mm is used for the simulation in this paper.

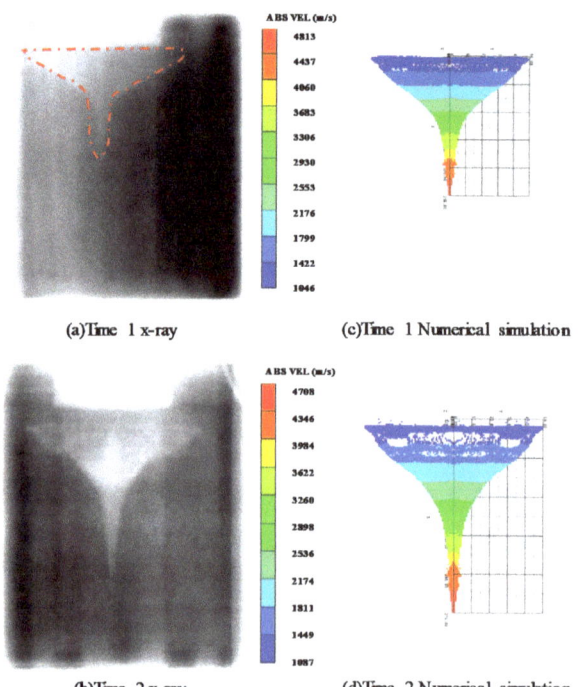

(a)Time 1 x-ray (c)Time 1 Numerical simulation

(b)Time 2 x-ray (d)Time 2 Numerical simulation

Figure 5. Comparison of experimental results [39] and numerical simulation.

Table 5. Simulated EFP head velocity under different grid sizes.

Grid Size (mm)	Grid Numbers	Head X-Velocity (m/s)
3.00 × 3.00	648	1204
2.00 × 2.00	1200	1362
1.00 × 1.00	4800	1568
0.40 × 0.40	30,000	1700
0.30 × 0.30	53,600	1713
0.20 × 0.20	120,000	1757
0.12 × 0.12	334,000	1784
0.10 × 0.10	480,000	1793

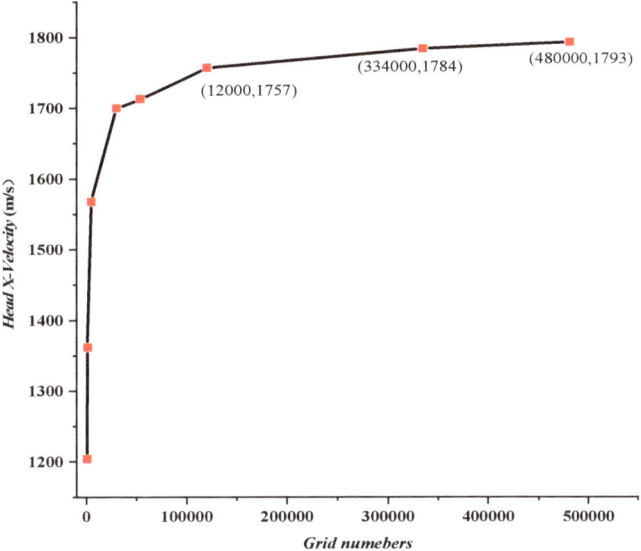

Figure 6. Head X-velocity of a projectile with different grid numbers.

3.3. Formation Process of Shaped Charge Projectile

Then, the formation processes of shaped charge projectiles in air and water with and without air cavities were further analyzed.

3.3.1. Case 1: Air Explosion

Firstly, the formation process of a shaped charge projectile in the air was analyzed. The velocity distribution of the projectile at different times is shown in Figure 7. At $t = 5$ μs, the detonation wave arrived at the top of the liner, with a plastic deformation caused. At $t = 10$ μs, with the shockwave effect, the liner was completely crushed, with an EFP initially formed. At $t = 15$ μs, an EFP was fully formed, and its head velocity peaked at approximately 1700 m/s. Due to the velocity gradient from the front to the back of the EFP, it was stretched, and its head and pestle could be distinguished. The EFP could fly smoothly in the air if its gravity and air resistance were ignored.

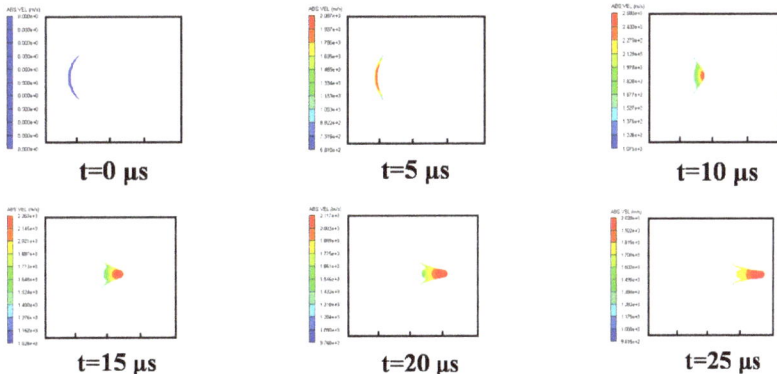

Figure 7. Velocity distribution of EFPs in a formation process in air.

3.3.2. Case 2: Water Explosion without Air Cavity

Then, the formation process of the shaped charge projectile in water was analyzed. The velocity distribution at different times is shown in Figure 8. At t = 5 μs, the detonation wave reached the liner top, with plastic deformation. At t = 10 μs, the liner began to turn over. Due to the great resistance effect of water, the shape of the EFP developed into a "crescent moon," which is different from that of the air in Figure 7. As the EFP moved in the water, its head shape kept stable at t = 25 μs. However, with the movement of the EFP in water, its head was worn, which led to mass loss and velocity decrease. Penetration performance decreased as a result.

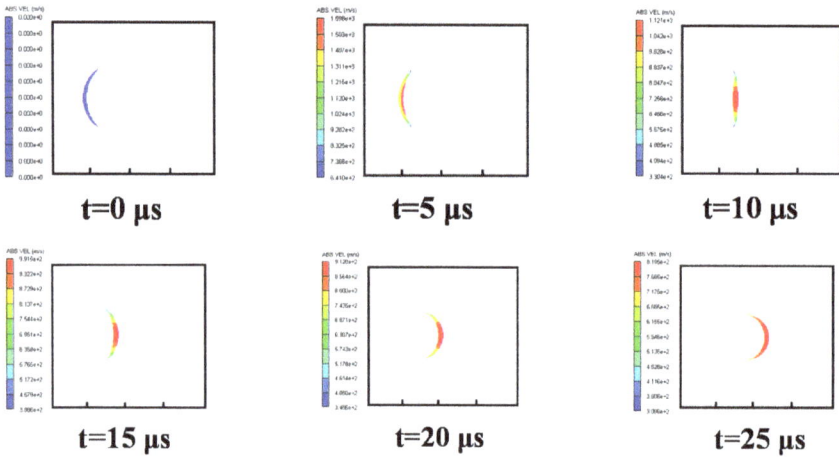

Figure 8. Velocity distribution of the EFP in a formation process in water.

3.3.3. Cases 3–5: Water Explosion with Air Cavity

Finally, the formation process of the shaped charge projectile, which moves from air to water, was analyzed. Three cases with lengths d of the air cavity of twice, three, and four times larger of charge radius are discussed in this section, namely Cases 3–5, respectively. Numerical results for velocity distributions of these three cases are illustrated in Figures 8–10, respectively.

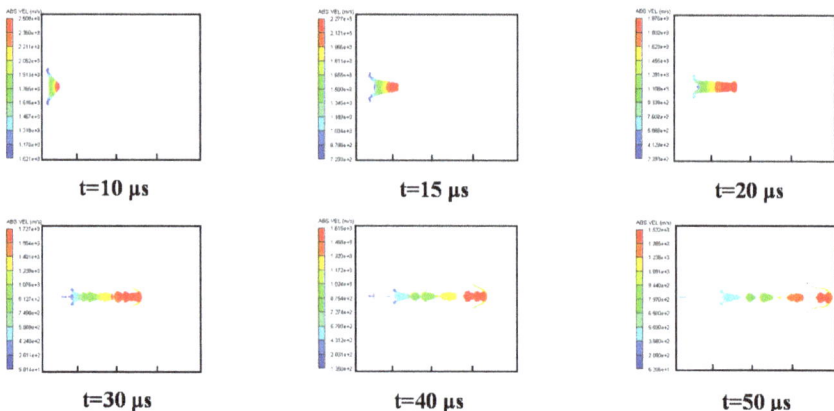

Figure 9. Velocity distribution of the EFP for Case 3.

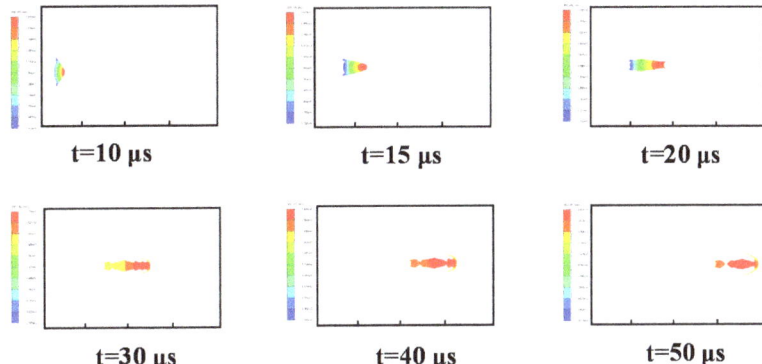

Figure 10. Velocity distribution of the EFP for Case 4.

The velocity distribution for Case 3 Is shown In Figure 9. At t = 15 μs, it can be seen that the EFP was not completely formed while it moved from air to water. Due to the water resistance, the head of EFP was worn. At t = 30 μs, the EFP began to break. Due to the large velocity gradient between the front and the rear of the EFP, it was overstretched, with multiple fractures formed. More fractures were found at t = 40 μs and 50 μs, which decreased the penetration performance of EFP.

The velocity distribution for Case 4 is shown in Figure 10. At t = 15 μs, a short EFP was formed stably. At t = 20 μs, the head of EFP entered the water and began to be worn, with a cavity generated around it in the fluid. Mass loss of the EFP is also found, and the head of the EFP is flattened. At t = 30 μs, the shape of the head of the EFP developed into a "mushroom." At t = 40 μs and 50 μs, the velocity gradient of the EFP was small so that fewer fractures were formed than that of Case 3.

The velocity distribution for Case 5 is shown in Figure 11. Although the EFP had been formed before it entered water, the velocity gradient of the EFP was more significant than that of Case 4, which also caused more fractures at t = 40 μs and 50 μs. In conclusion, when the length of the air cavity is three times of charge radius, a shaped charge projectile with better velocity and shape can be formed.

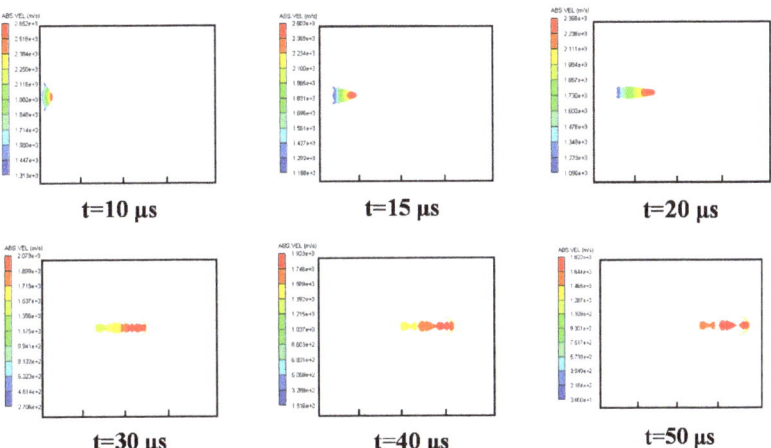

Figure 11. Velocity distribution of the EFP for Case 5.

3.3.4. Results Analysis and Discussion

After analyzing the formation processes of EFPs in different media, the comparison results show that the medium has an excellent effect on the formation of EFPs. A short and thick EFP is formed in air. The shaped charge projectile is turned over and develops into a "crescent moon" shape in the water. As for the case of a water explosion with an air cavity, the initial shape of an EFP before it arrives in water is similar to that in air. However, its shape gradually becomes a "mushroom" after its head arrives at the water. Due to the velocity gradient, the EFP breaks into many fractures. In addition, the effect of the length of the air cavity on the formation of EFPs is discussed. It can be found that when the length is three times the charge radius, such variables as tensile length, fracture, and water-entry velocity of the EFP are better than those of the other two cases.

4. Maximum Head Velocity of Projectile in Air and Water

4.1. Coefficient Modification of Head Velocity of Projectile in Air

According to the empirical formula in Section 2.1, the empirical coefficient of head velocity is set to 0.7 when the projectile forms in the air. Based on the air explosion model of a shaped charge in Section 3.3.1, the maximum head velocities of projectiles with four types of charge materials are discussed in this section. The empirical coefficients are numerically obtained in Table 6. Taking the average empirical coefficient of 0.647, the modified formula can be obtained as follows

$$V_{1-A} = 0.647 \times D \times u, \tag{17}$$

Table 6. Empirical coefficient of the head velocity of projectiles with different types of charge materials in air.

Material	D (m/s)	ρ_e(g/cm^3)	ρ_m(g/cm^3)	Simulation Velocity (m/s)	Empirical Coefficient	Average Empirical Coefficient
TNT	6930	1.630	8.960	1302.58	0.653	
COMP B	7980	1.717	8.960	1569.55	0.669	
C4	8193	1.601	8.960	1488.81	0.635	0.647
HMX	9110	1.891	8.960	1763.94	0.634	

Then, evolutions of velocity with different types of charge materials are further analyzed in Figure 12. The projectile is formed in the microsecond time scale, with its head velocity up to the peak value. After that, due to air resistance, the velocity slightly decreases. This decrease is affected by many factors, such as the windward area of the projectile, liner density, air density, etc. The detailed attenuation law of EFP flight in air was analyzed by Du et al. [17], which shall be given in Section 5.1 in detail.

4.2. Reduction Coefficient of Head Velocity of Projectile in Water

On the basis of the empirical formula of the maximum head velocity of the projectile in air, the formula in water is numerically deduced in this section. Evolutions of head velocity EFPs in water with different types of charges are shown in Figure 13. It can be seen that although their maximum velocities are so different, they share a similar attenuation law. Firstly, their velocity sharply decreases and then slowly declines. In this section, the maximum velocity of EFPs formed in water is studied, with an empirical coefficient for estimating the maximum velocity of EFPs given.

After validating the empirical coefficient of the head velocity of EFPs in air, similar numerical models of shaped charges subjected to underwater explosions with different types of charges are established. The obtained maximum velocities of EFPs and empirical coefficients are listed in Table 7. It is found that the empirical coefficient ranges from 0.455

to 0.476. Taking the average empirical constant of 0.462, the empirical formula of the initial head velocity of EFPs in water is obtained by:

$$V_{1-W} = 0.462 \times D \times u, \tag{18}$$

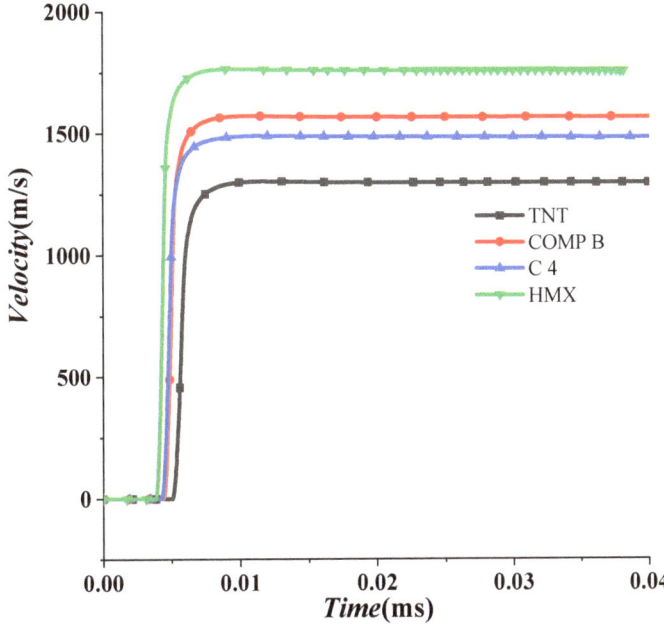

Figure 12. Evolutions of the maximum head velocity of projectiles with different types of charges in air.

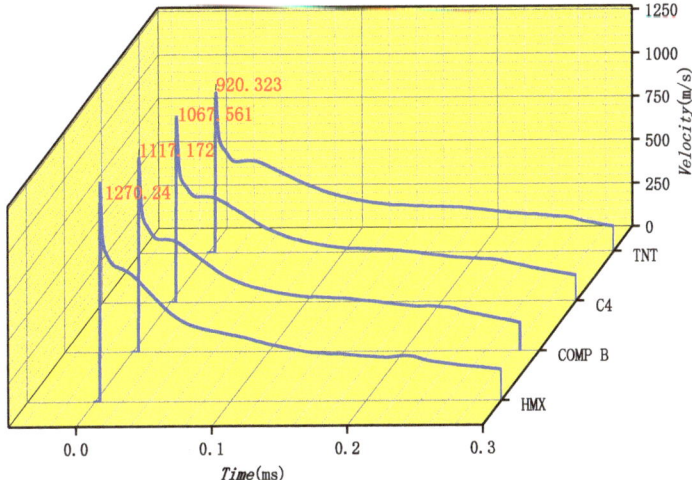

Figure 13. Evolution of the velocity of EFPs with different types of charge.

Table 7. Empirical coefficient of the head velocity of projectiles with different types of charge materials in water.

Material	D(m/s)	ρ_e(g/cm^3)	ρ_m(g/cm^3)	Simulation Velocity (m/s)	Empirical Coefficient	Average Empirical Coefficient
TNT	6930	1.630	8.960	920.428	0.461	
C4	8193	1.601	8.960	1067.892	0.456	
COMP B	7980	1.717	8.960	1118.419	0.477	0.462
HMX	9110	1.891	8.960	1270.936	0.457	

5. Effects of Media on the Evolution of Velocity

5.1. Velocity Attenuation Law of Projectiles in Air

It is found that such factors as flight distance, shape, the density of projectiles, etc., affect the residual velocity of projectiles [8]. Because the flight velocity of the projectile is much larger than the speed of sound, its weight is relatively small, and the air resistance is far greater than its weight, and the influence of gravity on the speed of EFP is ignored in the calculation, so the flight trajectory of the EFP can be regarded as a straight line, and its motion equation is [17]:

$$q_f = \frac{dV}{dt} = -C_D \frac{\rho_0}{2} A_s H(Y) V^2,$$ (19)

where q_f is the actual weight of the projectile; C_D is the air resistance coefficient; A_S is the windward area of the projectile; $H(Y)$ is the relative air density at height Y; ρ_0 is the ground air density, and V is the instantaneous flight speed of the projectile.

Among them, the resistance coefficient varies with the shape and flight velocity of the projectile. In order to obtain the analytical expression of residual velocity with distance, linear standardization is usually used for the solution of C_D, which is based on the measured results. At the second stage in air, the velocity attenuation formula of the projectile is given by [17]:

$$V_{2-A} = V_{1-A} \exp\left[-\frac{C_D H(Y) \rho A_s}{2q_f} r \right],$$ (20)

where V_{1-A} is the initial velocity in the first stage, r is the radius of the projectile, and ρ is the density of the air at the location.

5.2. Velocity Attenuation Law of Projectile in Water

HMX is found to work best in air and water. Therefore, it is used as an explosive in the following sections. Then, the velocity attenuation law of a projectile in water without an air cavity is analyzed. The formation and propagation process of a projectile can be divided into three stages: acceleration, rapid decay, and slow decay stages, as shown in Figure 14. In the first stage, the velocity increases linearly and peaks at 1300 m/s at about 0.1 μs. After that, the velocity sharply decreases in the second stage, with an attenuation coefficient. In the third stage, it slowly declines. Based on this, the velocity attenuation law of a projectile under different charges is further analyzed. According to Figure 13, it can be preliminarily judged that the underwater velocity attenuation is similar and has specific laws. It is found that the four cases with different types of charge share a similar evolution of velocity in Section 4.2. Next, the detailed velocity attenuation law in the second and third stages is analyzed.

5.2.1. Velocity Attenuation Law in the Second Stage

The velocity attenuation law in the second stage is first analyzed in Table 8. With the increase of explosive detonation velocity, both the maximum head velocity of the projectile and the attenuation coefficient increase. This indicates that the greater the initial velocity of the underwater projectile is, the greater its instantaneous attenuation velocity also is, which leads to a larger attenuation coefficient. Numerical results show that the attenuation coefficient ranges from 0.569 to 0.630, with an average attenuation percentage of about 58.8%. Therefore, the velocity attenuation formula in the second stage can be obtained by:

$$V_{2-W} = 0.588 \times V_{1-W}, \tag{21}$$

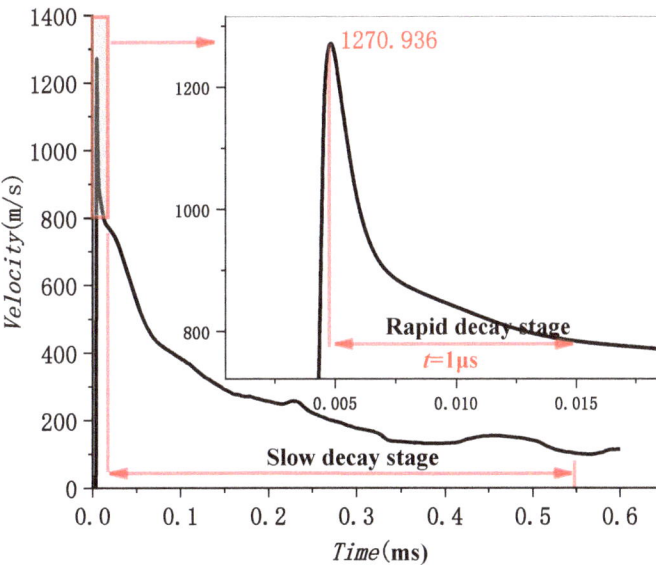

Figure 14. Evolution of velocity of projectile in water without air cavity.

Table 8. Velocity attenuation in the second stage.

Material	Simulation Maximum Velocity (m/s)	Sharply Decrease to Velocity (m/s)	Attenuation Factor	Average Attenuation Factor
TNT	920.428	524.312	0.570	
C4	1067.892	610.582	0.572	0.588
COMP B	1118.419	649.375	0.581	
HMX	1270.936	800.947	0.630	

5.2.2. Velocity Attenuation Law in the Third Stage

The velocity attenuation law in the third stage is analyzed in this section. Take (0.0128 ms, 801 m/s) as the starting point of the third stage in Figure 15. The numerical results for the evolution of velocity in the third stage are fitted according to Equation (4) in Section 2.2, given by:

$$V_{3-W} = \frac{V_{2-W}}{1 + 0.01486V_{2-W}t}, V_{2-W} = 800.947 \text{ m/s} \tag{22}$$

The theoretical formula fits well with the numerical simulation, but the numerical results fluctuate. The velocity attenuation coefficient β is set as a constant when fitting according to the theoretical formula. However, the velocity attenuation coefficient β varies in the actual process because of the fluctuation of numerical results. When a shaped charge projectile moves in water, the mass falls off. The head gradually became a "crescent moon" shape, and the projectile's head-on area changed. In the process of penetration, both of them changed simultaneously. According to Equation (5), the velocity attenuation coefficient β also changes. This paper selected three points, A, B, and C, with large fluctuations to further analyze the specific reasons, as shown in Figure 15 and Table 9.

There are some differences between theoretical fitting and numerical simulation in Figure 15. At the beginning of detonation, the projectile's mass does not change during 0~0.1 ms in Figure 16. However, the projectile head area is the main factor affecting the attenuation coefficient β. Detonation waves force the head area of the projectile to decrease during the extreme time of detonation of the explosive. Subsequently, the area increases due to the influence of water resistance. That is, the attenuation coefficient β decreases first and then increases. However, in this paper, the attenuation coefficient β is taken as a constant, which results in the theoretical velocity being small at first and then prominent in the range of 0~0.1 ms. The head area of the projectile is stable, and the head gradually becomes a "crescent moon" after $t = 0.1$ ms, as shown in Figure 8. At this time, the projectile mass is the main influencing factor of the attenuation coefficient β. The increase in attenuation coefficient β is caused by the shedding of projectile mass. However, this paper takes the attenuation coefficient β as a constant, which results in the theoretical velocity being less than the numerical simulation velocity. In this paper, three points with relatively large fluctuations are marked as A, B, and C, respectively, and the recorded data are shown in Table 9. First, the velocity fluctuation range of theoretical fitting is 20–30 m/s. Secondly, it has basically lost its penetration ability [31] when projectile velocity drops to approximately 200. Therefore, a 20–30 m/s velocity error does not affect the evaluation of the damage degree. The fitting formula of the third stage is reliable.

Table 9. Error analysis of head velocity.

	Point A	Point B	Point C
Time (ms)	0.218	0.350	0.450
Numerical simulation (m/s)	256.456	134.028	153.957
Theoretical equation (m/s)	222.448	155.046	126.015
Velocity error	−34.008	21.018	−27.941
Percentage error	−13.260%	15.682%	−18.149%

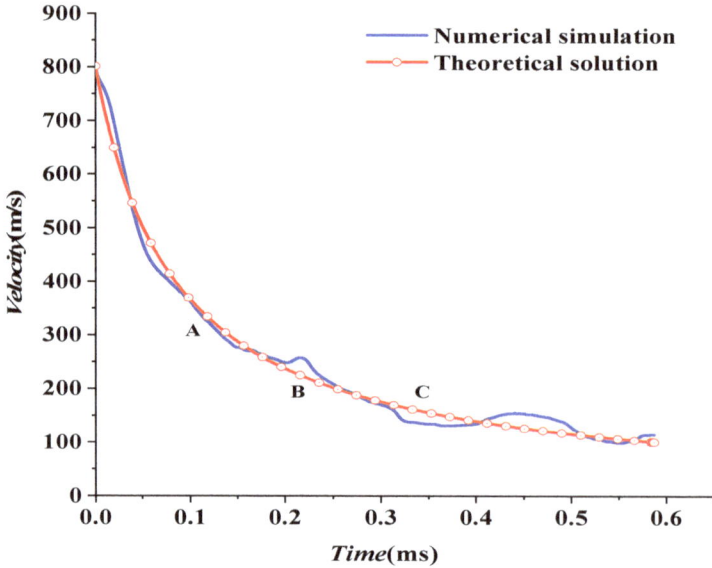

Figure 15. The fitting curve of underwater velocity attenuation of the EFP at the third stage.

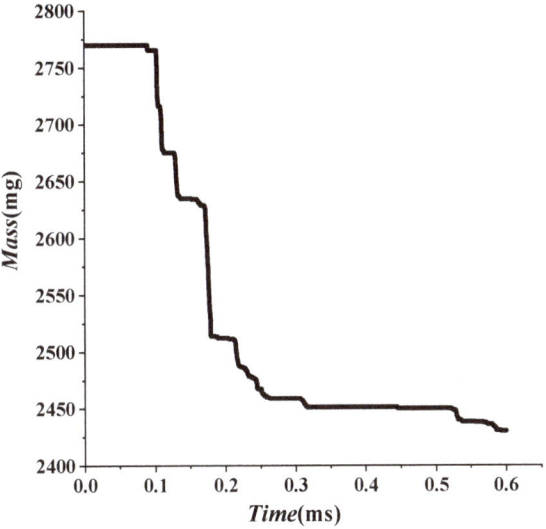

Figure 16. Evolution of mass of the shaped charge projectile in water.

5.3. Velocity Attenuation Law of Projectile from Air to Water

In order to obtain a better shape and velocity of the projectile, it is more suitable to add an air cavity inside the liner for the design of a lightweight torpedo rather than a formation in water. The water entry of the projectile should be investigated in this process. Therefore, the velocity attenuation law during the water entry process is further analyzed in this section. According to the results in Section 3.3.3, the better length of the air cavity is about three times larger than the charge radius. Thus, the evolution of the projectile's velocity for Case 3 is illustrated in Figure 17.

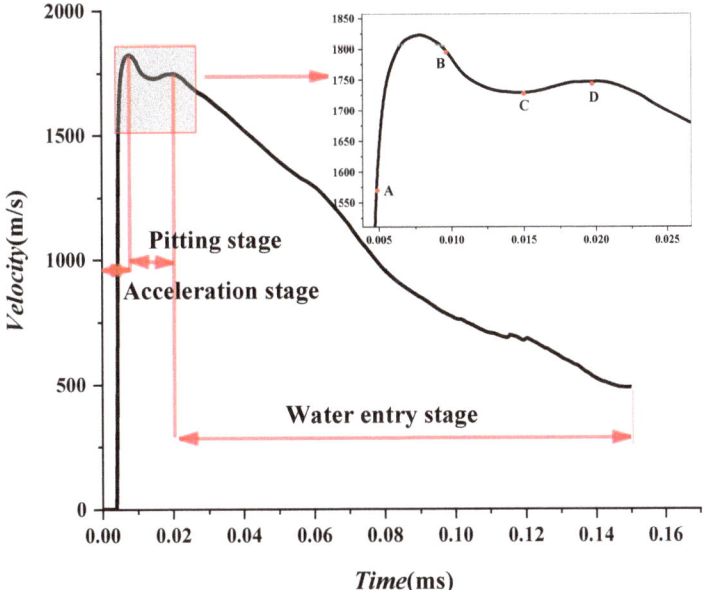

Figure 17. Evolution of the projectile entering water from air.

Three stages are included: acceleration, pitting, and water entry stages. Due to the fracture and collision of the projectile, the water entry stage is further subdivided into fracture and collision fluctuation stages, respectively. The projectile forms in the air in the first stage, and its velocity increases linearly. In the second pit stage in the BCD region in Figure 17, the projectile velocity first decreases and then climbs slightly. As for the third stage-water entry stage, the velocity decay is slow and fluctuates due to fracture and collision of the shaped charge projectile.

5.3.1. Velocity Analysis in Pit Stage

The pitting stage is a unique phenomenon of the projectile, which forms in water. Four points, A, B, C, and D, are marked in Figure 17, and their specific values of velocities are shown in Table 10. Meanwhile, pressure distributions when the projectile arrives at the above four points are shown in Figure 18. The wave load should be of concern because it is the main energy that overwhelms the liner at the moment of the burst. At $t = 5$ μs, a detonation wave is generated and propagates in water. Besides, with the effect of a detonation wave, the velocity of the liner peaks in a very short time. At $t = 10$ μs, the shockwave propagates from the water to the air cavity and begins to dissipate., so the head velocity of the projectile decreases slightly. However, At $t = 15$ μs, with the continuous effect on the projectile, its velocity increases slightly. At $t = 20$ μs, the projectile begins to enter the water. After that, the shockwave has little effect on the velocity of the projectile. The pit stage is basically over. Then, due to different media after the air cavity, the attenuation law is different. If the projectile moves in the air, it shall fly stably, with the velocity decreasing slightly, as in Figure 18. However, if it moves from air to water, the velocity rapidly declines. The effect of media on the velocity of the projectile in the pit stage is discussed in the next section. Finally, it is worth mentioning that the medium of wave load propagation in pure air and water (cases 1 and 2) remains unchanged, so the phenomenon of the pit stage does not occur.

Table 10. Parameters of data points in the pit stage.

	Point A	Point B	Point C	Point D
Time (μs)	5	10	15	20
Velocity (m/s)	1638.864	1786.006	1729.389	1747.401

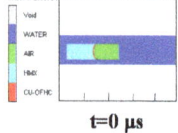 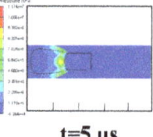 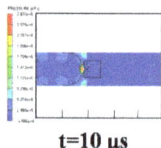 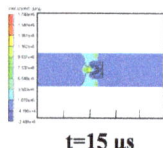 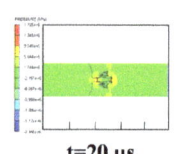

t=0 μs   t=5 μs   t=10 μs   t=15 μs   t=20 μs

Figure 18. Shock wave propagation in the pit stage.

Three cases are listed and discussed in Table 11. Evolutions of velocity for Cases 4 and 6 are first compared. Both of them showed a slight decline and climb, resulting in a concave phenomenon. Under the two working conditions, the time and speed are basically the same in Figure 19. This result indicates that if the length of the air cavity is sufficient to shape the projectile, then the water entry velocity of the projectile is essentially the same. Even if the length of the air cavity is increased further, the velocity of the projectile will not increase. Besides, evolutions of velocity for Cases 1 and 6 are compared. The maximum velocity of the projectile in water with an infinite air cavity is slightly higher than that only in air, as shown in Figure 19. The reason is that the shockwave dissipates quickly in the air, while the shockwave propagates faster in water, and the effect is more substantial than that in the air.

Table 11. Cases discussed in the pit stage.

Cases	1	4	6
Media	air	Water with air cavity	Water with infinite air cavity
Lengths of air cavity	-	Three times the charge radius	Infinite

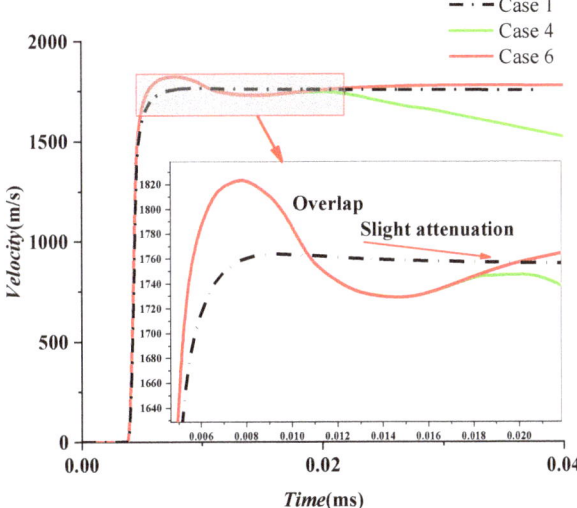

Figure 19. Comparison analysis diagram of the pitting stage and maximum velocity.

5.3.2. Velocity Analysis in Water Entry Stage

At the water entry stage, (0.019 ms, 1746.775 m/s) is taken as the initial point. After the time reaches zero, the velocity of the projectile in the water entry stage is fitted according to Equation (9) in Section 2.3, with $A = 3.880, B = 127.286, C = 652.968$. Therefore, the evolution of velocity can be obtained by:

$$u_t = 1746.775 \exp\left(-3.880t - 127.286t^2 + 652.968t^3\right) \tag{23}$$

Fluctuations are found in the numerical and fitting curves in the water entry stage. Three fluctuation points, D, E, and F, are chosen in Figure 20, with the fracture fluctuation stage of $t_D = 0.041$ ms and $t_E = 0.071$ ms collision fluctuation stages of $t_F = 0.085$ ms. It can be seen that the fracture begins to be caused at point D, and the curve fluctuates accordingly. The projectile during the phase between D and E breaks, with its head worn, and its head shape gradually develops into a "mushroom." However, its tail does not directly contact the water after the air cavity, with a higher velocity than the head. The tail catches up with the head at point F and begins to impact the head, with the velocity slightly increasing. After that, with the merge of the head and tail, the velocity gradually stays stable and drops to about 400 m/s. As a result, the projectile basically does not have penetration capability [31].

In order to further verify the reliability of the theoretical formula, three points, G, H, and I, in Figure 20, with large fluctuations, are selected for error analysis in Table 12. The maximum velocity fluctuation is 54 m/s, and the maximum error percentage is approximately 8%, validating the theoretical formula. After that, the shape of the projectile stays stable without fractures forming anymore, which corresponds to the velocity attenuation law in water. Finally, the residual velocity of the projectile decreases to approximately 400 m/s, and the projectile has basically lost its penetration ability.

Table 12. Velocity error analysis.

	Point G	Point H	Point I
Time (ms)	0.041	0.070	0.102
Numerical simulation (m/s)	1292.232	858.142	682.358
Theoretical equation (m/s)	1253.264	892.540	627.438
Velocity error	−38.968	34.398	−54.920
Percentage error	−3.02%	4.01%	−8.05%

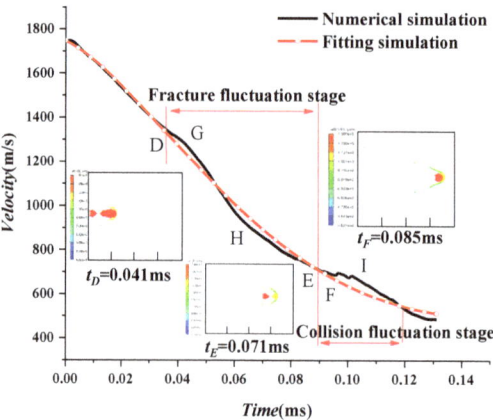

Figure 20. Attenuation curve of the shaped charge projectile entering water.

6. Conclusions

Based on the theoretical formula of the head velocity of the shaped charge projectile in the formation process, the Euler method was used to establish the air and underwater explosion models of a shaped charge with and without air cavity, with shapes of the projectile analyzed in different media. The empirical coefficient of head velocity attenuation in the formation process in water is given. The variation law of the head velocity of projectile in different media is discussed. The specific conclusions are given as follows:

1. A shaped charge projectile formed in air is short, thick, and dense while it turns over to be a "crescent moon" in water and develops into a "mushroom" shape from the air cavity to water. Due to the velocity gradient, fractures are found when the projectile enters and moves in the water. When the length of the air cavity is lower or larger than three times of charge radius, the projectile cannot be completely formed or easily fractured. Therefore, it is suggested to make the length of the air cavity three times larger than the charge radius;

2. Velocity attenuation laws of shaped charge projectiles with four types of explosives in air and water are discussed. Results show that the empirical coefficients of maximum velocity in air and water are 0.647 and 0.462, respectively. The head velocity of a projectile in water can be divided into three stages: acceleration, rapid decay, and slow decay. The higher the maximum head velocity of a projectile is, the greater the percentage of velocity attenuation is in the rapid decay stage. The residual velocity is about 60% of the maximum head velocity. The theoretical fitting formula is given in the slow decay stage, and its results agree well with the numerical ones. The maximum error of head velocity is only about 30 m/s, which proves the high reliability of the theoretical fitting formula;

3. The shaped charge projectile forms in the air cavity and then enters the water. Its head velocity includes acceleration, pit, and water entry stages. Because of the fracture and collision of the projectile, the water-entry stage is divided into fracture and collision stages. The pitting stage is a unique phenomenon of a projectile in water. Its velocity

tendency shows that the velocity first declines and then increases and eventually stays steady. The theoretical fitting formula of the head velocity of a projectile in the water-entry stage is given. The maximum error between the theoretical and numerical results for a projectile's head velocity is lower than 8.1%, which validates the theoretical fitting formula. Besides, the fluctuations are found in the numerical results caused by the fracture and the projectile collision.

Author Contributions: Formal analysis, H.L.; Methodology, G.Z.; Software, L.W.; Supervision, Z.Z. (Zhi Zong); Writing—original draft, Z.Z. (Zhifan Zhang); Writing—review & editing, Z.Z. (Zhi Zong). All authors have read and agreed to the published version of the manuscript.

Funding: The National Natural Science Foundation of China (52271307, 52061135107, 52192692, 11802025), the opening project of State Key Laboratory of Explosion Science and Technology (KFJJ21-09M), the Liao Ning Revitalization Talents Program (XLYC1908027) and the Fundamental Research Funds for the Central Universities (DUT20RC(3)025, DUT20TD108, DUT20LAB308).

Informed Consent Statement: Informed consent was obtained from all subjects involved in the study.

Acknowledgments: The authors wish to thank the National Natural Science Foundation of China (52271307, 52061135107, 52192692, 11802025), the opening project of State Key Laboratory of Explosion Science and Technology (KFJJ21-09M), the Liao Ning Revitalization Talents Program (XLYC1908027) and the Fundamental Research Funds for the Central Universities (DUT20RC(3)025, DUT20TD108, DUT20LAB308).

Conflicts of Interest: All authors declare that we have no financial and personal relationship with other people or organizations that have an interest in the submitted work. There are no other relationships or activities that could be construed as an influence of the submitted work.

References

1. Ma, J.X.; Wang, R.W.; Lu, S.Z.; Chen, W.D. Dynamic parameters of multi-cabin protective structure subjected to low-impact load -Numerical and experimental investigations. *Def. Technol.* **2020**, *16*, 988–1000. [CrossRef]
2. Jiang, X.W.; Zhang, W.; Li, D.C.; Chen, T.; Guo, Z.T. Experimental analysis on dynamic response of pre-cracked aluminum plate subjected to underwater explosion shock loadings. *Thin-Walled Struct.* **2021**, *159*, 107256. [CrossRef]
3. Ciepielewski, R.; Gieleta, R.; Miedzinska, D. Experimental Study on Static and Dynamic Response of Aluminum Honeycomb Sandwich Structures. *Materials* **2022**, *15*, 1793. [CrossRef] [PubMed]
4. Ji, L.; Wang, P.; Cai, Y.; Shang, W.; Zu, X. Blast Resistance of 240 mm Building Wall Coated with Polyurea Elastomer. *Materials* **2022**, *15*, 850. [CrossRef] [PubMed]
5. Si, P.; Liu, Y.; Yan, J.; Bai, F.; Huang, F. Ballistic Performance of Polyurea-Reinforced Ceramic/Metal Armor Subjected to Projectile Impact. *Materials* **2022**, *15*, 3918. [CrossRef]
6. Wan, M.; Hu, D.; Pei, B. Performance of 3D-Printed Bionic Conch-Like Composite Plate under Low-Velocity Impact. *Materials* **2022**, *15*, 5201. [CrossRef]
7. Yin, C.Y.; Jin, Z.Y.; Chen, Y.; Hua, H.X. Effects of sacrificial coatings on stiffened double cylindrical shells subjected to underwater blasts. *Int. J. Impact Eng.* **2020**, *136*, 103412. [CrossRef]
8. Wang, H.; Zhu, X.; Cheng, Y.S.; Liu, J. Experimental and numerical investigation of ship structure subjected to close-in underwater shock wave and following gas bubble pulse. *Mar. Struct.* **2014**, *39*, 90–117. [CrossRef]
9. Zhang, Z.H.; Chen, Y.; Huang, X.C.; Hua, H.X. Underwater explosion approximate method research on ship with polymer coating. *Proc. Inst. Mech. Eng. Part M J. Eng. Marit. Environ.* **2017**, *231*, 384–394. [CrossRef]
10. Liu, Y.; Yin, J.; Wang, Z.; Zhang, X.; Bi, G. The EFP Formation and Penetration Capability of Double-Layer Shaped Charge with Wave Shaper. *Materials* **2020**, *13*, 4519, Correction in *Materials* **2021**, *14*, 2210. [CrossRef]
11. Fu, H.; Jiang, J.; Men, J.; Gu, X. Microstructure Evolution and Deformation Mechanism of Tantalum–Tungsten Alloy Liner under Ultra-High Strain Rate by Explosive Detonation. *Materials* **2022**, *15*, 5252. [CrossRef] [PubMed]
12. Berner, C.; Fleck, V. Pleat and asymmetry effects on the aerodynamics of explosively formed penetrators. In Proceedings of the 18th International Symposium on Ballistics, San Antonio, TX, USA, 15–19 November 1999; CRC Press: Boca Raton, FL, USA, 1999; p. 11.
13. Li, Y.; Niu, S.J.; Shi, H.; Ji, X.S.; Hu, X.C.; Li, N.J. Effects of Environment Temperature on The Attenuation of Quasi-Spherical EFP Velocity. *J. Physics. Conf. Ser.* **2021**, *1855*, 12037. [CrossRef]
14. Liu, J.Q.; Gu, W.B.; Lu, M.; Xu, H.M.; Wu, S.H. Formation of explosively formed penetrator with fins and its flight characteristics. *Def. Technol.* **2014**, *10*, 119–123. [CrossRef]
15. Jeremić, O.; Milinović, M.; Marković, M.; Rašuo, B. Analytical and numerical method of velocity fields for the explosively formed projectiles. *FME Trans.* **2017**, *45*, 38–44. [CrossRef]

16. Wu, J.; Liu, J.B.; Du, Y.X. Experimental and numerical study on the flight and penetration properties of explosively-formed projectile. *Int. J. Impact Eng.* **2007**, *34*, 1147–1162. [CrossRef]
17. Ji, C.Q. *Missile Aerodynamics*; Aerospace Press: Beijing, China, 1996.
18. Zhang, A.M.; Cao, X.Y.; Ming, F.R.; Zhang, Z.F. Investigation on a damaged ship model sinking into water based on three dimensional SPH method. *Appl. Ocean. Res.* **2013**, *42*, 24–31. [CrossRef]
19. Zhang, Z.F.; Sun, L.; Yao, X.; Cao, Y. Smoothed particle hydrodynamics simulation of the submarine structure subjected to a contact underwater explosion. *Combust. Explos. Shock Waves* **2015**, *51*, 502–510. [CrossRef]
20. Zhang, Z.F.; Ming, F.R.; Zhang, A.M. Damage Characteristics of Coated Cylindrical Shells Subjected to Underwater Contact Explosion. *Shock Vib.* **2014**, *2014*, 763607. [CrossRef]
21. Zhang, Z.F.; Wang, C.; Xu, W.L.; Hu, H.L. Penetration of annular and general jets into underwater plates. *Comput. Part. Mech.* **2021**, *8*, 289–296. [CrossRef]
22. Zhang, Z.F.; Wang, C.; Xu, W.L.; Hu, H.L.; Guo, Y.C. Application of a new type of annular shaped charge in penetration into underwater double-hull structure. *Int. J. Impact Eng.* **2022**, *159*, 104057. [CrossRef]
23. Zhang, Z.F.; Wang, L.K.; Ming, F.R.; Silberschmidt, V.V.; Chen, H.L. Application of Smoothed Particle Hydrodynamics in analysis of shaped-charge jet penetration caused by underwater explosion. *Ocean. Eng.* **2017**, *145*, 177–187. [CrossRef]
24. Zhang, Z.F.; Wang, L.K.; Silberschmidt, V.V. Damage response of steel plate to underwater explosion: Effect of shaped charge liner. *Int. J. Impact Eng.* **2017**, *103*, 38–49. [CrossRef]
25. Zhang, Z.F.; Wang, L.K.; Silberschmidt, V.V.; Wang, S.P. SPH-FEM simulation of shaped-charge jet penetration into double hull: A comparison study for steel and SPS. *Compos. Struct.* **2016**, *155*, 135–144. [CrossRef]
26. Cao, J.Z.; Liang, L.H. Demonstration of numerically simulating figures for underwater explosions. In Proceedings of the 24th International Congress on High-Speed Photography and Photonics, Sendai, Japan, 17 April 2001.
27. Lee, M.; Longoria, R.G.; Wilson, D.E. Cavity dynamics in high-speed water entry. *Phys. Fluids* **1997**, *9*, 540–550. [CrossRef]
28. Chen, T.; Huang, W.; Zhang, W.; Qi, Y.F.; Guo, Z.T. Experimental investigation on trajectory stability of high-speed water entry projectiles. *Ocean Eng.* **2019**, *175*, 16–24. [CrossRef]
29. Li, Y.; Sun, T.Z.; Zong, Z.; Li, H.T.; Zhao, Y.G. Dynamic crushing of a dedicated buffer during the high-speed vertical water entry process. *Ocean Eng.* **2021**, *236*, 109526. [CrossRef]
30. Zhang, G.Y.; You, C.; Wei, H.P.; Sun, T.Z.; Yang, B.Y. Experimental study on the effects of brash ice on the water-exit dynamics of an underwater vehicle. *Appl. Ocean Res.* **2021**, *117*, 102948. [CrossRef]
31. Wang, Y.J.; Li, W.B.; Wang, X.M.; Li, W.B. Numerical Simulation and Experimental Study on Flight Characteristics and Penetration against Spaced Targets of EFP in Water. *Chin. J. Energetic Mater.* **2017**, *25*, 459–465. (In Chinese)
32. Sun, Y.X.; Hu, H.L.; Zhang, Z.F. Simulation Study on Influential Factors of EFP Underwater Forming. *Chin. J. High Press. Phys.* **2020**. (In Chinese) [CrossRef]
33. Zhou, F.Y.; Jiang, T.; Wang, W.L.; Zhan, F.M.; Zhang, K.Y.; Huang, Z.X. Simulation Study on Tapered and Spherical Shaped Charge under Underwater Explosion. *Appl. Mech. Mater.* **2012**, *157–158*, 852–855. [CrossRef]
34. Ahmed, M.; Malik, A.Q.; Rofi, S.A.; Huang, Z.X. Penetration Evaluation of Explosively Formed Projectiles Through Air and Water Using Insensitive Munition: Simulative and Experimental Studies. *Eng. Technol. Appl. Sci. Res.* **2016**, *6*, 913–916. [CrossRef]
35. Liu, F. *The Explosively Formed Penetrator and Its Engineering Damage Effects Research*; University of Science and Technology of China: Hefei, China, 2006. Available online: https://kns.cnki.net/KCMS/detail/detail.aspx?dbname=CDFD9908&filename=2007020858.nh (accessed on 23 September 2022). (In Chinese)
36. Li, Y.; Zhang, L.; Zhu, H.Q.; Zhang, W.; Zhao, P.D. Velocity Attenuation of Blast Fragments in Water Tank. *Shipbuild. China* **2016**, *57*, 127–137. (In Chinese)
37. Yang, L.; Zhang, Q.M.; Shi, D.Y. Numerical Simulation for the Penetration of Explosively Formed Projectile into Water. *J. Proj. Rocket. Missiles Guid.* **2009**, *29*, 117–119. (In Chinese)
38. Century Dynamics Inc. *Interactive Non-Linear Dynamic Analysis Software AUTODYNTM User Manual*; Century Dynamics Inc.: Houston, TX, USA, 2003.
39. Zhang, X.W.; Duan, Z.P.; Zhang, Q.M. Experimental Study on the Jet Formation and Penetration of Conical Shaped Charges with Titanium Alloy Liner. *Trans. Beijing Inst. Technol.* **2014**, *34*, 1229–1233. (In Chinese)

materials

MDPI

Article

The Mechanical and Energy Release Performance of THV-Based Reactive Materials

Mengmeng Guo [1], Yanxin Wang [2], Haifu Wang [1],* and Jianguang Xiao [2,3,*]

[1] State Key Laboratory of Explosion Science and Technology, Beijing Institute of Technology, Beijing 100081, China
[2] College of Mechatronic Engineering, North University of China, Taiyuan 030051, China
[3] Science and Technology on Transient Impact Laboratory, No. 208 Research Institute of China Ordnance Industries, Beijing 102202, China
* Correspondence: wanghf@bit.edu.cn (H.W.); xiaojg@nuc.edu.cn (J.X.)

Abstract: A polymer of tetrafluoroethylene, hexafluoropropylene, and vinylidene fluoride- (THV) based reactive materials (RMs) was designed to improve their density and energy release efficiency. The mechanical performances, fracture mechanisms, thermal behavior, energy release behavior, and reaction energy of four types of RMs (26.5% Al/73.5% PTFE, 5.29% Al/80% W/14.71% PTFE, 62% Hf/38% THV, 88% Hf/12% THV) were systematically researched by conducting compressive tests, scanning electron microscope (SEM), differential scanning calorimeter, thermogravimetric (DSC/TG) tests and ballistic experiments. The results show that the THV-based RMs have a unique strain softening effect, whereas the PTFE-based RMs have a remarkable strain strengthening effect, which is mainly caused by the different glass transition temperatures. Thermal analysis indicates that the THV-based RMs have more than one exothermic peak because of the complex component in THV. The energy release behavior of RMs is closely related to their mechanical properties, which could dominate the fragmentation behavior of materials. The introduction of tungsten (W) particles to PTFE RMs could not only enhance the density but also elevate the reaction threshold of RMs, whereas the reaction threshold of THV-based RMs is decreased when increasing Hf particles content. As such, under current conditions, the THV-based RMs (88% Hf/12% THV) with a high density of 7.83 g/cm^3 are adapted to release a lot of energy in thin, confined spaces.

Keywords: THV-based reactive materials; mechanical performances; thermal analysis; reaction threshold; energy release behavior

Citation: Guo, M.; Wang, Y.; Wang, H.; Xiao, J. The Mechanical and Energy Release Performance of THV-Based Reactive Materials. *Materials* **2022**, *15*, 5975. https://doi.org/10.3390/ma15175975

Academic Editor: Davide Palumbo

Received: 13 July 2022
Accepted: 19 August 2022
Published: 29 August 2022

Publisher's Note: MDPI stays neutral with regard to jurisdictional claims in published maps and institutional affiliations.

1. Introduction

In order to effectively attack light armored targets, armor-piercing projectiles are usually filled with incendiary agents or explosives of a certain quality, forming armor-piercing incendiary projectiles, armor-piercing incendiary explosive projectiles, and other highly effective damage munitions [1,2]. An important feature of this kind of ammunition is that there is generally no fuze inside the warhead, which can provide more space for high-energy materials or fragile metals with high damage ability. Therefore, more efficient damage elements can be loaded to achieve a more pronounced damage effect. However, these projectiles usually explode first and then inert armor-piercing, when they impact on the target. The problem is that the incendiary explosive reaction with high damage ability only works in front of the target and could not produce a large damage effect inside the target. Therefore, people began to explore new ways of damaging light armored targets. In 2005, Daniel B. Nielson et al. [3,4] proposed, in their patent, the idea of using reactive materials enhanced projectile to achieve efficient damage to thin wall armor. Dozens of RMs formulations were designed and the performances such as overpressure, perforation, and firelight size of reactive material enhanced projectiles in spacer plates were tested. Significantly different from traditional projectiles that use inert metal to penetrate, this enhanced projectile uses reactive damage elements to pierce armor, which could be initiated and releases a large amount of chemical energy during or after its penetration process, that is, it realizes the combined effect of projectile penetration and internal explosion, thus, greatly improving the damage effect inside the target [5–7].

As the basic formula of fluorine polymer-based reactive material, the reactive damage element prepared from aluminum and polytetrafluoroethylene powder (Al/PTFE) through mechanical mixing, molding, and sintering process has been extensively studied by scholars and has achieved certain application results [8–13]. For example, in 2002, E. L. Baker et al. [14,15] proposed the technical concept of a unitary terminal chemical energetic blasting warhead and designed four different formulations of Al/PTFE reactive liners (inert metal type, oxygen-rich type, oxygen balance type, and oxygen deficiency type) to verify the damage effect of this unitary chemical energetic blasting warhead. The results show that, as compared to the aluminum liner, the RMs-shaped charge can not only penetrate the concrete target but also produces a stronger demolition effect due to the deflagration reaction in the concrete target, which greatly improves the damage efficiency of the shaped charge. Another example is its application in warhead fragments. Zhao Hongwei et al. studied the terminal demolition lethality of reactive fragments through experimental methods, and the results show that the reactive fragments have greatly improved the penetration of warheads, the ignition of the fuel, and the ability to detonate explosives [16].

However, the disadvantages of low Al/PTFE density (about 2.27 g/cm^3) were exposed when the reactive material was used to replace the steel core in traditional armor-penetrating combustion projectiles to form reactive enhanced projectiles. Generally speaking, high density is an important factor to ensure the depth of armor piercing. However, the density of Al/PTFE reactive material is much lower than that of traditional steel core, resulting that its armor-piercing ability is greatly reduced, which seriously restricts the successful application of the reactive material in anti-light armor ammunition. In order to effectively improve the density of RMs, one method is to add high-density inert metal powder (such as tungsten and tantalum, etc.), into the RMs of Al/PTFE [17–20]. Since the inert metal hardly releases energy due to chemical reaction during the impact, this method reduces the release of chemical energy while improving the density. The introduction of inert metal also significantly reduces the energy release efficiency of the reactive material under the same condition, which is not conducive to the realization of efficient damage. Another method is to introduce metal oxides (such as CuO, MoO_3, etc.), to the basic formula [21,22], which can form thermite with the Al component. Although thermite can react chemically and release a large amount of heat energy, almost no gaseous products are produced after the thermite reaction, which is not beneficial to achieving efficient damage. The ideal method is to replace the relatively light aluminum powder with higher density reactive metals, such as titanium (Ti), zirconium (Zr), hafnium (Hf), tantalum (Ta), uranium (U), etc. The reactive material composed of metal hafnium and fluoropolymer has the highest formation enthalpy, but metal hafnium has a higher activity. It is easy to react with oxygen in the air near the conventional sintering temperature of PTFE (about 380 °C), which may lead to sintering accidents. To improve the safety of Hf powder in the sintering process, fluoropolymers with lower melting and crystallization temperature can be selected, such as polyvinylidene fluoride (PVDF), tetrafluoroethylene-hexafluoropropylene copolymer (FEP), tetrafluoroethylene-hexafluoropropylene-vinylidene terpolymer (THV), etc. The fluorine content of THV 220 is 72.61%, which is the closest to the fluorine content of PTFE (76%), and it can complete melting and crystallization at about 120 °C, making it an ideal replacement material for PTFE [23]. Therefore, the RMs composed of high-density active metal powder and THV are an ideal formula for both the energy release efficiency and the density of the damage element.

Traditional PTFE-based RMs are usually prepared by a mechanical mixing–molding–sintering process, while THV products are generally large particles rather than ultra-fine powder form, so it is impossible to obtain a uniform mixture of metal powder and THV particles by traditional mechanical mixing means, the traditional preparation process is obviously not suitable for THV-based RMs. In this paper, the reactive material samples composed of metal Hf and THV were prepared by solvent evaporation and hot pressing sintering method, and the microstructure of the samples was analyzed by scanning electron microscopy. In addition, many studies have shown that the energy release of RMs is related to the fracture process of materials [24,25]. The formation of crack is related to the strain energy absorbed by the material before fracture. The larger the strain energy is, the more energy released by the material during fracture, and the easier it is to form hot spots to ignite near the crack. The strain energy absorbed by materials before fracture is closely related to their mechanical properties. Therefore, this paper intends to study the mechanical properties and fracture mechanism of THV-based RMs, and the results are conducive to a deeper understanding of the energy release mechanism of RMs.

2. Sample Preparation

The Hf/THV specimens were prepared by the process of solvent loss and hot pressing. The raw Hf is particle powders with the size of 10 μm, and THV 220 granules are a flexible, transparent fluoroplastic composed of tetrafluoroethylene, hexafluoropropylene, and vinylidene fluoride in the form of melt pellets, as shown in Figure 1. The preparation steps were as follows: (1) the Hf powders and THV 220 granules were poured into ethyl acetate solvent, and the concentration of THV220 should be controlled at less than 0.1 g/mL. Then, the formed solution was heated to 70–90 °C and stirred continuously until THV220 dissolved completely. (2) The solution temperature was kept at about 80 °C for distillation. To make the metal powder uniformly dispersed, continuous stirring is required during distillation until the solvent completely evaporated, resulting in a dry solid block consisting of Hf and THV 220. (3) The solid block was put into the heating mold that was placed on a universal testing machine, and the mold was heated directly to 120–160 °C in the vacuum drying box and kept warm for 30 min. Then, the pressure was loaded on the mold to 15 MPa, at a speed of 0.05 MPa/s. Finally, the sample of desired shape was obtained after 30 min of holding pressure. The density of the RMs samples fabricated by this method could reach more than 92% of theoretical maximum density (TMD). In addition, the basic formula Al/PTFE specimens were also prepared for comparison, the average size of Al powder is 10 μm as shown in Figure 1. The previous method was employed to fabricate these Al/PTFE specimens [26]. The specimen formulas and density of PTFE, THV-based RMs are listed in Table 1. The specimen density is the average density of five specimens.

(a)

(b)

Figure 1. *Cont.*

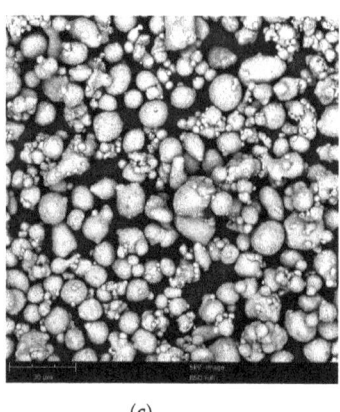

(c)

(d)

Figure 1. The scanning electron microscope (SEM) images of the raw materials: (**a**) Hf powder; (**b**) THV 220 granules; (**c**) Al powder; (**d**) PTFE powder.

Table 1. The specimen formulas and theoretical energy content.

NO.	Specimen Formula	Theoretical Density (g/cm^3)	Specimen Density (g/cm^3)	Compactness
1	26.5% Al/73.5% PTFE	2.31	2.31	100%
2	5.29% Al/80% W/14.71% PTFE	7.73	7.73	100%
3	62% Hf/38% THV	4.14	3.93	94.93%
4	88% Hf/12% THV	7.83	7.28	92.98%

Field emission scanning electron microscope (Phenom Pure, The Netherlands, desktop scanning electron microscope) was employed to investigate the microstructures of the specimens. As shown in Figure 2, Hf powders are uniformly scattered in the THV 220 matrix, but some cavities are observed in THV-based specimens, resulting in relatively low compactness for them. This may be caused by the poor compatibility between Hf and THV 220. In contrast, PTFE and Al are tightly aggregated in PTFE-based specimens, leading to high compactness of nearly 100%.

(a)

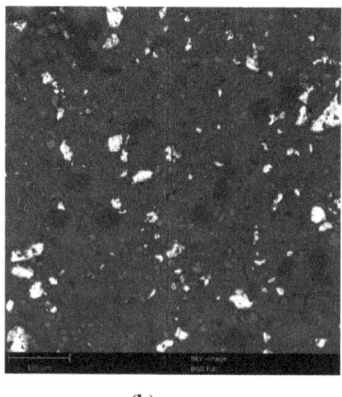

(b)

Figure 2. *Cont.*

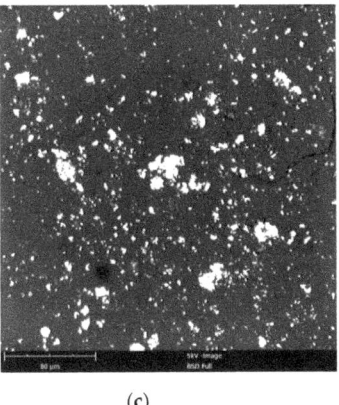
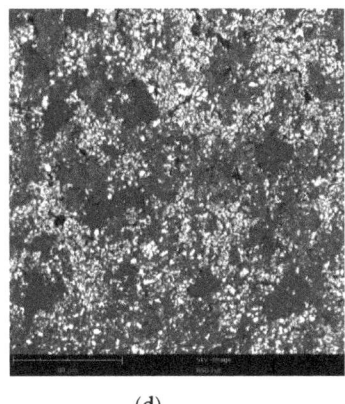

(c)   (d)

Figure 2. Microstructures of RMs specimens: (**a**) 26.5% Al/73.5% PTFE; (**b**) 5.29% Al/80% W/14.71% PTFE; (**c**) 62% Hf/38% THV; (**d**) 88% Hf/12% THV.

3. Experiment

3.1. Quasi-Static Compressive Test

The cylindrical specimens with a size of Φ12 mm × 12 mm were fabricated using the methods mentioned above. The static compressive test was carried out on the CMT 5105 electronic universal mechanical testing machine (MTS Industrial Systems Co., Ltd., Eden Prairie, MN, USA) at room temperature, with an initial loading strain rate of $0.8 \times 10^{-2} \text{ s}^{-1}$. The end surfaces of the specimens were polished with lubricating oil to reduce the friction with the punch of the testing machine. Under constant strain rate, three specimens were tested for each type of RMs to obtain quasi-static compressive performance curves, which are illustrated in Table 2 and Figure 6. The data in Table 2 are their average value.

Table 2. Quasi-static compressive performance of RMs.

No.	Elasticity Modulus (MPa)	Yield Stress (MPa)	Yield Strain	Compressive Strength (MPa)	Failure Strain
1	728	13.30	0.0322	72.57	2.14
2	661	16.89	0.0308	68.60	2.05
3	665	28.85	0.1205	55.12	2.18
4	3550	79.39	0.0257	98.17	0.06

3.2. SHPB Test

The split Hopkinson pressure bars (SHPB) system was used to investigate the dynamic compressive mechanical performance of RMs. The test system is mainly composed of one stage light gas gun subsystem, the bullet with the size of 250 × 12 mm, the compressive bar and signal collecting subsystem, etc. The different strain rates were achieved by changing the charge pressure of light gas. After the signals in the incident and transient bar are collected, the incident, reflected, and transmitted waves (ε_i, ε_r, and ε_t, respectively), are obtained when divided by an amplification that is defined as follows:

$$\varepsilon_i = \frac{U_i}{C_{amp}}, \ \varepsilon_r = \frac{U_r}{C_{amp}}, \ \varepsilon_i = \frac{U_t}{C_{amp}}, \ where \ C_{amp} = \frac{k_1 k_2 U_0}{k_3} \tag{1}$$

where k_1, k_2, k_3 are sensitivity coefficient of strain gage, amplification of dynamic strain meter, and the amplification related to the bridge circuit. U_0 is the bridge voltage. Then the strain, stress, and strain rate in the specimen could be calculated as,

$$\varepsilon_s = \frac{c_0}{l_s} \int (\varepsilon_i - \varepsilon_r - \varepsilon_t)dt = -\frac{2c_0}{l_s} \int \varepsilon_r dt$$
$$\sigma_s = \frac{EA}{2A_s}(\varepsilon_i + \varepsilon_r + \varepsilon_t) = \frac{EA}{A_s}\varepsilon_t \tag{2}$$

where ε_s, σ_s are the strain and stress in the specimen, c_0 is the material sound speed of the bars, l_s is the length of the specimen. The dynamic mechanical performance of RMs is shown in Table 3 and Figure 7.

Table 3. Dynamic compressive performance of RMs.

Formula	Strain Rate (s⁻¹)	Yield Stress (MPa)	Yield Strain	Compressive Strength (MPa)	Failure Strain
26.5% Al/73.5% PTFE	2269	30.14	0.0173	47.21	0.20
	3905	21.44	0.0122	57.88	0.36
	5685	23.04	0.0165	82.91	0.54
	8081	34.65	0.0161	128.15	0.69
5.29% Al/80% W/14.71% PTFE	2725	36.05	0.0154	53.84	0.23
	4534	49.30	0.0337	75.90	0.40
	6318	53.52	0.0322	80.22	0.47
	9069	76.12	0.0479	101.65	0.66
62% Hf/38% THV220	4325	74.03	0.0374	86.85	0.35
	5353	85.56	0.0486	85.56	0.39
	6645	86.49	0.0302	86.49	0.49
	7201	84.32	0.0561	85.79	0.56
88% Hf/12% THV220	2411	103.49	0.0241	164.71	0.11
	4633	184.20	0.0321	184.20	0.03
	5610	211.79	0.0301	211.79	0.03
	7535	231.94	0.0318	231.94	0.03

3.3. DSC/TG Test

The thermal decomposition behavior of the samples was investigated using a Mettler Toledo TGA/DSC 3+ differential scanning calorimeter and thermogravimetric analyzer. The system was programmed to heat the samples at a rate of 20 K/min from room temperature to 800 K. Sample masses of 2 mg were loaded into the sample crucible of the TGA/DSC 3+ and the sample was slightly compacted to obtain good thermal contact between the sample and the crucible. The DSC column was first evacuated using a turbo molecular drag pump and then backfilled with nitrogen. The DSC/TG experimental results of four samples are displayed in Figure 9, in which an exotherm appears as a peak while an endotherm will appear as a valley.

3.4. Ballistic Experiment

The closed bomb vessel with the size of $\Phi100$ mm $\times$ 500 mm was employed to examine the energy release performance of RMs, as shown in Figure 3. The test system is mainly composed of light gas gun, speed network target, high-speed camera, pyrometer, overpressure transducer, and closed bomb vessel. RMs projectile was launched by light gas gun, and the projectile velocity was adjusted by changing the charge pressure in the gas chamber. Then, the velocity could be calculated based on the on–off signal of speed network target. The on–off signal is also the trigger signal for high-speed camera, overpressure transducer, and pyrometer. When impacting on the aluminum plate in the closed bomb vessel, the RMs projectile will be broken to be debris, by which the energy release phenomenon was induced, resulting in the abundant gas with high temperature and pressure. The released energy by the chemical reaction of RMs is also known as reaction enthalpy including the internal energy and pressure potential energy of the deflagration product. When the temperature and pressure of gas products are detected by pyrometer and overpressure transducer, respectively, the assessment of the energy release performance could be conducted.

$$H = m_p C_{vp} \Delta T + PV \tag{3}$$

where H is the released energy by the chemical reaction of RMs, m_p, C_{vp}, and T are the mass, specific heat and temperature of gas product, P is overpressure, V is the volume of the closed bomb vessel.

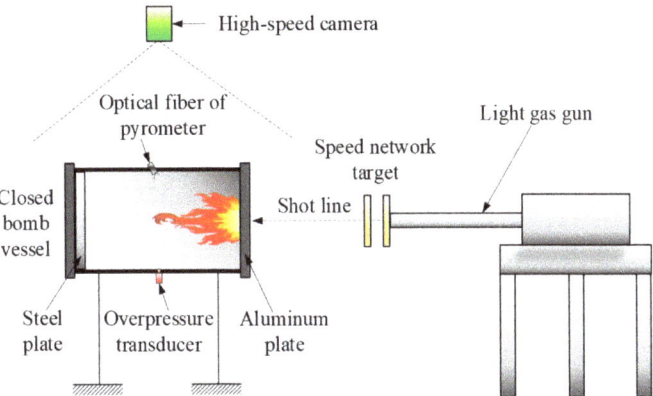

Figure 3. Test system of the energy release performance of RMs.

3.4.1. Temperature of the Gas-Phase Product

After initiation, chemical reaction occurs in RMs sample, resulting in abundant energy release in the form of internal energy of deflagration product and luminous emission that could be sensed by optical fiber probe. The optical fiber probe should be calibrated by standard light source. If the calibrated values of some wavelength were h_c, then the flash intensity could be calculated as [27,28],

$$I = \frac{h_{\exp}}{h_c l^2} \cdot \frac{N_r(\lambda)}{2\pi(1 - \cos\theta)} \tag{4}$$

where $h_{\exp}$ is the signal obtained in experiments, l is the distance between the object and optical fiber, $N_r(\lambda)$ is the illuminance of the standard light source, θ is fiber aperture angle.

After the flash intensity is achieved, the temperature of the deflagration product could be deduced based on Planck's blackbody radiation law [29].

$$I(\lambda, T) = \varepsilon \cdot c_1 \cdot \lambda^{-5} \cdot [\exp(c_2/\lambda T) - 1]^{-1} \tag{5}$$

where c_1 and c_2 are Planck constants, ε is radiation coefficient, λ is the wavelength, T is the temperature of blackbody.

Because the radiation coefficient is variational in different environments, the flash intensity results of at least two wavelengths are required in general to achieve the deflagration product, by the least square method. However, the disadvantage of this method is that it requires higher experimental equipment and measurement accuracy because small changes in single-channel observations can cause large changes in radiation temperature, and the obtained radiation temperature curve is not completely reasonable. As such, a ratio method is developed to obtain more accurate results. In the current experiments, the flash intensity of four wavelengths (400 nm, 500 nm, 600 nm, 700 nm) was recorded, as illustrated in Figure 4. Then, an expression is introduced to find the temperature,

$$S = I_1 I_3 / (I_2 I_4) = \frac{\left(\lambda_1\lambda_3 \Big/ \lambda_2\lambda_4\right)^{-5} [\exp(^{c_2} / \lambda_2 T) - 1] \cdot [\exp(^{c_2} / \lambda_4 T) - 1]}{[\exp(^{c_2} / \lambda_1 T) - 1] \cdot [\exp(^{c_2} / \lambda_3 T) - 1]} \quad (6)$$

where I_1, I_2, I_3, I_4 are flash intensity of the four wavelengths, respectively. The flash intensity has been obtained in the experiments, then the only unknown variable in Equation (6) is temperature T. From Equation (6) one can obtain the temperature of the deflagration product by numerical method. The typical temperature curves are shown in Figure 4, the results of low-velocity impact experiments could be found in Table 4.

Table 4. The performance of energy release by RMs projectile.

Formula	Density (g/cm³)	Velocity (m/s)	Over-Pressure (Mpa)	Temperature (K)	Pressure Potential Energy (kJ/g)	Internal Energy (kJ/g)	Energy Content (kJ/g)	Efficiency (%)	Duration (ms)
1	2.42	635	0.128	2514	0.50	3.41	14.64	26.69	89.28
1	2.27	1880	0.0289	2169	0.50	3.15	14.64	24.90	9.44
2	7.92	464	0.089	2929	0.11	0.71	4.34	18.86	172.21
2	7.87	1150	0.106	2261	0.53	5.09	4.34	129.33	92.06
3	3.94	450	0.031	-	0.08	-	10.03	-	4.92
3	3.97	1500	0.216	4721	1.77	11.80	10.03	135.32	79.63
4	7.45	480	0.108	4664	0.13	1.99	6.97	30.42	211.30
4	7.24	1100	0.179	3140	0.88	7.57	6.97	121.15	100.67

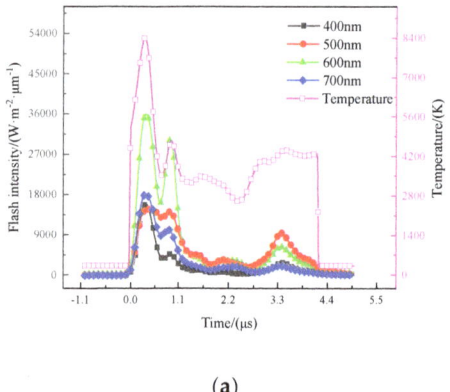

(a)

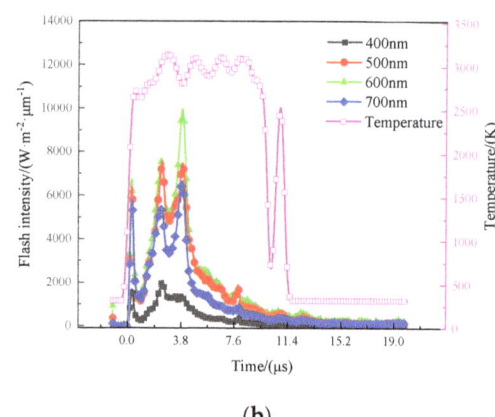

(b)

Figure 4. *Cont.*

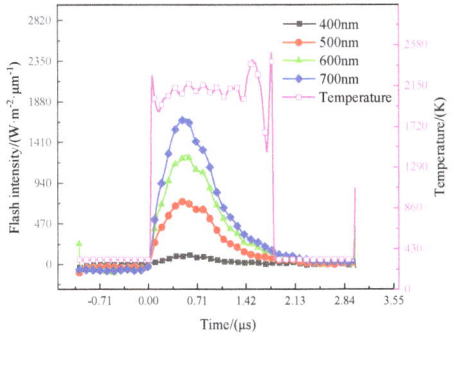

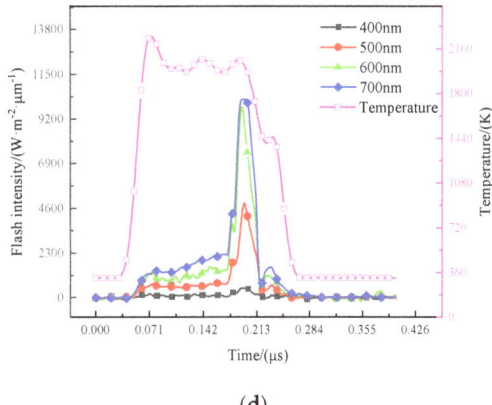

(c) (d)

Figure 4. Electrical signal recorded by transient pyrometer and temperature of deflagration flame: (**a**) 62% Hf/38% THV−1500 m/s (**b**) 88% Hf/12% THV−1100 m/s (**c**) 26.5% Al/73.5% PTFE−1880 m/s (**d**) 5.29% Al/80% W/14.71% PTFE−1150 m/s.

3.4.2. Overpressure in the Closed Bomb Vessel

Figure 5 shows the typical overpressure curve of RMs projectile in the closed bomb vessel. It can be seen from the figure that a pressure peak will be generated quickly after the deflagration reaction occurs, because the abundant gaseous products generated at the initial stage of the reaction sharply compress the air in the environment, resulting in a strong shock wave. As the reaction continues, the gas volume continues to increase, and shock waves continue to be generated. These shock waves undergo multiple reflections and superpositions in the closed bomb vessel, and the gas pressure tends to equalize, eventually forming the quasi-static pressure, which is the peak of the red dotted line in the figure. The impact of the projectile against the front aluminum plate of the container creates a pressure relief hole, and the pressure gradually decreases to the ambient pressure. The duration of overpressure can reach hundred milliseconds, while the temperature rise only occurs in tens of microseconds, which is mainly caused by the shielding of reaction products. On the other hand, the good sealing of closed space is also the reason for the longer duration of overpressure.

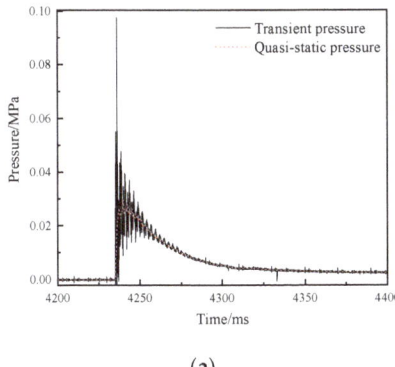

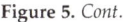

(a)

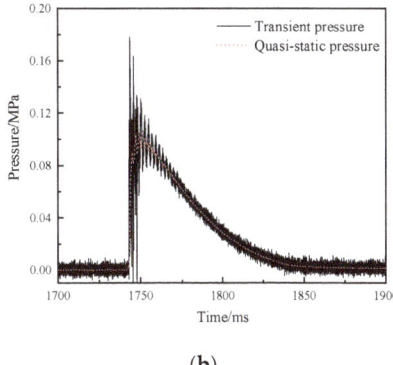

(b)

Figure 5. *Cont.*

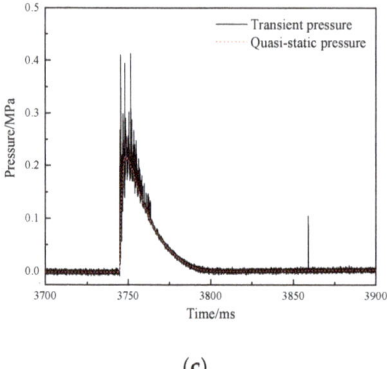
(c)

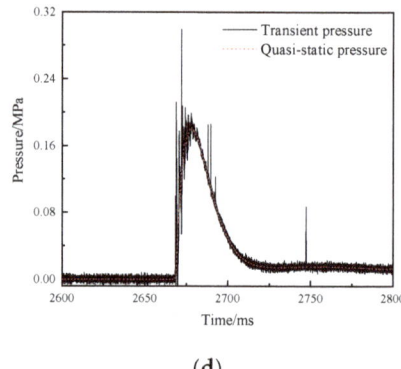
(d)

Figure 5. Typical overpressure versus time in the closed bomb vessel: (**a**) 26.5% Al/73.5% PTFE−1880 m/s (**b**) 5.29% Al/80% W/14.71% PTFE−1150 m/s (**c**) 62% Hf/38% THV−1500 m/s (**d**) 88% Hf/12% THV−1100 m/s.

4. Result and Discussion

4.1. Mechanical Performance of RMs

From Figure 6, one can see that the fracture compression strain of fluorine polymer-based RMs is more than 2.05, except for 88% Hf/12% THV specimens. This may be attribute to the low volume fraction of THV matrix in this formula. However, the fracture strength of this formula is the highest among them, which is about 93.5 MPa. In addition, there is no distinct yield stage in the loading process for this formula, it is to say, the 88% Hf/12% THV specimens have the highest brittleness. In addition, significantly different from PTFE-based RMs, the strain softening effect is observed in THV-based RMs (Figure 6c,d) immediately after the materials yield. When the specimens are further compressed, the stress is increases with strain again, until the final rupture. The mechanisms will be discussed in detail in the following section.

Figure 7 shows the dynamic compressive curves of RMs. All specimens exhibit a remarkable strain rate enhancement effect. The higher the strain rate, the stronger the materials. The stress of PTFE-based RMs is generally increasing with a strain before failure. However, the stress of THV-based RMs shows a downward trend with strain after the yield point, indicating that the THV-based RMs are strain-softening materials, whereas the PTFE-based RMs are strain-hardening materials.

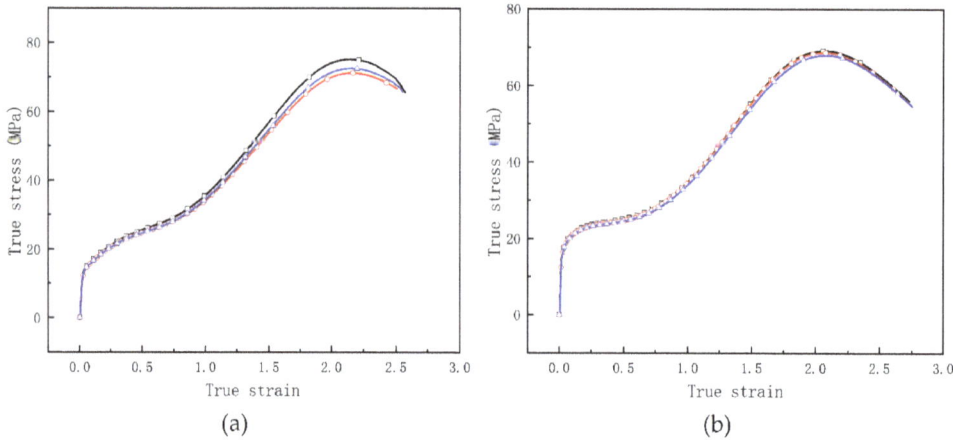

(a)   (b)

Figure 6. *Cont.*

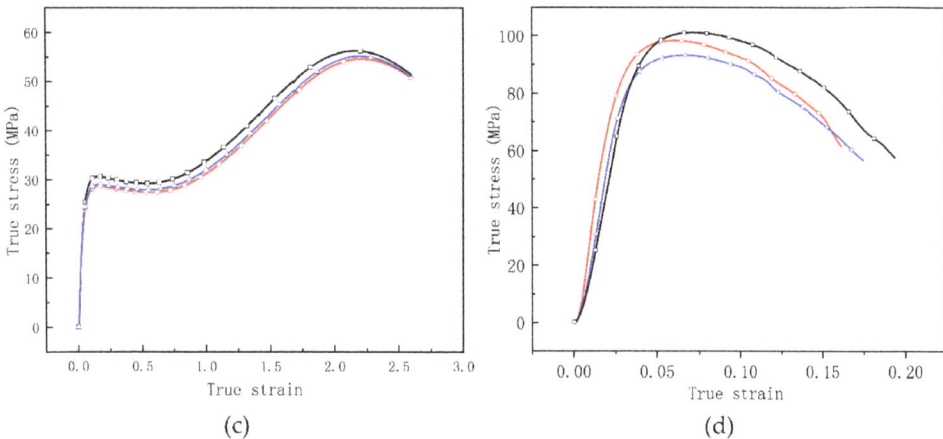

Figure 6. Quasi-static compressive performance curves of RMs: (**a**) 26.5% Al/73.5% PTFE (**b**) 5.29% Al/80% W/14.71% PTFE (**c**) 62% Hf/38% THV (**d**) 88% Hf/12% THV.

(a)

(b)

(c)

(d)

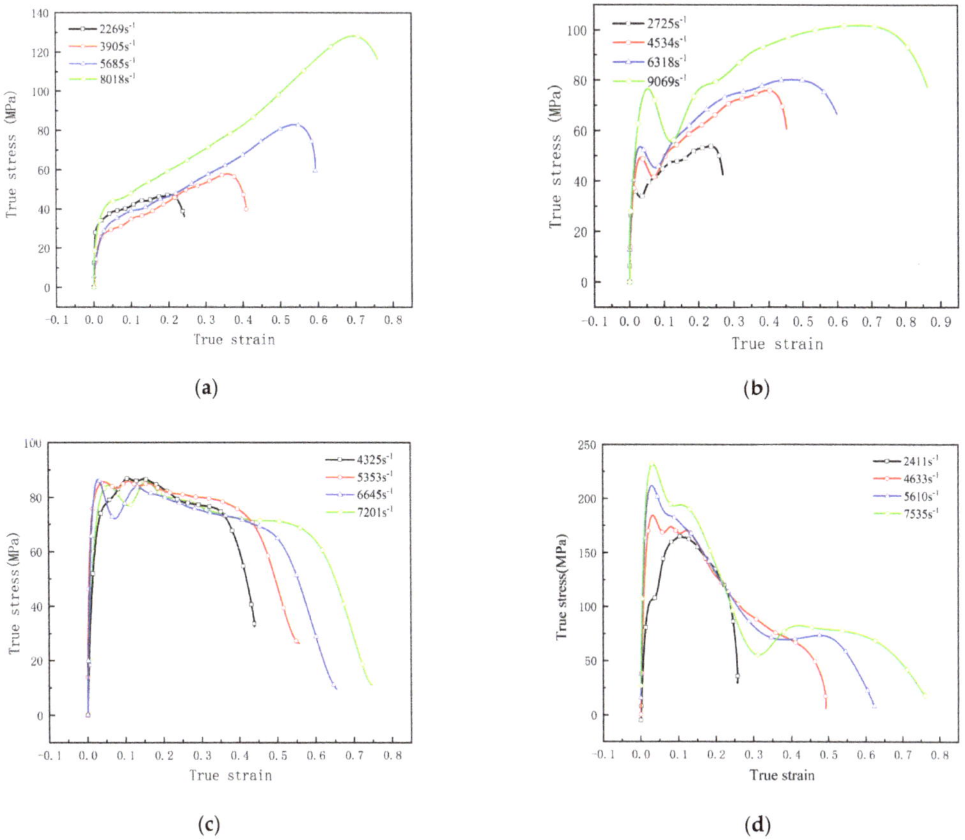

Figure 7. Dynamic compressive curves of RMs: (**a**) 26.5% Al/73.5% PTFE (**b**) 5.29% Al/80% W/14.71% PTFE (**c**) 62% Hf/38% THV (**d**) 88% Hf/12% THV.

4.2. The Fracture Mechanism

Field emission scanning electron microscope (Phenom Pure, desktop scanning electron microscope) was employed to examine the micromorphology of the surface of the specimen after compression of these RMs, as shown in Figure 8. When compression was loaded in the vertical direction, the materials would expand in the horizontal direction because of the Poisson effect, leading to the metal particles being separated from the fluorine polymer matrix (Figure 8a,b). From Figure 8, one can see that the THV presents a more remarkable orientation near the microcosmic interface with metal than PTFE. This may be mainly caused by the different glass transition temperatures of them, which has a significant effect on the movement of disordered chains of molecules in RMs. If the ambient temperature remains under the glass transition temperature, the movement of the disordered chains of molecules is restricted, and relative motion between them could be only found in a small range (Figure 8a). In this condition, the stress is increasing with the deformation of the fluorine polymer matrix until the fracture. The ambient temperature of the compression test is about 25 °C, which is far below the glass transition temperature of PTFE (115 °C). Therefore, the PTFE-based RMs are "frozen" in this case. However, the glass transition temperature of THV is 5 °C, which is below the test temperature. In this case, the disordered chains of molecules in RMs will be "unfrozen" so that the relative motion between them is promoted to a great extent under loading. For the deformation of RMs during the period of movement of disordered chains of molecules, the materials are apt to present the characteristics of the fluid, resulting in the non-continuously increased stress suffered by them.

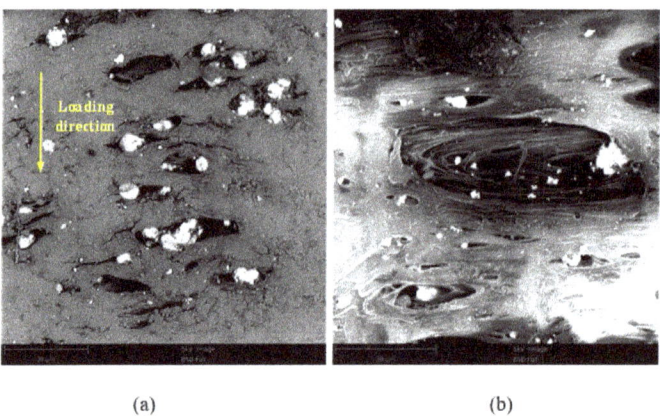

(a)   (b)

Figure 8. Micromorphology of the surface of specimen after compression: (**a**) PTFE-based RMs; (**b**) THV-based RMs. The surface is the lateral surface of the sample.

4.3. Thermal Behavior under DSC/TG Tests

The TG/DSC curves of Al/PTFE are depicted in Figure 9a. It can be determined that the endothermic peak A is the melting temperature of PTFE, and the endothermic peak C is the melting temperature of Al. The endothermic peak B covers a temperature range from 530 °C to 601 °C, at the same time, the mass of the samples decreases according to the TG curve, indicating that the small gas molecule is produced from the decomposition of the PTFE matrix. Generally, violent exothermic reactions between the small gas molecules and Al particles will immediately occur after the decomposition of PTFE [5,30]. However, violent exothermic reactions were not found before the Al particles are melted in this experiment. For mechanism consideration, the Al particles are often coated with a hard layer of alumina, which could be broken by volume expansion from the melted Al [31]. Then, a violent exothermic reaction occurs when small gas molecules meet fine aluminum, leading to the formation of the exothermic peak D in the DSC curve. When W particles are introduced to the formula, violent exothermic peak B occurs immediately after the decomposition of PTFE, which could be corresponding to the reaction between W particles and small gas molecules, as shown in Figure 9b. As such, it could be inferred that the rupture of the alumina layer outside the Al core plays an important part in the onset of the whole chemical reaction of Al/PTFE RMs. If the alumina layer is ruptured under the impact load, the chemical reaction will bring forward the decomposition temperature of PTFE.

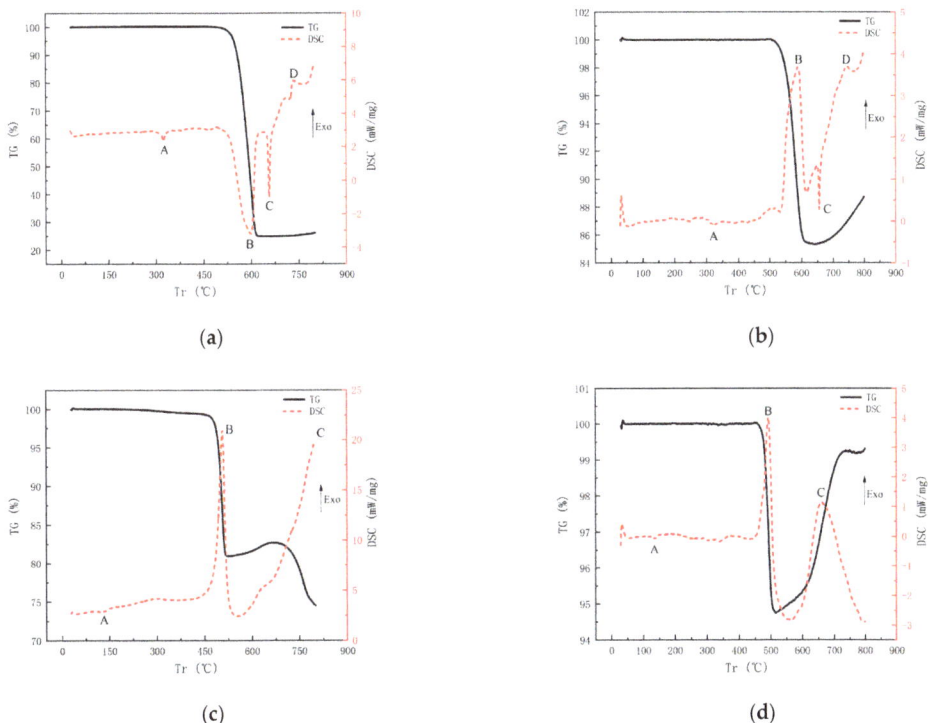

Figure 9. DSC/TG results of RMs: (**a**) 26.5% Al/73.5% PTFE (**b**) 5.29% Al/80% W/14.71% PTFE (**c**) 62% Hf/38% THV (**d**) 88% Hf/12% THV.

The TG/DSC curves of Hf/THV are shown in Figure 9c,d. The endothermic peak A is the melting endothermic peak of THV. Significantly different from Al/PTFE, two exothermic peaks (B and C) are observed in the DSC curves, both of which are accompanied by a decrease in mass (Figure 9c). The first exothermic peak B leads to a 17.9% drop in mass, which is consistent with the content of TFE (17.40%) in THV. The subsequent exothermic peak may be associated with the reaction caused by the HFP and VDF. In addition, from Figure 9b,d, one can find that the weight increase in the TG curve may be caused by the reaction product's desublimation in the crucible lid. A closer look shows that there is also an uptrend in the late TG curve in Figure 9a and the middle part of the TG curve in Figure 9c. However, due to the higher content of PTFE or THV in a and c, the weight increased by the reaction product desublimation was much smaller than the weight lost by the reaction, so the uptrend of the TG curve was not obvious. For Figure 9b,d, the content of PTFE or THV was less, and the end of the reaction was earlier, so the weight increase caused by the desublimation of reaction products was more obvious. Secondly, due to the desublimation of exothermic heat, the formation of exothermic peaks was accompanied by the rise of the TG curve.

The chemical reaction processes could be summarized as follows. For Al/PTFE and Al/W/PTFE, the possible chemical reaction could be:

$$(-C_2F_4-)n \rightarrow nC_2F_4$$
$$4Al + 3C_2F_4 \rightarrow 4AlF_3 + 6C$$
$$W + C_2F_4 \rightarrow WF_4 + 2C$$

For Hf/THV, it could be:

$$(-C_2F_4-)_n(-C_3F_6-)_m(-CH_2CF_2-)_l \rightarrow nC_2F_4 + mC_3F_6 + lCH_2CF$$
$$Hf + C_2F_4 \rightarrow HfF_4 + 2C$$
$$3Hf + 2C_3F_6 \rightarrow 3HfF_4 + 6C$$
$$Hf + 2CH_2CF_2 \rightarrow HfF_4 + 4C + 2H_2$$

4.4. The Energy Release Behavior of RMs

The energy release process in the closed vessel with the front aluminum plate of 2 mm and the back steel plate of 10 mm captured by the high-speed camera is shown in Figure 10. For the basic formula with the impact velocity of 635 m/s, as illustrated in Figure 10a, violent chemical reactions in the 26.5% Al/73.5% PTFE RMs projectile were observed in the front of the aluminum plate, whereas the chemical reactions became a little weaker in the back of the aluminum plate. Subsequently, more violent chemical reactions were induced by the rigid steel plate, which did not stop until 89.28 ms. When the impact velocity was increased to 1880 m/s (Figure 10b), the basic formula RMs projectile released much more energy at the front and back of the first aluminum plate as compared to that with the velocity of 635 m/s, and the chemical reaction induced by the steel plate became less. This is to say, a large part of the energy is released outside the closed vessel with a higher impact velocity. In addition, the whole reaction time became only 9.44 ms, indicating that the rate of a chemical reaction is highly dependent on the impact conditions, including projectile material and velocity.

When introducing tungsten particles to the basic formula, a little chemical reaction of RMs projectile with the impact velocity of 464 m/s was found at the front or back of the aluminum plate, and the obvious chemical reaction was not found until it impacted the second steel plate. Significantly different from the basic formula, the chemical reaction duration was prolonged to 172.21 ms, which may be caused by the chemical reaction with a low rate between tungsten or aluminum and oxygen in the air. When the impact velocity was increased to 1150 m/s, a violent chemical reaction was observed at the front and back of the first aluminum plate. Subsequently, the residual penetrator after the first plate produced more reaction when impacting the second steel plate, the duration was reduced to 92 ms.

For the 62% Hf/38% THV RMs projectile, the energy release pictures are illustrated in Figure 10e,f. When the impact velocity is 450 m/s, a chemical reaction only occurred before the second steel plate. When the impact velocity is increased to 1600 m/s, the chemical reactions were at a competitive level before the aluminum and steel plate, and the reactions were fast, with a duration of about 6.64 ms. When increasing the Hf content to 88%, the RMs produced a remarkable chemical reaction at the back of the aluminum plate. Extremely fragmentation of RMs projectile could be concluded according to the flame pattern in Figure 10g. When impacting the second steel plate, a more violent chemical reaction occurred, and the duration was about 209.86 ms, which could be caused by the chemical reaction with a low rate between the hafnium and oxygen in the air. When the impact velocity was increased to 1100 m/s, a certain chemical reaction occurred before the first aluminum plate, whereas most of the energy was released inside the closed vessel, which met the original requirement of the RMs projectile.

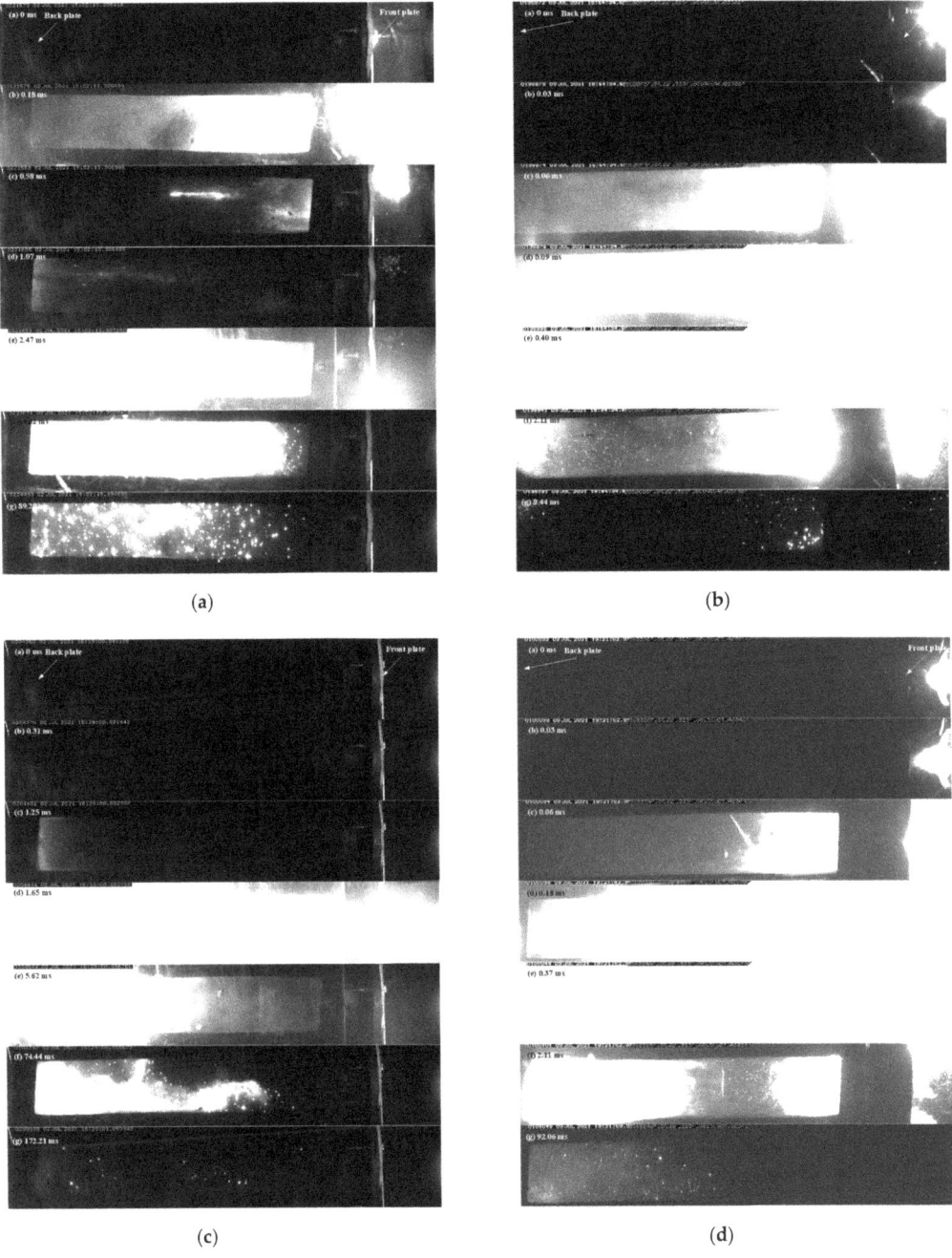

Figure 10. *Cont.*

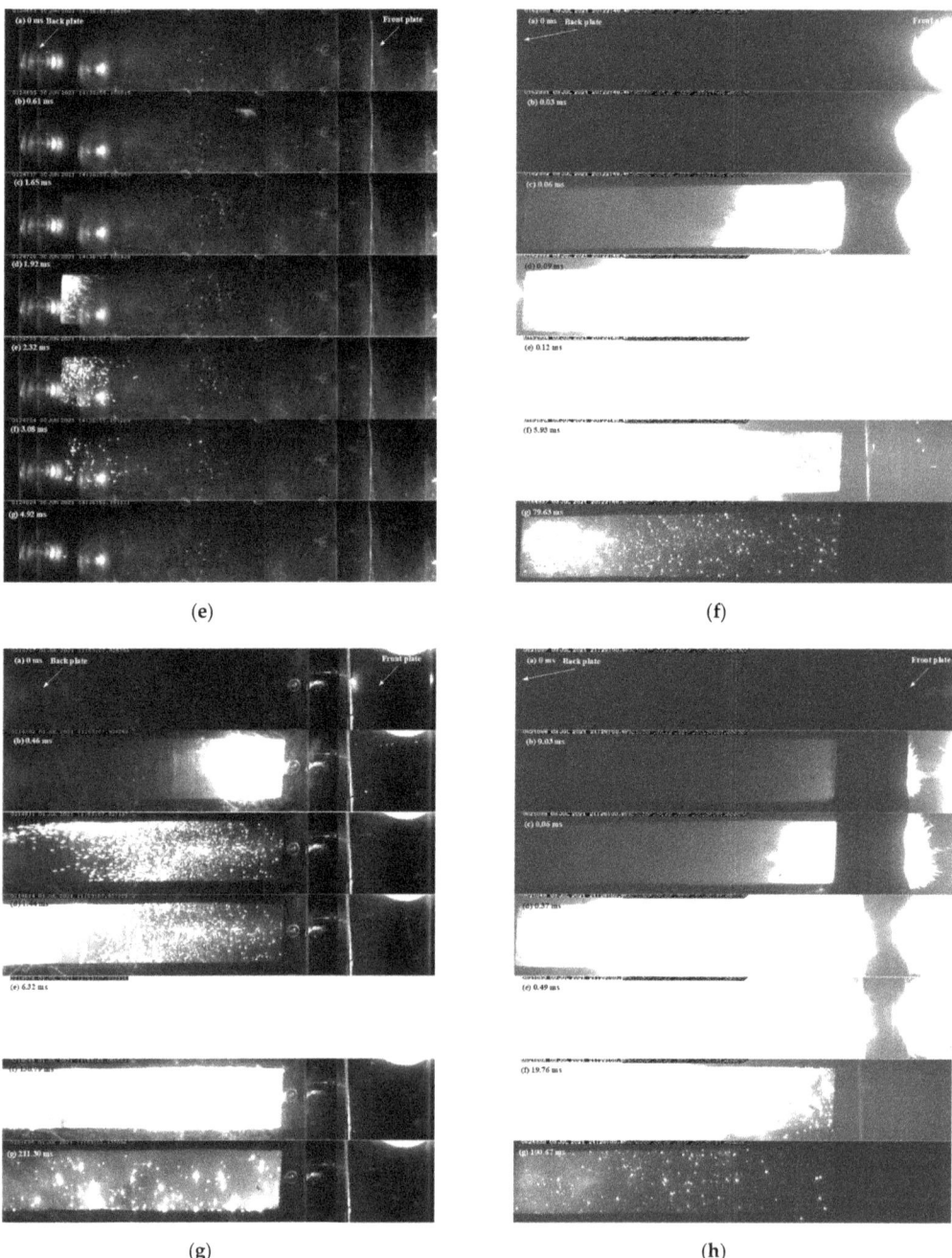

Figure 10. Energy release process of RMs: (**a**) 26.5% Al/73.5% PTFE-635 m/s (**b**) 26.5% Al/73.5% PTFE-1880 m/s (**c**) 5.29% Al/80% W/14.71%PTFE-464 m/s (**d**) 5.29% Al/80% W/14.71% PTFE-1150 m/s (**e**) 62% Hf/38% THV-450 m/s (**f**) 62% Hf/38% THV-1500 m/s (**g**) 88% Hf/12% THV-480 m/s (**h**) 88% Hf/12% THV-1100 m/s.

From the above analysis, one can conclude that the energy release process of RMs projectile is mainly dependent on the mechanical characteristics of RMs and the impact condition. The mechanical characteristics of RMs are mainly reflected in the stress–strain curve (Figures 6 and 7). The strain energy of the RMs can be obtained by integrating the corresponding stress–strain curve. Integrating the function of curves in Figure 7, the result of the integration with a strain rate of about 4000 S^{-1} is as follows: (a) 14.8588 KJ, (b) 23.8584 KJ, (c) 32.2153 KJ, (d) 4.40197 KJ. As can be seen from the results of the integration, when increasing the Hf content to 88%, the THV-based RMs become easier to be fragmented. The introduction of tungsten (W) particles to PTFE RMs make RMs not easy to be fragmented. In fact, fragmentation of RMs projectile is likely the prerequisite condition for ignition because of the general non-self-sustaining reaction in RMs. If it is the fact, the reaction process is a particle burning process in essence. The addition of W particles increases the density of the material and thus improves its penetration ability so that the strain is smaller during perforating the first Al plate than that suffered by the basic formula, leading to a smaller amount of fragmentation to release energy before and after the first Al plate (Figure 10c). The reaction threshold of THV-based RMs is higher because of their unique strain softening effect, which is not conducive to breakage. However, the reaction threshold of THV-based RMs is decreased when increasing the Hf content to 88%. This is because the high content of Hf powder improves the fracture strength of the material and increases the energy released near the crack after the fracture of the material to generate a hot spot at a higher temperature, thus making the RMs more prone to ignition reaction. In conclusion, the introduction of tungsten (W) particles to PTFE RMs could not only enhance the density but also elevate the reaction threshold of RMs, whereas the reaction threshold of THV-based RMs is decreased when increasing Hf particle content to achieve an equivalent density of RMs projectile. As such, the THV-based RMs with high density is adapted to release a lot of energy in thin, confined spaces. This has important reference significance to the design of the RMs warhead.

4.5. The Total Energy Release of THV-Based RMs

Since deflagration reaction time is very short, the heat radiated by deflagration gas products to the environment is ignored, and it is considered that there is no heat exchange between deflagration products and the external environment. Therefore, the temperature of the gas in the closed bomb vessel can be considered as its radiation temperature. Then, the final released energy by the reactive material projectile can be calculated according to Equation (3). As listed in Table 4, for the basic formula, the energy release efficiency is 26.69% when the impact velocity is 635 m/s. When the speed was increased to 1880 m/s, the energy release efficiency decreased to 24.90%. On one hand, the increase in velocity will increase the friction between the projectile and the gun barrel, resulting in the obvious temperature rising phenomenon of the projectile after leaving the muzzle. The temperature rising phenomenon under high strain rate load will make part of the RMs react outside the closed bomb vessel. On the other hand, when the velocity is increased, the impact stress between the reactive projectile and the first aluminum plate is higher, so that more RMs are broken and react in front of the aluminum plate to release energy. The result of both actions is that the RMs release a large amount of energy outside the closed container. So, even though the velocity is higher, the energy released inside the container is lower. When tungsten powder is introduced to the basic formula, the density of the projectile is close to that of steel, so the degree of breakage of the reactive projectile before the first aluminum plate is reduced, and the energy release efficiency of the reactive projectile in the container is improved. When the velocity level is the same, the energy release efficiency of THV-based RMs is lower than that of PTFE-based RMs, indicating that the reaction threshold of THV-based RMs is higher. The main reason is that the toughness of THV-based RMs is better than that of PTFE-based RMs. Under the same impact conditions, the degree of breakage of THV-based RMs is lower than that of PTFE-based RMs, and the reaction degree is lower, resulting in lower energy release efficiency. After further increasing the proportion of hafnium powder, the strength and brittleness of THV-based RMs are improved, and the difficulty of crushing is significantly reduced. Therefore, under the same impact conditions, the fragmentation degree and energy release efficiency are also improved, reaching 121.15% at high speed. At high speed, the energy release efficiency of the reactive projectile of various formulations is more than 100%, and the highest is 135.32. The reasons for this overestimation are twofold: (1) the kinetic energy of the projectile is not considered; (2) the temperature in the closed bomb vessel is constantly changing as the reaction progresses; however, the highest temperature is used to calculate the internal energy, resulting in the overestimation for the internal energy listed in Table 4.

5. Conclusions

In this work, four types of PTFE- and THV-based RMs were prepared. The mechanical performance and reactive characteristics of the materials were systematically investigated through quasi-static compressive tests, SHPB tests, SEM investigations, DSC/TG tests, and ballistic experiments. Combined with the above analysis, the impact-induced energy release process of RMs could be summarized as follows: (1) the fragmentation of RMs samples; (2) the product of small gas molecules by the decomposition of fluoropolymer matrix under impact loading (impact-induced hot spot or impact-induced fracture); (3) exposure of reactive metal to small gas molecule atmospheres; (4) burning process of the fragmentized composite particle. The main conclusions can be drawn as follows:

(1) For the compression tests with quasi-static strain rate, the THV-based RMs have a unique strain softening effect whereas the PTFE-based RMs have a remarkable strain strengthening effect, that is, the stress decreases with the increase in strain after the materials yield. This phenomenon is mainly caused by the different glass transition temperatures of them. The glass transition temperature of THV is 5 °C, which is below the test temperature. In this case, the disordered chains of molecules in RMs will be "unfrozen" so that the relative motion between them is promoted to a great extent under loading, leading to their unique strain softening effect.

(2) Thermal analysis indicates that the THV-based RMs have more than one exothermic peak because of the complex component in THV. The first exothermic peak leads to a 17.9% drop in mass, which is consistent with the content of TFE (17.40%) in THV. The subsequent exothermic peak may be associated with exothermic reaction caused by the HFP and VDF. In addition, the rupture of the alumina layer outside the Al core plays an important part in the onset of the whole chemical reaction of Al/PTFE RMs. If the alumina layer is ruptured under the impact load, a chemical reaction will bring forward the decomposition temperature of PTFE. Additionally, the increase in the TG curve is caused by the reaction product's desublimation in the crucible lid.

(3) The reaction threshold is closely related to the mechanical characteristics of RMs. The introduction of tungsten (W) particles to PTFE RMs could not only enhance the density but also elevate the reaction threshold of RMs, whereas the reaction threshold of THV-based RMs is decreased when increasing Hf particles content to achieve an equivalent density of RMs projectile. This is because the high content of Hf powder makes it easier for the RMs to be fragmented and it increases the energy released near the crack after the fracture of the material to generate a hot spot at a higher temperature, thus making the RMs more prone to ignition reaction. However, the introduction of tungsten (W) particles to PTFE RMs make RMs not easy to be fragmented. As such, under current conditions, the THV-based RMs (88% Hf/12% THV) with a high density of 7.83 g/cm^3 are adapted to release a lot of energy in thin, confined spaces.

Author Contributions: Conceptualization, J.X.; Methodology, H.W.; Supervision, H.W. and J.X.; Writing—original draft, M.G.; Writing—review & editing, Y.W.; Funding acquisition, J.X. All authors have read and agreed to the published version of the manuscript.

Funding: The research was funded by the National Natural Science Foundation of China (Grant No.11702256), Natural Science Foundation of Shanxi Province (Grant No.20210302124214), Scientific and Technological Innovation Programs of Higher Education Institutions in Shanxi (Grant No.201802071), and Scientific and Technological Innovation Team Programs of North University of China (Grant No.TD201903).

Institutional Review Board Statement: Not applicable.

Informed Consent Statement: Informed consent was obtained from all subjects involved in the study.

Data Availability Statement: The data that support the findings of this study are available from the corresponding author upon reasonable request.

Conflicts of Interest: The authors declare no conflict of interest.

References

1. Eryong, H. *Investigation of Mechanism and Performance of Spaced Ceramic Target under Impact of 12.7 mm Armor Piercing Projectile*; National University of Defense Technology: Changsha, China, 2008.
2. Shengcai, Z. *Research on Damage Mechanism of Light Armored Target by PELE*; Nanjing University of Science and Technology: Nanjing, China, 2010.
3. Nielson, D.; Ashcroft, B.; Doll, D. Reactive Material Enhanced Munition Compositions and Projectiles Containing Same. U.S. Patent 10/801,948, 15 September 2005.
4. Nielson, D.B.; Truitt, R.M.; Ashcroft, B.N. Reactive Material Enhanced Projectiles and Related Methods. U.S. Patent 9,103,641, 11 August 2015.
5. Wu, J.X.; Liu, Q.; Feng, B.; Yin, Q.; Li, Y.C.; Wu, S.Z.; Huang, J.I.; Ren, X.X. Improving the energy release characteristics of PTFE/Al by doping magnesium hydride. *Def. Technol.* **2022**, *18*, 219–228. [CrossRef]
6. Zhang, H.; Zheng, Y.F.; Yu, Q.B.; Ge, C.; Su, C.H.; Wang, H.F. Penetration and internal blast behavior of reactive liner enhanced shaped charge against concrete space. *Def. Technol.* **2022**, *18*, 952–962. [CrossRef]
7. Zhang, H.; Zheng, Y.F.; Yu, Q.B.; Ge, C.; Su, C.H.; Wang, H.F. Energetic Materials Based on W/PTFE/Al: Thermal and Shock-Wave Initiation of Exothermic Reactions. *Metals* **2021**, *11*, 1355.
8. Saikov, I.; Seropyan, S.; Malakhov, A.; Saikova, G.; Denisov, I.; Petrov, E. An effective way to enhance energy output and combustion characteristics of Al/PTFE. *Combust. Flame* **2020**, *214*, 419–425.
9. Wang, J.; Zhang, L.; Mao, Y.; Gong, F. Sensitivity of Al-PTFE upon Low-Speed Impact. *Propellants Explos. Pyrotech.* **2019**, *44*, 630–636.
10. Feng, B.; Qiu, C.L.; Zhang, T.H.; Hu, Y.F.; Li, H.G.; Xu, B.C. Investigation on mechanical properties and reaction characteristics of Al-PTFE composites with different Al particle size. *Adv. Mater. Sci. Eng.* **2019**, *44*, 630–636.
11. Wu, J.X.; Fang, X.; Gao, Z.R.; Wang, H.X.; Huang, J.Y.; Wu, S.Z.; Li, Y.C. Study on initiation mechanism of reactive fragment to covered explosive. *Trans. Beijing Inst. Technol.* **2012**, *32*, 786–789.
12. Wang, H.F.; Zheng, Y.F.; Yu, Q.B.; Liu, Z.W.; Yu, W. Damage effect of energetic fragment warhead. *Chin. J. Energ. Mater.* **2012**, *19*, 450–453.
13. Ge, C.; Dong, Y.; Maimaitituersun, W. Microscale simulation on mechanical properties of Al/PTFE composite based on real microstructures. *Materials* **2016**, *9*, 590. [CrossRef]
14. Baker, E.L.; Daniels, A.S.; Ng, K.W.; Martin, V.O.; Orosz, J.P. Barnie: A unitary demolition warhead. In Proceedings of the 19th International Symposium of Ballistics, Interlaken, Switzerland, 7–11 May 2001.
15. Daniels, A.S.; Baker, E.L.; DeFisher, S.E.; Ng, K.W.; Pham, J. BAM BAM: Large Scale Unitary Demolition Warheads. In Proceedings of the 23rd International Symposium on Ballistics, Tarragona, Spain, 16–20 April 2007.
16. Zhao, H.; Yu, Q.; Deng, B.; Cun, H. Experimental Study on Terminal Demolition Lethality of Reactive Fragments. *Trans. Beijing Inst. Technol.* **2020**, *40*, 375–381. [CrossRef]
17. Zhang, X.F.; Zhang, J.; Qiao, L.; Shi, A.S.; Zhang, Y.G.; H, Y.; Guan, Z.W. Experimental study of the compression properties of Al/W/PTFE granular composites under elevated strain rates. *Mater. Sci. Eng. A* **2013**, *581*, 48–55. [CrossRef]
18. Zhang, X.F.; Zhang, J.; Qiao, L.; Shi, A.S.; Zhang, Y.G.; He, Y.; Guan, Z.W. Influence of particle size grading on strength of Al/W/PTFE composite. *Ordnance Mater. Sci. Eng.* **2014**, *6*, 17–21.
19. Ren, K.; Chen, J.; Qing, H.; Chen, R.; Chen, P.; Lin, Y.; Guo, B. Study on Shock-Induced Chemical Energy Release Behavior of Al/W/PTFE Reactive Material with Mechanical-Thermal-Chemical Coupling SPH Approach. *Propellants Explos. Pyrotech.* **2020**, *45*, 1937–1948. [CrossRef]
20. Ren, K.; Chen, J.; Qing, H.; Chen, R.; Chen, P.; Lin, Y.; Guo, B. Investigation on the thermal behavior, mechanical properties and reaction characteristics of Al-PTFE composites enhanced by Ni particle. *Materials* **2018**, *11*, 1741.
21. Wang, R.; Huang, J.; Liu, Q.; Wu, S.; Wu, J.; Ren, X.; Li, Y. Impact energy release characteristics of PTFE/Al/CuO reactive materials measured by a new energy release testing device. *Polymers* **2019**, *11*, 149.
22. Huang, J.; Fang, X.; Wu, S.; Yang, L.; Yu, Z.; Li, Y. Mechanical Response and Shear-Induced Initiation Properties of PTFE/Al/MoO$_3$ Reactive Composites. *Materials* **2018**, *11*, 1200. [CrossRef]

23. Aboud, N.; Ferraro, D.; Taverna, M.; Descroix, S.; Smadja, C.; Tran, N.T. Dyneon THV, a fluorinated thermoplastic as a novel material for microchip capillary electrophoresis. *Analyst* **2016**, *141*, 5776–5783. [CrossRef]
24. Feng, B.; Li, Y.C.; Hao, H.; Wang, H.X.; Hao, Y.F.; Fang, X. A Mechanism of Hot-spots Formation at the Crack Tip of Al-PTFE under Quasi-static Compression. *Propellants Explos. Pyrotech.* **2017**, *42*, 1366–1372. [CrossRef]
25. Ge, C.; Yu, Q.; Zhang, H.; Qu, Z.; Wang, H.; Zheng, Y. On dynamic response and fracture-induced initiation characteristics of aluminum particle filled PTFE reactive material using hat-shaped specimens. *Mater. Des.* **2020**, *188*, 108472. [CrossRef]
26. Xiao, J.; Nie, Z.; Wang, Z.; Du, Y.; Tang, E. Energy release behavior of Al/PTFE reactive materials powder in a closed chamber. *J. Appl. Phys.* **2020**, *127*, 165106. [CrossRef]
27. Xiao, J. *Experimenl Study on the Physical Characteristics of Flash Produced by Hypervelocity Impact with LY12 Aluminum Plate*; Shenyang Ligong University: Shenyang, China, 2012.
28. Yafei, H.; Enling, T.; Liping, H.; Meng, W.; Kai, G.; Jin, X.; Ruizhi, W.; Zhenbo, L. Evolutionary characteristics of thermal radiation induced by 2A12 aluminum plate under hypervelocity impact loading. *Int. J. Impact Eng.* **2019**, *125*, 173–179. [CrossRef]
29. Thornhill, T.F.; Reinhart, W.D.; Chhabildas, L.C. Characterization of prompt flash sig-natures using high-speed broadband diode detectors. *Int. J. Impact Eng.* **2008**, *35*, 827–835. [CrossRef]
30. Osborne, D.T.; Pantoya, M.L. Effect of Al particle size on the thermal degradation of Al/Teflon mixtures. *Combust. Sci. Technol.* **2007**, *179*, 1467–1480. [CrossRef]
31. Dreizin, E.L.; Schoenitz, M. Correlating ignition mechanisms of aluminum-based reactive materials with thermoanalytical measurements. *Prog. Energy Combust. Sci.* **2015**, *50*, 81–105. [CrossRef]

Article

Controlling Shock-Induced Energy Release Characteristics of PTFE/Al by Adding Oxides

Ying Yuan, Yiqiang Cai, Dongfang Shi, Pengwan Chen, Rui Liu and Haifu Wang *

State Key Laboratory of Explosion Science and Technology, Beijing Institute of Technology, Beijing 100811, China
* Correspondence: wanghf@bit.edu.cn; Tel.: +86-010-6891-5848

Abstract: Polytetrafluoroethylene (PTFE)/aluminum (Al)-based energetic material is a kind of energetic material with great application potential. In this research, the control of the shock-induced energy release characteristics of PTFE/Al-based energetic material by adding oxides (bismuth trioxide, copper oxide, molybdenum trioxide, and iron trioxide) was studied by experimentation and theoretical analysis. Ballistic impact experiments with impact velocity of 735~1290 m/s showed that the oxides controlled the energy release characteristics by the coupling of impact velocities and oxide characteristics. In these experiments, the overpressure characteristics, including the quasi-static overpressure peak, duration, and impulse, were used to characterize the energy release characteristics. It turned out that when the nominal impact velocity was 735 m/s, the quasi-static overpressure peak of $PTFE/Al/MoO_3$ (0.1190 MPa) was 1.99 times higher than that of PTFE/Al (0.0598 MPa). Based on these experimental results, an analytical model was developed indicating that the apparent activation energy and impact shock pressure dominated the energy release characteristic of PTFE/Al/oxide. This controlling mechanism indicated that oxides enhanced the reaction after shock wave unloading, and the chemical and physical properties of the corresponding thermites also affected the energy release characteristics. These conclusions can guide the design of PTFE-based energetic materials, especially the application of oxides in PTFE-based reactive materials.

Keywords: PTFE/Al/oxide; shock-induced; energy release characteristic; controlling effect; shock wave

Citation: Yuan, Y.; Cai, Y.; Shi, D.; Chen, P.; Liu, R.; Wang, H. Controlling Shock-Induced Energy Release Characteristics of PTFE/Al by Adding Oxides. *Materials* **2022**, *15*, 5502. https://doi.org/10.3390/ma15165502

Academic Editors: Chuanting Wang, Yong He, Wenhui Tang, Shuhai Zhang, Yuanfeng Zheng and Xiaoguang Qiao

Received: 25 June 2022
Accepted: 8 August 2022
Published: 10 August 2022

Publisher's Note: MDPI stays neutral with regard to jurisdictional claims in published maps and institutional affiliations.

1. Introduction

Polytetrafluoroethylene (PTFE)/Aluminum (Al), as a novel energetic material, is extensively utilized in explosion and warhead terminal damage due to its unique impact reaction characteristics and high energy density (21 kJ/cm^3) [1,2].

In recent years, a lot of studies have been conducted on the chemical reaction of PTFE/Al. The noticeable decomposition of PTFE/Al occurs at temperature above 673 K, and the main reaction products involve AlF_3, CO, and CO_2 [3]. A standardized evaluation technique for characterizing the energy release of PTFE/Al, which offers a feasible pathway to present the energy release of PTFE/Al quantitatively, has been developed [4–6]. Furthermore, considering the energy consumption of the test chamber and the energy of leakage gas, a more perfect method for calculating and measuring the impact energy release of active materials has been developed [7].

However, its applications are restricted by its low mechanical strength and low reaction efficiency due to non-self-sustaining reactions. Many energetic components, such as hydrides [8–11], active metals [12–14], and oxides, have been introduced to PTFE/Al to enhance its energy release characteristics. Among them, the effects of adding oxides on the energy release characteristics of PTFE/Al has received much attention from scholars due to the excellent reaction performance and various reaction characteristics of thermite (Al/oxide). Experiments have been conducted by self-designed energy release testing devices and the results have shown that CuO promotes the energy release efficiency of PTFE/Al [15]. Drop-weight tests have been conducted, indicating that Bi_2O_3 improves the impact sensitivity of PTFE/Al [16]. In addition, the burning speed, specific volume, and mechanical properties of PTFE/Al/Fe$_2$O$_3$ [17], and the mechanical and reaction properties and thermal decomposition of PTFE/Al/MnO$_2$ [18] have also been studied.

These pioneering works indeed have demonstrated the potential of oxides in adjusting the energy release characteristics of PTFE/Al energetic material. However, the lack of a systematic study on how the oxides control the energy release characteristics of PTFE/Al-based energetic material seriously restricts further application of PTFE/Al/oxide in weapons.

In this work, PTFE/Al and four kinds of PTFE/Al/oxide, including bismuth trioxide (Bi$_2$O$_3$), copper oxide (CuO), molybdenum trioxide (MoO$_3$), and iron trioxide (Fe$_2$O$_3$), were fabricated to investigate the shock-induced energy release characteristics by vented-chamber tests. An analytical model was developed to discuss how the oxides control the shock-induced energy release characteristics of PTFE/Al-based energetic material. The results revealed the mechanism of oxides controlling shock-induced reactions and can guide the design and application of reactive materials. In the Section 1, the development of PTFE-based reactive materials was introduced, and the studies on PTFE/Al/oxide were summarized. Imperfections in published studies were pointed out. In the Section 2, the sample preparation and energy release test setup is introduced in detail. In the Section 3, the shock-induced energy release behavior of the samples is introduced, and an analytical model is established to quantitatively describe the shock-induced energy release of reactive materials. Combined with the analytical model, the energy release characteristics of different types of reactive materials were analyzed when they impacted with 735~1290 m/s. In the Section 4, the results, analysis, and discussions are concluded.

2. Materials and Methods

2.1. Sample Preparation

There were five kinds of energetic materials fabricated in this work. The raw powders were: Al (2.78 g/cm^3, FLQT2, from Xingrongyuan, Beijing, China), PTFE (2.20 g/cm^3, MP1300, from dongfu, Shanghai, China), Bi$_2$O$_3$ (8.90 g/cm^3, 325 mesh, from Xingrongyuan, Beijing, China), CuO (6.50 g/cm^3, 325 mesh, from Xingrongyuan, Beijing, China), Fe$_2$O$_3$ (5.24 g/cm^3, 325 mesh, from Xingrongyuan, Beijing, China), and MoO$_3$ (4.69 g/cm^3, 325 mesh, from Xingrongyuan, Beijing, China). The chemical reaction information of the involved energetic mixtures is listed in Table 1.

Table 1. Summary of chemical reaction information for mixtures [19].

Mixture	Chemical Reaction Equation	Stoichiometric Ratio	Theoretical ΔH (J/g)
Al/PTFE	$4Al + 3C_2F_4 \rightarrow 4AlF_3 + 6C$	26.5/73.5	8530
Al/Bi$_2$O$_3$	$2Al + Bi_2O_3 \rightarrow Al_2O_3 + 3Bi$	10.4/89.6	2115
Al/CuO	$2Al + 3CuO \rightarrow Al_2O_3 + 3Cu$	18.4/81.6	4072
Al/MoO$_3$	$2Al + MoO_3 \rightarrow Al_2O_3 + Mo$	27.3/72.7	4698
Al/Fe$_2$O$_3$	$2Al + Fe_2O_3 \rightarrow Al_2O_3 + 2Fe$	25.3/74.7	3156

According to the stoichiometric ratio of each reaction, the PTFE/Al-based energetic materials were mixed with 20 wt.% oxide to meet the oxygen equilibrium. The specific information of samples is listed in Table 2. The actual density in Table 2 was calculated according to the actual size and mass of the sample after preparation, in which PTFE/Al/Fe$_2$O$_3$ had the lowest relative density and PTFE/Al/Bi$_2$O$_3$ had the highest relative density.

Table 2. Specific information of PTFE/Al/oxide.

Type	Oxide	PTFE/Al/Oxide [a]	ρ_{TMD}/ρ_a [b] (g/cm^3)	Relative Density	E_t [c] (kJ/g)
A	/	73.5/26.5	2.33/2.28	97.6%	8.528
B	Bi_2O_3	57.1/22.9/20.0	2.74/2.73	99.6%	7.098
C	CuO	55.5/24.5/20.0	2.69/2.64	98.1%	7.418
M	MoO_3	53.3/26.7/20.0	2.63/2.53	96.2%	7.467
F	Fe_2O_3	53.8/26.2/20.0	2.65/2.48	93.6%	7.586

[a] The mass fraction; [b] theoretical maximum density/actual density; [c] theoretical total energy.

The preparation process mainly included mixing, cold isostatic pressing, and high-temperature sintering. Firstly, the raw powders of a certain mass were added to the anhydrous ethanol solution and mixed by a blender for about 60 min, followed by a drying process at room temperature lasting 48 h. Then, the mixed powder was filled into a mold with an inner diameter of 10 mm and uniaxially cold-pressed at about 250 MPa. Finally, the cold isostatic pressing samples were placed in a vacuum sintering oven. The oven temperature rose to 370 °C at a rate of 60 °C/h, then held at 370 °C for 4.5 h, and finally brought down to room temperature at a rate of 60 °C/h.

The typical prepared PTFE/Al/oxide samples with Φ10 mm × 10 mm are shown in Figure 1. As shown in Figure 1, the PTFE/Al-based energetic materials differed in color, presenting gray-green (Bi_2O_3), black (CuO), gray-white (MoO_3), and red (Fe_2O_3). The microstructure of the PTFE/Al/oxides is shown in Figure 2. Al particles and oxide particles were wrapped in PTFE matrix, and there were a few pores between the matrix and particles and in the matrix itself. Among them, the PTFE/Al, PTFE/Al/Bi_2O_3, and PTFE/Al/CuO particles were evenly distributed and closely bonded with the matrix. The MoO_3 in the PTFE/Al/MoO_3 had an agglomeration phenomenon, and the Fe_2O_3 showed obvious porous characteristics in the PTFE/Al/Fe_2O_3.

Figure 1. Morphology of PTFE/Al/oxides: A, B, C, M, and F are PTFE/Al, PTFE/Al/Bi_2O_3, PTFE/Al/ CuO, PTFE/Al/ MoO_3, and PTFE/Al/ Fe_2O_3, respectively.

2.2. Experimental Setup

Figure 3 presents the experimental setup used to investigate the shock-induced energy release characteristics by the quasi-vented-chamber calorimetry technique. The test system mainly included a ballistic gun, chamber, pressure sensors (AK-1, measuring range from 0 to 1 MPa, sampling frequency 1 MHz), data acquisition system (TST3206), and velocity measuring instrumentation. The samples, encapsulated in nylon sabots, as shown in Figure 3a, were launched from the ballistic gun with a diameter of 12.7 mm. The velocity of the sample was controlled by adjusting the mass of gunpowder loaded into the cartridge. However, the combustion of gunpowder is complicated and affected by many factors, so the projectile velocity fluctuated within a certain range when the same charge was filled.

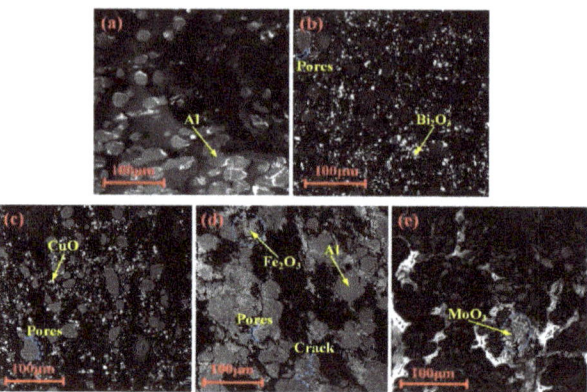

Figure 2. Microscopic characteristics of PTFE/Al/oxides: (**a**) PTFE/Al; (**b**) PTFE/Al/Bi$_2$O$_3$; (**c**) PTFE/Al/CuO; (**d**) PTFE/Al/MoO$_3$; (**e**) PTFE/Al/Fe$_2$O$_3$.

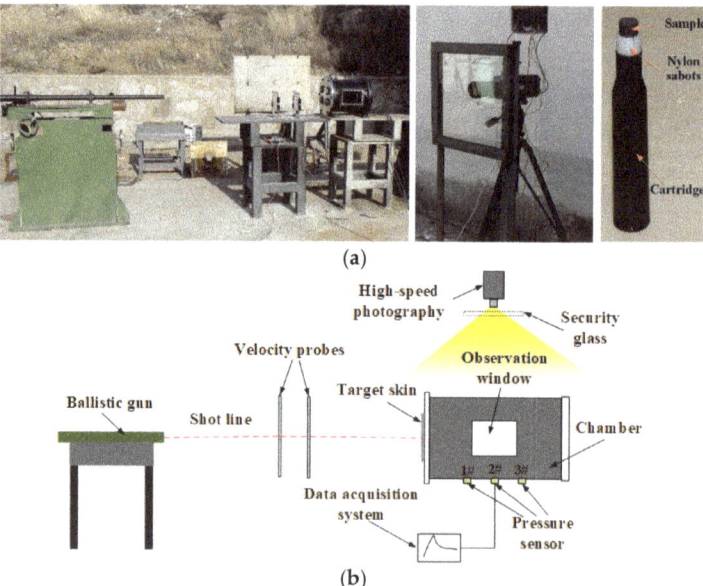

Figure 3. Experimental setup: (**a**) physical; (**b**) schematic.

The chamber with a volume of 27.35 L was sealed initially with a thin-skin plate (2024-T3 aluminum, thicknesses of 3 mm) at one end. In the interior of the chamber there was a hardened steel anvil on the other end for the energetic material to impact after passing through the target skin. In the experiment, it was considered that the test tank was a rigid body, and that no deformation occurred during the reaction of the energetic material. Three sensors were arranged in an equidistant sequence parallel with the axis of the chamber to record the overpressure characteristics. The pressure sensor was close to the inner wall of the chamber, and the sensor data was transmitted to the data processing system through signal lines. When the pressure in the chamber increased, the sensor acted as a pressure-sensitive material, and its resistance changed with the pressure change to correspond to the electrical signal change and recorded the pressure change in the chamber. The sensor began to record the experimental data when the pressure in the tank exceeded

3% of the medium range. The experiment under the same conditions was carried out three times to exclude accidental errors.

The Phantom V710 high-speed photography camera (Vision Research, Inc., Wayne, NJ, USA) was used to record the shock-induced energy release characteristics of the PTFE/Al/oxide materials. The selected frame rate was 20,000 fps so that a frame was taken every 50 µs. The resolution was 640×480 pixels and the exposure time was set to 10 µs. These settings were selected based on early testing and represent an optimal tradeoff between available lighting and the minimization of blur in the images.

3. Results and Discussion

3.1. Typical Shock-Induced Energy Release Characteristics

The energetic materials launched by the ballistic gun perforated the target skin of the chamber, and entered the test chamber with a violent reaction. The typical shock-induced reaction phenomena of the PTFE/Al/oxide energetic materials are shown in Figure 4. As shown in Figure 4, when the energetic materials impacted the skin plate, some debris of the energetic materials were formed and reacted outside the chamber (as shown in sequence 0.1 ms). It can be observed that energetic material started to react to an extent and continued on to the impact anvil (as shown in sequence 0.2 ms). When the energetic material impacted the anvil inside the test chamber, the energetic material had a more violent reaction (as shown in sequence 0.4 ms). The reaction in the test chamber lasted for tens of microseconds and then stopped gradually.

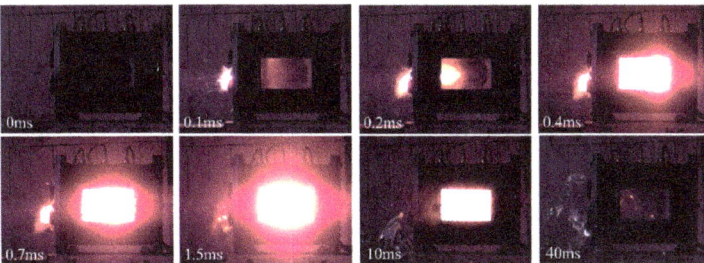

Figure 4. Typical shock-induced reaction phenomena (PTFE/Al/MoO$_3$, 910.45 m/s).

The violent exothermic reaction triggered overpressure in the test chamber, and the typical variation in overpressure with time is shown in Figure 5. As shown in Figure 5, the overpressure firstly went through a very high peak, followed by a relatively high quasi-static overpressure peak ($\Delta P_{\max}$). The first extremely high peak was caused by the initial blast of the energetic material, and the subsequent quasi-static overpressure peak was caused by the heat release by the energetic material reaction. After that, the high-pressure gas in the chamber leaked out, and the pressure in the chamber decreased. An analysis that combines the overpressure variation with high-speed photographic frames indicates that the peak of overpressure lagged behind, in time scale, the most intense reaction of energetic materials. This strongly suggests that the quasi-static overpressure in the test chamber characterized the accumulated energy released by the energetic material in the test chamber, not the reaction intensity instantaneously. Because the blast peak pressure had the characteristics of short-time and rapid attenuation, the measured peak pressure was greatly affected by the sampling frequency and sensor position. So, quasi-static overpressure was used to characterize the energy release of the shock-induced reaction.

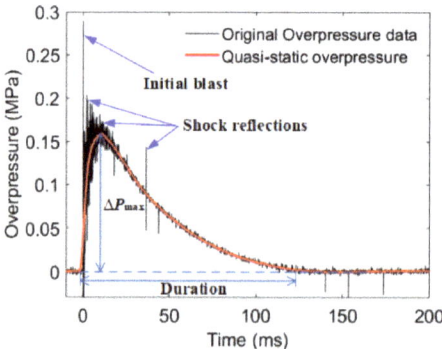

Figure 5. The typical overpressure in the chamber varied with time.

The energy released by the energetic material was adopted to characterize the quasi-static overpressure peak ($\Delta P_{\max}$). Ignoring the influence of the reaction products in the gas, which is very little, and the gas leakage through perforation, which is negligible during such a short time, it was assumed that the heat released by the energetic material is all used to heat the initial gas in the chamber. The relationship between the quasi-static overpressure peak in the chamber and the energy released by the energetic material in the chamber can be expressed as [7]

$$\Delta P_{\max} = \frac{\gamma_a - 1}{V} \Delta E,$$ (1)

where ΔE is the released energy, V is the volume of the test chamber and γ_a is the ratio of the specific heat of the gas. The reaction efficiency of the PTFE/Al/oxide can be expressed as

$$\eta = \frac{\Delta E}{\Delta E_t},$$ (2)

where η is the reaction efficiency of energetic materials, and ΔE_t is the theoretical total energy of the sample, which ignores the heat released by further side reaction.

3.2. Analytical Model of Shock-Induced Energy Release Characteristics

The shock-induced energy release mechanism of PTFE/Al/oxide is complex due to the combination effect of mechanics-thematic chemistry. When PTFE/Al/oxide energetic material impacts the skin plate, the shock wave is generated and propagates within the energetic material. Upon the adiabatic compression of the shock wave, the temperature of the energetic material increases, triggering the chemical reaction of the PTFE/Al/oxide. Based on the one-dimensional shock wave theory and the conservation of mass and momentum at the impact interface, the initial shock wave induced by the impact within the PTFE/Al/oxide can be expressed as

$$P_0 = v_0 \frac{\rho_{p0}\rho_{t0}U_pU_t}{\rho_{p0}U_p + \rho_{t0}U_t},$$ (3)

where P_0 is the initial impact shock pressure, v_0 is the impact velocity of the energetic projectile, U is the shock wave velocity, and ρ_0 is the initial density. Subscripts p and t represent energetic projectile and plate, respectively. The relationship between the shock velocity and particle velocity can be expressed as

$$\begin{cases} U_p = C_p + S_p u_p, \\ U_t = C_t + S_t u_t, \end{cases}$$ (4)

where C is the sound speed of the material, S_p is the Hugoniot parameter, and u is the particle velocity. The temperature of the energetic material rises under the compression of the shock wave, which induces a chemical reaction. The relationship between the shock wave pressure and temperature can be expressed as [20]

$$T = T_0 \exp\left[\left(\tfrac{\gamma_0}{V_0}\right)(V_0 - V_1)\right] + \tfrac{V_0 - V_1}{2C_V} P \\ + \tfrac{\exp[(-\gamma_0/V_0)V_1]}{2C_V} \int_{V_0}^{V_1} P \exp\left(\tfrac{\gamma_0}{V_0} V\right)\left[2 - \tfrac{\gamma_0}{V_0}(V_0 - V)\right] dV \tag{5}$$

where T is the temperature, γ is the Gruneisen coefficient, V is the specific volume, C_V is the heat capacity at constant volume, and P is the shock pressure. Subscript 0 represents the material parameters in the initial state, and subscript 1 represents the parameters after the shock wave compression state.

It is assumed that the PTFE/Al/oxide chemical reaction efficiency varies linearly with time. According to the research of Ortega [21], the relationship between the reaction efficiency η and temperature T of the energetic material can be expressed as

$$\frac{dT}{d\eta} = \frac{R_u T^2}{E_a}\left[\frac{1}{2\eta} - \frac{n\ln(1-\eta) + n - 1}{n(1-\eta)[-\ln(1-\eta)]}\right], \tag{6}$$

where R_u is the universal gas constant, which is 8.314 J/(mol K), E_a is the apparent activation energy, and n is the chemical coefficient related to boundary conditions and reaction mechanisms. As a typical composite energetic material, the material parameters of PTFE/Al/oxide can be estimated by its composition and content as follows [22]

$$\chi = \sum_{i=1}^{n} \chi_i m_i, \tag{7}$$

where χ is the material parameter, such as the sound velocity C, Hugoniot parameter S, heat capacity C_V, Gruneisen coefficient γ, chemical coefficient related to boundary conditions and reaction mechanisms n, and the apparent activation energy E_a, χ_i is the material parameter of each specific composition, and m_i is the mass ratio of each specific composition.

The reaction parameters of PTFE/Al/oxide, such as the chemical reaction coefficient and apparent activation energy, are approximated by considering the two kinds of reaction parameters. The chemical reaction coefficient of PTFE/Al and Al/oxide are 0.625 [23] and 0.1 [24], respectively. The apparent activation energy of PTFE/Al is 50.836 kJ mol^{-1} [23]. The activation energy of Al/oxide can be calculated by the Arrhenius kinetic model approach of the Flynn–Wall–Ozawa isoconversion method [25]. Some materials and reaction parameters involved in the calculation of the model are listed in Table 3. The effects of some properties of oxides (S, C, and γ) on the PTFE/Al/oxide properties are not considered temporarily and are replaced by those of Bi$_2$O$_3$ [26].

Table 3. Material parameter of oxide [27].

Oxide Type	Bi$_2$O$_3$	CuO	Fe$_2$O$_3$	MoO$_3$
C_v (J/(mol K))	236	530	662	521
E_a [1] (kJ/mol)	201.5	349.5	425.4	252.3

[1] represents the apparent activation energy of the corresponding thermite.

3.3. Overpressure Characteristics

Figure 6 presents the quasi-static pressure vs. time induced by the PTFE/Al/oxide with impacting at different velocities. The quasi-static overpressure characteristics of the PTFE/Al/oxide with different impact velocities are listed in Table 4. The experimental results indicate that the controlling effect of oxides on overpressures depended on the oxide type and impact velocity.

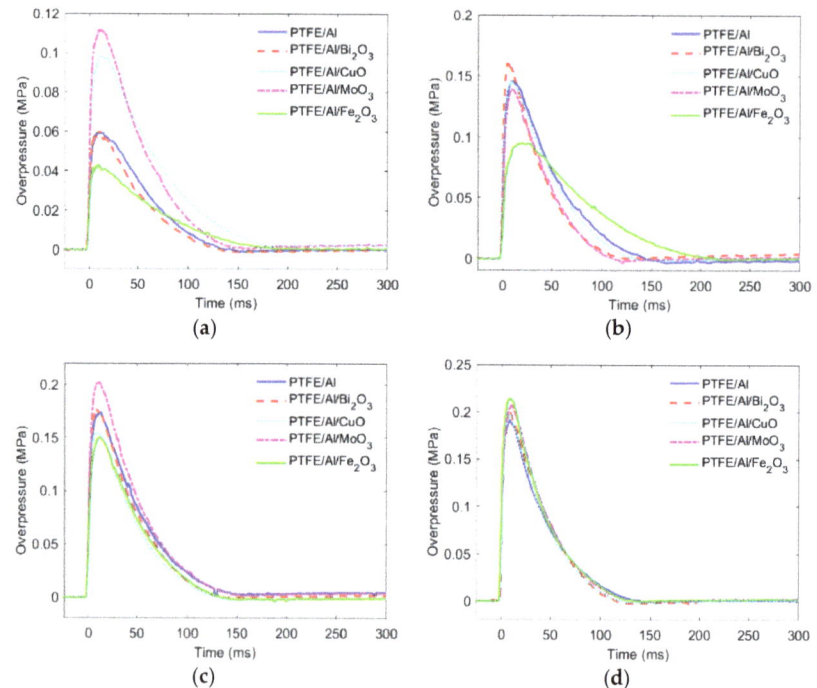

Figure 6. The quasi-static overpressure varying time: (**a**) nominal velocity 735 m/s; (**b**) nominal velocity 920 m/s; (**c**) nominal velocity 1127 m/s; (**d**) nominal velocity 1290 m/s.

Table 4. The quasi-static overpressure characteristic.

Sample	Impact Velocity (m/s)	ΔP_{max} (MPa)	Duration (ms)	Impulse (s kPa)
A	726.80	0.0598	126.76	3.6795
B	741.48	0.0602	118.88	3.2596
C	739.97	0.0987	182.44	7.1655
M	723.98	0.1190	138.96	6.5227
F	741.07	0.0434	166.72	3.1393
A	910.54	0.1466	139.74	8.2577
B	949.80	0.1619	117.04	6.7877
C	930.28	0.1456	126.40	7.6622
M	915.29	0.1397	106.38	6.4186
F	911.87	0.0953	198.96	8.5054
A	1070.43	0.1745	127.42	9.5816
B	1157.47	0.1767	125.70	8.9589
C	1098.18	0.1544	123.02	7.7877
M	1149.03	0.2023	152.02	10.883
F	1165.57	0.1503	129.08	8.1513
A	1345.80	0.1917	131.30	9.1400
B	1270.97	0.2029	114.77	9.3705
C	1309.50	0.1978	115.84	8.7886
M	1235.94	0.2078	127.32	9.7313
F	1299.46	0.2150	124.12	9.8345

As shown in Figure 6 and Table 4, the impact velocity was the most primary factor affecting the overpressure characteristics. In general, the quasi-static overpressure peak increased monotonically with the increase in impact velocity. For PTFE/Al, with the impact velocity increasing for 726.8 m/s to 1345.8 m/s, ΔP_{max} increased from 0.0598 MPa to 0.1917 MPa. Taking PTFE/Al/Fe$_2$O$_3$ for example, which was most affected by the velocity in the PTFE/Al/oxide reactive materials, the overpressure increased from 0.0434 MPa to 0.2150 MPa as the velocity increased from 741.07 m/s to 1299.46 m/s. This is because the intensity of the shock wave in the energetic material increased and the energy release efficiency of the energetic material increased. The variation trend of the overpressure duration was not obvious, and the variation rule of each type of energetic material was different, ranging from 100 ms and 200 ms. With the increase in impact velocity, the impulse generally increased. Impulse, a parameter that comprehensively considers the overpressure intensity and duration, can more reasonably characterize the performance of the energy release of energetic materials.

Obviously, different types of PTFE/Al/oxide showed different energy release characteristics under different impact velocities. When the nominal velocity was 735 m/s, the quasi-static overpressure induced by MoO$_3$ was the highest, and the impulse effect induced by copper oxide was the strongest. When the nominal velocity was 920 m/s, the quasi-static overpressure induced by Fe$_2$O$_3$ was significantly lower than that of the other four kinds of PTFE/Al/oxide energetic materials, with little difference in impulse. When the nominal velocity was 1127 m/s, the quasi-static overpressure peak and impulse induced by molybdenum oxide were the highest. When the nominal velocity was 1290 m/s, the quasi-static overpressure peak and impulse of the five kinds of energetic materials were basically the same.

3.4. Energy Release Efficiency of PTFE/Al/Oxide

Based on the shock-induced reaction model of PTFE/Al/oxide, the initial impact shock pressures of PTFE/Al/oxide with different impact velocities are presented in Figure 7. As shown in Figure 7, under the same impact velocity, the initial pressure of all kinds of PTFE/Al/oxide were higher than that of PTFE/Al, which can be attributed to the high shock impedance of the oxide. Among them, under the same nominal velocity, the initial impact pressure within PTFE/Al/Bi$_2$O$_3$ was highest. This is because Bi$_2$O$_3$ has the highest density, which is conducive to promoting the impact impedance of energetic material. In addition, with the increase in impact velocity, the increase extent of pressure of PTFE/Al/oxide increased compared with that of PTFE/Al. The analysis indicates that the high-density oxides controlled the shock-induced energy release characteristics of PTFE/Al-based energetic materials by increasing the initial impact pressure.

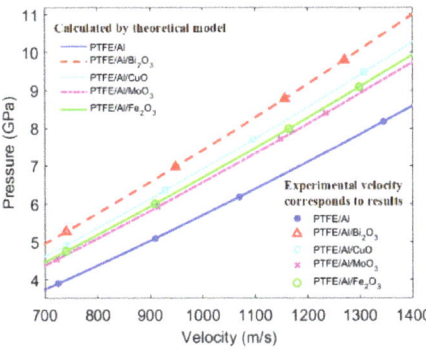

Figure 7. Pressure varying with impact velocity.

Based on the analytical model, considering the influence of the chemical reaction characteristics and apparent activation energy of energetic materials with different oxides, the energetic release efficiency of the PTFE/Al/oxide varying with impact pressure is shown in Figure 8. As shown in Figure 8, in terms of the analytical model predictions, with the increase in impact pressure, the energetic release efficiency of the PTFE/Al/oxide increased in an S-shaped tendency [28]. When the impact pressure was low (<2 GPa) or high (>10 GPa) enough, the energetic release efficiency increased slowly with impact pressure increasing. However, there are some differences in the specific change law of PTFE/Al/oxide, mainly due to the difference in activation energy and reaction coefficient of the different PTFE/Al/oxide energetic materials. On the whole, with impact pressure increasing, the energy release efficiency of PTFE/Al with MoO_3 increased at the fastest rate, and the energetic release efficiency of PTFE/Al with Fe_2O_3 increased at the slowest rate.

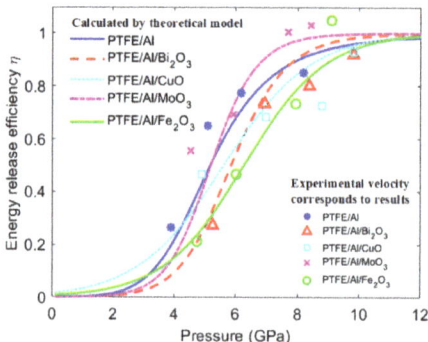

Figure 8. Energy release efficiency of PTFE/Al/oxide varying with impact pressure.

In the above analysis, it was assumed that all the energetic materials reacted in the test chamber without considering the mass of backsplash debris, and the attenuation of the shock wave during the propagation was not considered. Such an assumption will result in the energy release efficiency by calculation being higher than the actual energy release efficiency but will not affect the relative law. The shock-induced energy release characteristics have also been studied by experiments in recent research [29]. The predictions of the analytical model are also in good agreement with the literature [29].

3.5. Controlling Mechanism of Oxides on Energy Release Characteristics

In this section, the comprehensive effect of oxides on the energy release characteristics of PTFE/Al energetic materials is discussed based on the reaction mechanism. The response behaviors of PTFE/Al-based energetic materials can be distinguished as four classes, reacting from weak to strong. For Type I, no chemical reaction occurs, and the energetic materials only become densified and homogenized; for Type II, partial chemical reaction occurs in the energetic material, and the reaction stops when the pressure decays; for Type III, chemical reaction occurs in the energetic material, and the reaction continues as a self-sustaining chemical reaction after pressure unloading; for Type IV, complete chemical reaction occurs.

Under the experimental conditions in this study, as the impact velocity increased, the shock-induced reaction of energetic materials underwent Type II, Type III, and Type IV, gradually. This means that, when impact velocity was high, the energetic materials reacted completely (Type IV). In this case, the energy released by energetic materials was mainly determined by the total energy content. The energy released by the energetic materials was determined by the mass and energy release efficiency of the energetic materials involved in the reaction at other impact loads.

The control effect of the oxides on the shock-induced energy release of the energetic materials mainly reflected in controlling the Type II or Type III energetic material response. Thermites showed more significant self-sustaining property than Al/PTFE, and the addition of oxides made the reaction type of energetic materials evolve from Type II to Type

III, which improvex the energy release efficiency of energetic materials at a relatively low velocity. It is worth noting that the addition of different oxides also lead to some differences in the controlling effect, which was determined by the specific properties of the various thermites.

When the impact velocity ranged from 723.98 m/s to 1345.80 m/s, MoO_3 presented the best optimization enhancement effect on the reaction performance of PTFE/Al because of the comprehensive effect of the highest heat of Al/MoO_3 per unit mass (4.698 kJ/g), the lower ignition temperature of Al/MoO_3 (~880 K) [26], and the further reaction between reaction products (MoC_2).

Note that the control effect of Fe_2O_3 strongly depended on the impact velocity. It was attributed that as Al/Fe_2O_3 has a high onset reaction temperature (~937 K) [26], it was difficult to start its reaction under low-velocity impact, leading to the decrease in the overall energy release efficiency of the energetic material. The other reason is that the energy release per unit mass of Al/Fe_2O_3 is lower than that of Al/PTFE, and the additional Fe_2O_3 reduced the total energy content of energetic material.

The addition of Bi_2O_3 in PTFE/Al improved ΔP_{max} of the energetic materials modestly. Since the reaction temperature of the energetic materials exceeded 3000 K, the reaction product Bi formed vapor, which increased the amount of gas produced in the reaction of the energetic materials and raised the overpressure. However, the increase in gaseous product volume had a limited contribution to the increase in overpressure due to the large volume of the test chamber used in the experiments. It can be inferred that if $PTFE/Al/Bi_2O_3$ were to react in a relatively narrow space, its overpressure peak would be higher than that of other PTFE/Al/oxide energetic materials.

The analyses above indicate that the overpressure of the PTFE/Al/oxide was controlled by many factors, such as the specific heat capacity of the oxide, the reaction onset temperature of the thermite, the gas product volume of reaction, and so on.

4. Conclusions

The shock-induced energy release characteristics of PTFE/Al-based energetic material with oxides (Bi_2O_3, CuO, MoO_3, and Fe_2O_3) were studied by vented-chamber tests and by theoretical analysis. The overpressure characteristics were analyzed with consideration of the shock wave and activation energy. Furthermore, the controlling effect of oxides on PTFE/Al shock-induced energy release characteristics was analyzed and discussed. The main conclusions are drawn as follows:

(a) The experimental results indicate that the oxides controlled the shock-induced energy release characteristics, and this controlling effect was affected by the impact velocity. With a lower impact velocity (usually lower than 750 m/s), the energy release characteristics of the PTFE/Al was significantly enhanced by MoO_3, by 1.99 times. The oxides also presented a significant influence on the overpressure duration of the PTFE/Al-based energetic materials.

(b) The analytical model for PTFE/Al/oxide shock-induced energy release indicated that the oxides dominated the energy release characteristics by affecting the apparent activation energy and impact shock pressure of the energetic materials. Oxides with a high-sensitivity corresponding thermite, or with a high density could enhance the energy release performance of PTFE/Al/oxide.

(c) The mechanism of oxides controlling the shock-induced energetic behaviors of PTFE/Al energetic materials was revealed. It indicated that oxides improved the continuous reaction ability of energetic materials after shock wave unloading. The controlling effects of different oxides was determined by the chemical and physical properties of the corresponding thermites.

(d) This study fills the gap in the theoretical study of PTFE/Al shock-induced energy release behaviors and has great guiding significance for the design and application of energetic materials.

Author Contributions: Conceptualization, H.W. and Y.Y.; methodology, H.W. and Y.Y.; software, Y.Y. and Y.C.; validation, H.W. and Y.Y.; formal analysis, Y.Y., Y.C. and D.S.; investigation, Y.Y., Y.C. and D.S.; resources, H.W., P.C. and R.L.; data curation, Y.Y.; writing—original draft preparation, Y.Y.; writing—review and editing, H.W.; visualization, Y.C. and D.S.; supervision, P.C. and H.W.; project administration, H.W.; funding acquisition, P.C. and R.L. All authors have read and agreed to the published version of the manuscript.

Funding: This research was funded by the State Key Program of the National Natural Science Foundation of China, grant number 12132003. This research was funded by the State Key Laboratory of Explosion Science and Technology, grant number QNKT20-07.

Institutional Review Board Statement: Not applicable.

Informed Consent Statement: Not applicable.

Data Availability Statement: Not applicable.

Conflicts of Interest: The authors declare no conflict of interest.

References

1. Granier, J.J.; Pantoya, M.L. Laser ignition of nanocomposite thermites. *Combust. Flame* **2004**, *138*, 373–383. [CrossRef]
2. Wang, H.; Zheng, Y.; Yu, Q.; Liu, Z.; Yu, W. Impact-induced initiation and energy release behavior of reactive materials. *J. Appl. Phys.* **2011**, *110*, 074904. [CrossRef]
3. Martin, L.; Santanu, C. Theoretical Study of Elementary Steps in the Reactions between Aluminum and Teflon Fragments under Combustive Environments. *J. Phys. Chem. A* **2009**, *113*, 5933–5941. [CrossRef]
4. Ames, R.G. Energy Release Characteristics of Impact-Initiated Energetic Materials. *Mater. Res. Soc. Symp. Proc.* **2006**, *896*, 321–333. [CrossRef]
5. Ames, R.G. A Standardized Evaluation Technique for Reactive Warhead Fragments. *Int. Symp. Ballist* **2007**, *23*, 49–58.
6. Ames, R.G. Vented chamber calorimetry for impact-initiated energetic materials. In Proceedings of the 43rd AIAA Aerospace Sciences Meeting and Exhibit, Reno, Nevada, 10–13 January 2005; p. 279. [CrossRef]
7. Chen, C.; Tang, E.; Zhu, W.; Han, Y.; Gao, Q. Modified model of Al/PTFE projectile impact reaction energy release considering energy loss. *Exp. Therm. Fluid Sci.* **2020**, *116*, 110132. [CrossRef]
8. Yu, Z.S.; Fang, X.; Li, Y.; Wu, J.X.; Wu, S.Z.; Zhang, J.; Ren, J.K.; Zhong, M.S.; Chen, L.P.; Yao, M. Investigation on the Reaction Energy, Dynamic Mechanical Behaviors, and Impact-Induced Reaction Characteristics of PTFE/Al with Different TiH$_2$ Percentages. *Materials* **2018**, *11*, 2008. [CrossRef]
9. Ren, X.X.; Li, Y.C.; Huang, J.Y.; Wu, J.X.; Wu, S.Z.; Liu, Q.; Wang, R.Q.; Feng, B. Effect of addition of HTa to Al/PTFE under quasi-static compression on the properties of the developed energetic composite material. *RSC Adv.* **2021**, *11*, 8540–8545. [CrossRef]
10. Zhang, J.; Li, Y.C.; Huang, J.Y.; Wu, J.X.; Liu, Q.; Wu, S.Z.; Gao, Z.R.; Zhang, S.; Yang, L. The effect of al particle size on thermal decomposition, mechanical strength and sensitivity of Al/ZrH$_2$/PTFE composite. *Def. Technol.* **2021**, *17*, 829–835. [CrossRef]
11. Wu, J.X.; Liu, Q.; Feng, B.; Yin, Q.; Li, Y.C.; Wu, S.Z.; Yu, Z.S.; Huang, J.Y.; Ren, X.X. Improving the energy release characteristics of PTFE/Al by doping magnesium hydride. *Def. Technol.* **2022**, *18*, 219–228. [CrossRef]
12. Wu, J.X.; Wang, H.X.; Fang, X.; Li, Y.C.; Mao, Y.M.; Yang, L.; Yin, Q.; Wu, S.Z.; Yao, M.; Song, J.X. Investigation on the Thermal Behavior, Mechanical Properties and Reaction Characteristics of Al-PTFE Composites Enhanced by Ni Particle. *Materials* **2018**, *11*, 1741. [CrossRef]
13. Mozaffari, A.; Manesh, H.D.; Janghorban, K. Evaluation of mechanical properties and structure of multilayered Al/Ni composites produced by accumulative roll bonding (ARB) process. *J. Alloys Compd.* **2010**, *489*, 103–109. [CrossRef]
14. Patselov, A.; Greenberg, B.; Gladkovskii, S.; Lavrikov, R.; Borodin, E. Layered metal-intermetallic composites in Ti-Al system: Strength under static and dynamic load. *AASRI Procedia* **2012**, *3*, 107–112. [CrossRef]
15. Ding, L.L.; Zhou, J.Y.; Tang, W.H.; Ran, X.W.; Hu, Y.X. Impact Energy Release Characteristics of PTFE/Al/CuO Reactive Materials Measured by a New Energy Release Testing Device. *Polymers* **2019**, *11*, 149. [CrossRef]
16. Yuan, Y.; Geng, B.Q.; Sun, T.; Yu, Q.B.; Wang, H.F. Impact-Induced Reaction Characteristic and the Enhanced Sensitivity of PTFE/Al/Bi$_2$O$_3$ Composites. *Polymers* **2019**, *11*, 2049. [CrossRef]
17. Lan, J.; Liu, J.X.; Zhang, S.; Xue, X.Y.; He, C.; Wu, Z.Y.; Yang, M.; Li, S.K. Influence of multi-oxidants on reaction characteristics of PTFE-Al-X$_m$O$_Y$ reactive material. *Mater. Design.* **2020**, *186*, 108325. [CrossRef]
18. Zhang, J.; Huang, J.Y.; Fang, X.; Li, Y.C.; Yu, Z.S.; Gao, Z.R.; Wu, S.Z.; Yang, L.; Wu, J.X.; Kui, J.Y. Thermal Decomposition and Thermal Reaction Process of PTFE/Al/MnO$_2$ Fluorinated Thermite. *Materials* **2018**, *11*, 2451. [CrossRef]
19. Glavier, L.; Taton, G.; Ducere, J.M.; Baijot, V.; Pinon, S.; Calais, T.; Esteve, A.; Rouhani, M.D.; Rossi, C. Nanoenergetics as pressure generator for nontoxic impact primers: Comparison of Al/Bi$_2$O$_3$, Al/CuO, Al/MoO$_3$ nanothermites and Al/PTFE. *Combust. Flame* **2015**, *162*, 1813–1820. [CrossRef]

20. Feng, S.S.; Wang, C.L.; Huang, G.Y. Experimental Study on the Reaction Zone Distribution of Impact-Induced Reactive Materials. *Propellants Explos. Pyrotech.* **2017**, *42*, 896–905. [CrossRef]
21. Ortega, A.; Maqueda, L.P.; Criado, J.M. The problem of discerning Avrami-Erofeev kinetic models from the new controlled rate thermal analysis with constant acceleration of the transformation. *Thermochim. Acta* **1995**, *254*, 147–152. [CrossRef]
22. Meyers, M.A. *Dynamic Behavior of Materials*; University of California: San Diego, CA, USA, 1994.
23. Zhang, X.F.; Shi, A.S.; Qiao, L.; Zhang, J.; Zhang, Y.G.; Guan, Z.W. Experimental study on impact-initiated characters of multifunctional energetic structural materials. *J. Appl. Phys.* **2013**, *113*, 083508. [CrossRef]
24. Umbrajkar, S.M.; Schoenitz, M.; Dreizin, E.L. Exothermic Reactions in Al-CuO Nanocomposites. *Thermochim. Acta* **2006**, *451*, 34–43. [CrossRef]
25. Rehwoldt, M.C.; Yang, Y.; Wang, H.Y.; Holdren, S.; Zachariah, M.R. Ignition of Nanoscale Titanium/Potassium Perchlorate Pyrotechnic Powder: Reaction Mechanism Study. *J. Phys. Chem. C* **2018**, *20*, 10792–10800. [CrossRef]
26. Fredenburg, D.A.; Thadhani, N.N. High-pressure equation of state properties of bismuth oxide. *J. Appl. Phys.* **2011**, *110*, 063510. [CrossRef]
27. Williams, R.A.; Patel, J.V.; Ermoline, A.; Schoenitz, M.; Dreizin, E.L. Correlation of optical emission and pressure generated upon ignition of fully-dense nanocomposite thermite powders. *Combust. Flame* **2013**, *160*, 734–741. [CrossRef]
28. Xiong, W.; Zhang, X.F.; Tan, M.T.; Liu, C.; Wu, X. The Energy Release Characteristics of Shock-Induced Chemical Reaction of Al/Ni Composites. *J. Phys. Chem. C* **2016**, *120*, 24551–24559. [CrossRef]
29. Yuan, Y.; Shi, D.; He, S.; Guo, H.; Zheng, Y.; Zhang, Y.; Wang, H. Shock-Induced Energy Release Performances of PTFE/Al/Oxide. *Materials* **2022**, *15*, 3042. [CrossRef]

Article

Mechanical Properties, Constitutive Behaviors and Failure Criteria of Al-PTFE-W Reactive Materials with Broad Density

Tao Sun, Aoxin Liu, Chao Ge 🄳, Ying Yuan and Haifu Wang *🄳

Beijing Institute of Technology, 5 South Zhongguancun Street, Beijing 100081, China;
3120185143@bit.edu.cn (T.S.); 1120183024@bit.edu.cn (A.L.); gechao@bit.edu.cn (C.G.);
3120185181@bit.edu.cn (Y.Y.)
* Correspondence: wanghf@bit.edu.cn; Tel.: +86-10-68915676

Abstract: Quasi-static tension tests, quasi-static compression tests and dynamic compression tests were conducted to investigate the mechanical properties, constitutive behaviors and failure criteria of aluminum-polytetrafluoroethylene-tungsten (Al-PTFE-W) reactive materials with W content from 20% to 80%. The analysis of the quasi-static test results indicated that the strength of the materials may be independent of the stress state and W content. However, the compression plasticity of the materials is significantly superior to its tension plasticity. W content has no obvious influence on the compression plasticity, while tension plasticity is extremely sensitive to W content. Dynamic compression test results demonstrated the strain rate strengthening effect and the thermal softening effect of the materials, yet the dynamic compression strengths and the strain rate sensitivities of the materials with different W content show no obvious difference. Based on the experimental results and numerical iteration, the Johnson–Cook constitutive (A, B, n, C and m) and failure parameters ($D_1 \sim D_5$) were well determined. The research results will be useful for the numerical studies, design and application of reactive materials.

Keywords: reactive materials; Al-PTFE-W composites; tension failure; mechanical properties; Johnson–Cook modeling

Citation: Sun, T.; Liu, A.; Ge, C.; Yuan, Y.; Wang, H. Mechanical Properties, Constitutive Behaviors and Failure Criteria of Al-PTFE-W Reactive Materials with Broad Density. *Materials* 2022, *15*, 5167. https://doi.org/10.3390/ ma15155167

Academic Editor: Tomasz Trzepieciński

Received: 29 June 2022
Accepted: 21 July 2022
Published: 26 July 2022

Publisher's Note: MDPI stays neutral with regard to jurisdictional claims in published maps and institutional affiliations.

1. Introduction

Reactive materials, which not only meet a certain mechanical strength but also have chemical energy release characteristics, are classified as a kind of energetic structural material [1]. Different from metals or explosives, such material remains inert under ambient conditions but reacts violently with a large amount of chemical energy release under dynamic loading. The materials have potential applications in aerospace, fortifications and efficient damage. For example, the shield structures made of reactive materials defend against space debris more effectively than those made of inert materials [2]. Additionally, the reactive material projectile shows a combined damage of kinetic energy and chemical energy when it impacts an aluminum plate, and the average diameter of perforation is 4–19 times the diameter of the projectile [3].

Al-PTFE composites are a typical representative of reactive materials. Much research has been conducted on fabrications, mechanical property, constitutive behavior and energy release characteristics of Al-PTFE reactive materials. Joshi [4] patented a preparation process of Al-PTFE reactive materials, which mainly includes powder mixing, molding and sintering. The samples fabricated by the method can withstand a larger overload without breaking. Based on quasi-static and dynamic compression tests, Ge [5] studied the mechanical property of Al-PTFE (26.5 wt. %/73.5 wt. %) reactive materials. The results demonstrated that the materials show typical elasto-plastic behavior with prominent strain hardening, strain rate strengthening and thermal softening effect. Raftenberg [6] used the Johnson–Cook constitutive parameters and simulated the impact deformation of Al-PTFE rods. Wang [7] found a mechanical property transforming response from brittle to ductile by comparing the effects of sintering temperature, cooling rate and initial and final cooling temperature on the properties of Al-PTFE reactive materials. Through numerical simulation, Tang [8] found that with Al content increasing, the ultimate compressive strength of Al-PTFE reactive materials increased first and then decreased, and failure mode evolved from the shear failure of matrix to debonding failure of particles. Feng [9,10] found that Al-PTFE composites would react under quasi-static compression when fabricated at a pressure of 60 MPa and a sintering temperature of 350 °C; he then proposed a crack-induced initiation mechanism. Osborne [11] revealed the thermal reaction mechanism of Al-PTFE reactive materials. Mock and Holt [12] investigated the shock initiation of Al-PTFE rods using a gas gun and then obtained the relationship between the initiation pressure threshold and initiation time. Ames [13] proposed a vented chamber calorimetry for measuring and evaluating the impact energy release; the measured blast pressure reached almost 110 psig.

Higher density, greater strength, more reaction energy and greater insensitivity of reactive materials are expected for engineering applications. Although Al-PTFE composites have high energy release rates, their density and strength are relatively low compared to metals such as steel or tungsten, resulting in poor kinetic energy and penetration ability. At present, one of the most effective methods is adding high-density metals (e.g., tungsten) into the mixture of Al and PTFE powders. The density of Al-PTFE-W reactive materials can be improved to steel-like density (7.8 g/cm^3) or even higher by controlling W content. Xu [14] reported that the average yield stress of Al-PTFE-W reactive materials, fabricated at a pressure of 200 MPa and a sintering temperature of 380 °C, increased from 9.2 MPa to 23.5 MPa as the density increased from 2.61 g/cm^3 to 9.28 g/cm^3 under quasi-static compression. In addition, he found that the sintered Al-PTFE-W reactive materials showed higher strength and greater fracture toughness compared with the pressed-only materials. Herbold [15] demonstrated that materials with fine W particles have a higher ultimate compressive strength compared to those with coarse W particles. Cai [16] revealed that the failure of Al-PTFE-W reactive materials was concentrated primarily in the PTFE matrix, and that the debonding between W particles and the matrix provided the initiation and propagation of cracks. Zhou [17] found that W content influences the trend of reaction efficiency by affecting the shock temperature and the initial shock pressure of Al-PTFE-W reactive materials. Ren [18] revealed that the impact reaction process of Al-PTFE-W reactive materials includes deformation, failure and combustion reaction. Moreover, she found that the reaction threshold of the materials increases with the increase in W content. However, Ge [19] obtained an opposite conclusion, in which the impact-initiation sensitivity increased with the increased W content when the materials were fabricated at a much lower sintering temperature and shorter duration. Xu [20,21] conducted ballistic experiments and indicated that the damage of the aluminum plate impacted by an Al-PTFE-W reactive material projectile was determined directly by the mechanical strength and fragmentation degree of the materials. When describing the mechanical behavior of materials subjected to significant strain, a high stain rate and temperature, constitutive model and failure criteria play an important role in simulating the deformation and fragmentation of materials under different loadings, such as impact and blast. Zhang [22] obtained Johnson–Cook constitutive parameters of Al-PTFE-W reactive materials with W content from 0% to 50% based on quasi-static and dynamic compression tests. However, the dynamic mechanical properties and constitutive behaviors of Al-PTFE-W reactive materials with higher density are rarely reported. Moreover, the failure criteria of Al-PTFE-W reactive materials remain to be explored.

For this paper, quasi-static compression tests, quasi-static tension tests and dynamic compression tests were conducted to investigate the mechanical properties, constitutive behaviors and failure criteria of Al-PTFE-W reactive materials with W content from 20% to 80%. The effects of W content, stress state, strain rate and temperature on the mechanical properties of the materials were analyzed and discussed. Finally, based on the experimental results and numerical iteration, Johnson–Cook constitutive parameters and failure parameters were obtained.

2. Experimental

2.1. Specimen Fabrication

In this work, three different types of specimens, i.e., quasi-static compression specimens, quasi-static tensile specimens and dynamic compression specimens, were prepared to investigate the constitutive behaviors and failure criteria of Al-PTFE-W reactive materials. All specimens were designed based on the Chinese standard GB/T 7314-2017 and GB/T 228.1-2010, as shown in Figure 1. *R* is the notched radium. The preparation of the specimens mainly includes component mixing, molding and vacuum sintering. Firstly, powders of PTFE (100 nm), Al (24 μm) and W (44 μm) with a certain mass ratio were mixed uniformly by a V-shaped mild machine (Zhongcheng Pharmacy Machine Co., Ltd., Hunan, China, VH-50) for 40 min. Keeping the mass ratio between PTFE and Al at about 73.5:26.5, Al-PTFE-W reactive materials with three different W contents were prepared in this research, as tabulated in Table 1. TMD is the theoretical maximum density. After this, the uniform mixtures were isostatically pressed at a pressure of 200 MPa in a self-designed steel model with an inner diameter of 10 mm. Finally, the pressed samples were sintered in an oven filled with argon. The temperature history of the sintering cycle is shown in Figure 2a. It is noted that the specimen expands slightly after sintering. The actual density is generally lower than the theoretical maximum density. For the tensile samples in particular, additional machining was required to form the smooth round bar specimens and the notched specimens. Typical specimens are presented in Figure 2b.

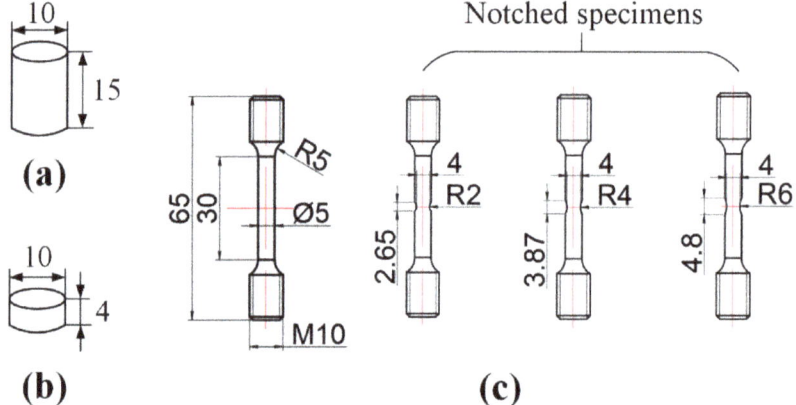

Figure 1. Schematics of specimens prepared: (**a**) quasi-static compression specimens, (**b**) dynamic compression specimens and (**c**) quasi-static tensile specimens.

Table 1. Compositions and densities of the Al-PTFE-W reactive materials.

Material Types	Component Mass Ratios (wt. %)			TMD (g·cm^{-3})	Actual Density (g·cm^{-3})
	Al	PTFE	W		
M20	21.2	58.8	20	2.78	2.65
M50	13.2	36.8	50	4.09	3.97
M80	5.3	14.7	80	7.8	7.66

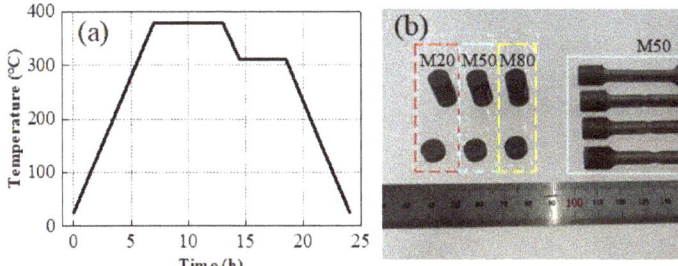

Figure 2. (a) Temperature history of sintering cycle and (b) typical specimens.

2.2. Quasi-Static and Dynamic Test Methods

1. Quasi-static compression and tension

The quasi-static compression and tension tests were conducted using the MTS universal material testing machine. For the quasi-static compression specimens and the smooth bound bar specimens, the strain rate, the true stress and the true strain can be expressed as

$$\dot{\varepsilon}_s(t) = \frac{v(t)}{l_0} \tag{1}$$

$$\sigma_s(t) = \left(1 \pm \frac{\Delta l}{l_0}\right) \frac{F(t)}{A_0} \tag{2}$$

$$\varepsilon_s = \pm \ln\left(1 \pm \frac{\Delta l}{l_0}\right) \tag{3}$$

where $\dot{\varepsilon}_s(t)$ is the strain rate, $v(t)$ is the loading speed, l_0 is the effective length of the specimen and $\sigma_s(t)$ and $\varepsilon_s(t)$ are the true stress and the true strain, respectively. $F(t)$ is the loading force, A_0 is the initial cross-sectional area of the specimen. '+' and '−' correspond to tension and compression, respectively. To correspond to an equal strain rate of 0.001 s^{-1}, the loading speeds $v(t)$ were set to 0.9 mm/min and 1.8 mm/min for compression and tensile condition, respectively.

2. Dynamic compression at elevated temperatures

The dynamic compression tests were performed using the Split Hopkinson Pressure Bar (SHPB) system equipped with 16 mm diameter 7075 T6 aluminum bars, as shown in Figure 3. Prior to the experiment, the specimen is set between the incidence bar and the transmitted bar. The striker bar is driven by the gas chamber to impact the incidence bar, which results in a compression stress wave that propagates through the bar and provides impact loading on the specimen. The impact velocity can be controlled by the gas chamber. When the compression stress wave propagates to the interface of the incidence bar and the specimen, the reflected wave and the transmitted wave are formed and then propagate toward the incidence bar and the transmitted bar, respectively. The strain gauges glued on the bars record the strain signals. Combined with the one-dimensional stress wave theory and the strain signals, the true stress–strain curves of the specimen can be calculated [5], as presented by Equations (4)–(6):

$$\dot{\varepsilon}_d(t) = \frac{C_0}{l_0}[\varepsilon_i(t) - \varepsilon_r(t) - \varepsilon_t(t)] \tag{4}$$

$$\sigma_d(t) = \frac{E_b A_b}{2A_0}[\varepsilon_i(t) + \varepsilon_r(t) + \varepsilon_t(t)] \tag{5}$$

$$\varepsilon_d(t) = -\frac{C_0}{l_0}\int_0^t [\varepsilon_i(t) - \varepsilon_r(t) - \varepsilon_t(t)]dt \tag{6}$$

where $\dot{\varepsilon}_d(t)$ is the strain rate and C_0 is the sound speed in the bar. $\varepsilon_i(t)$, $\varepsilon_r(t)$ and $\varepsilon_t(t)$ are the incidence, reflected and transmitted strain pulse, respectively. $\sigma_d(t)$ and $\varepsilon_d(t)$ are true stress and strain of the specimen, respectively. E_b and A_b are the Young's modulus and the cross-sectional area of the bar, respectively. Note that the dynamic loading of the specimens at elevated temperatures requires the employment of a furnace. The temperature of the furnace was set to 25 °C, 100 °C, 150 °C and 200 °C, respectively.

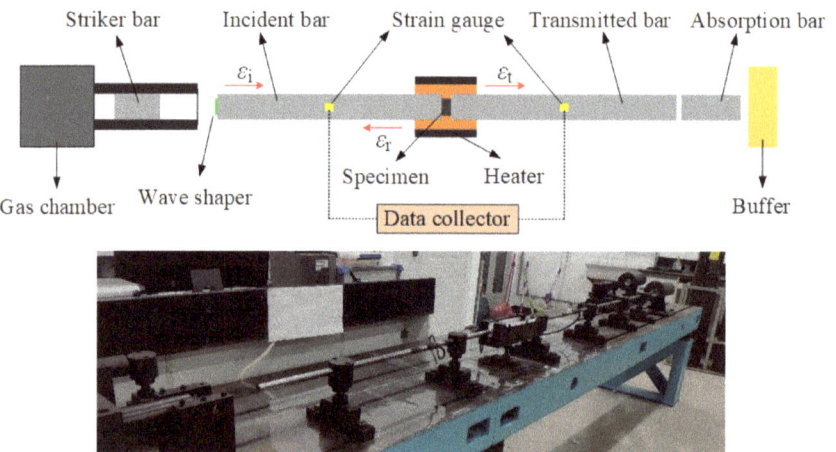

Figure 3. The SHPB system.

3. Results and Discussion

3.1. Stress State Analysis under Quasi-Static Condition

The true stress–strain curves of quasi-static compression and smooth round bar specimens are shown in Figure 4. From the stress–strain curves, typical elastic–plastic properties could be observed. When the ultimate strength is attained, the materials fail and the stress drops rapidly. Relevant material characteristic parameters were obtained according to the curves, as listed in Table 2. Herein, the elastic modulus is defined as the slope of the straight elastic stage. For the curves with obvious yield process, the yield strength is defined by the lower yield point. For the curves without obvious yield, the yield strength is determined by a 0.2% plastic offset [5,22].

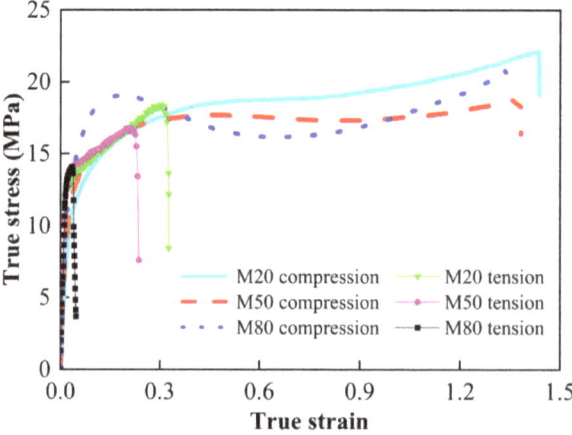

Figure 4. True stress–strain curves in quasi-static tests.

Table 2. Characteristic parameters of the Al-PTFE-W materials under quasi-static condition.

	Material Type	Elastic Modulus (MPa)	Yield Strength (MPa)	Ultimate Strength (MPa)	Critical Failure Strain
Compression	M20	508.5	10.6	22.1	1.43
	M50	628.8	11.4	19.1	1.38
	M80	734.6	16.5	20.9	1.34
Tension	M20	806	12.2	18.4	0.3
	M50	831	13.2	16.7	0.22
	M80	874	13.4	14.1	0.036

From Table 2, it is found that the ultimate strength under compression is slightly higher than that under tension, while the critical failure strain under compression is significantly higher than that under tension. The elastic modulus and the yield strength both show an increasing tendency with the increased W content, indicating the W particle strengthening effect on the deformation resistance of the materials. However, the critical failure strain shows a decreasing trend with the increased W content. Generally, the failure modes of the materials mainly include the shear failure of matrix and the debonding failure of particles [8]. The increased W particles break the continuity of the PTFE matrix and accelerate the formation of cracks, thus resulting in the premature failure of the materials. It should be noted that as the W content increases from 20% to 80%, the critical failure strain decreases by 6.2% and 88% under compression and tension, respectively. The significant difference indicates that the materials' failure is more sensitive to W content under tension than under compression. This may be attributed to force chain effect [8]. Under compression, the metal particles are linked with each other from the top to the bottom in the matrix, forming several force chains which bear and transmit the main compressive load. The force chains effectively prevent further deformation of the PTFE matrix. Furthermore, the force chain effect is enhanced as W content increases. However, no force chain is formed when the materials are tensioned. The addition of W particles only accelerates the materials' failure.

Figure 5 presents the samples after quasi-static loading. For the smooth round bar specimens, no obvious necking is observed, while the compressed specimens assume the shape of a round cake. An obvious crack appears in the M80, indicating again that the increased W particles break the continuity of the PTFE matrix. Furthermore, this may provide a failure mechanism induced by internal damage accumulation. It should be noted that in the quasi-static tensile test, both ends of the sample are connected to the MTS universal material testing machine by fixtures. Due to the low strength of the materials, the clamped thread section seriously deforms after the test.

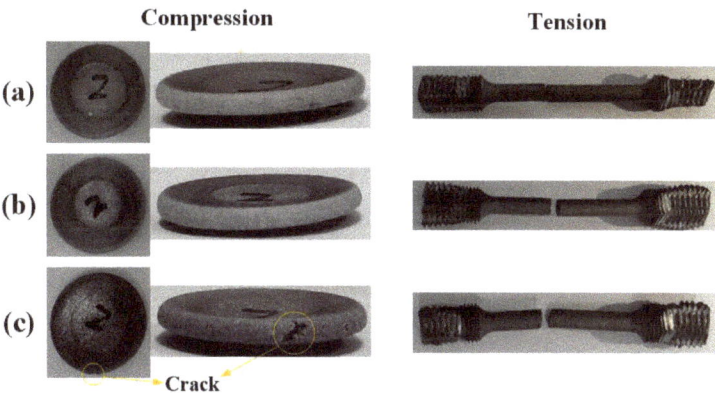

Figure 5. Samples after quasi-static loading. (**a**) M20, (**b**) M50 and (**c**) M80.

The load–displacement curves of smooth and notched bound bar specimens under quasi-static tension, as well as the photographs of the specimens after loading, are shown in Figure 6. The smooth bound bar is regarded as $R = \infty$. It is evident in Figure 6 that the elongation at break increases with the increased notch radius, which is consistent with the properties of metal materials [23]. In addition, it is found in Figure 6 that the elongation at break of the notched specimen decreases with the increased W content, which is consistent with the variation in the smooth bound bar specimen.

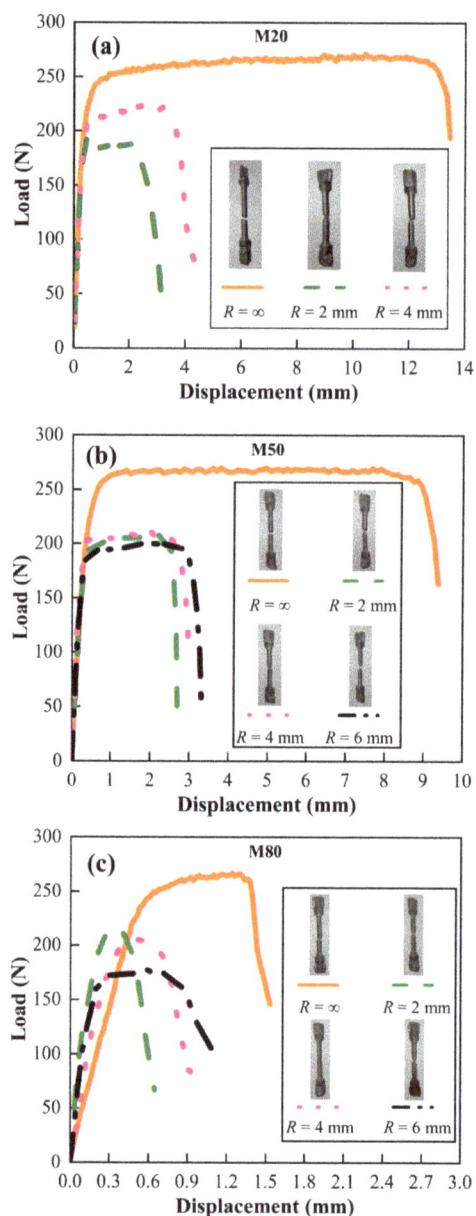

Figure 6. Load–displacement curves in quasi-static tensile tests for (**a**) M20, (**b**) M50 and (**c**) M80.

In order to obtain the fracture strain of tensile specimens, the method described in reference [24] was adopted. The method is thus briefly described here. Firstly, the finite element software Abaqus/Standard is used to establish the numerical model of the tensile specimen. Then, all elements on the center section of the sample are taken as the research object, and the simulated load–displacement curve is modified through iteration until it corresponds to the experimental load–displacement curve. Finally, when the simulated load–displacement curve reaches the experimental fracture point, the simulated equivalent plastic strain is regarded as the fracture strain of the specimen. Figures 7 and 8 show the numerical model and the equivalent plastic strain cloud diagram at the fracture time of the smooth bound bar specimen, respectively. It is observed that the simulated curves are in good agreement with the experimental curves before specimen fractures. Furthermore, the stress triaxiality $\sigma*$ was also obtained by the simulation. The stress triaxiality $\sigma*$ is defined as the ratio of hydrostatic pressure σ_m to von Mises equivalent stress σ_{eq}, which can be expressed as

$$\sigma* = \frac{\sigma_m}{\sigma_{eq}} = \frac{\sigma_{11} + \sigma_{22} + \sigma_{33}}{3\sigma_{eq}} \tag{7}$$

where σ_{11}, σ_{22} and σ_{33} are the three normal stresses.

Figure 7. Numerical model of smooth bound bar specimen.

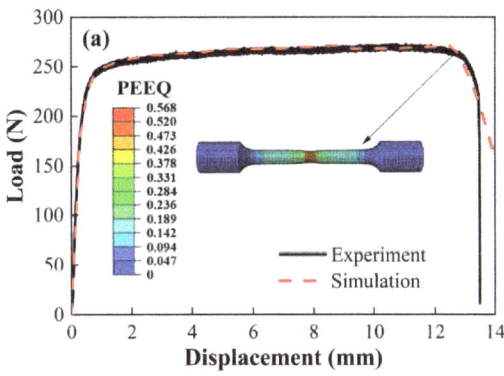

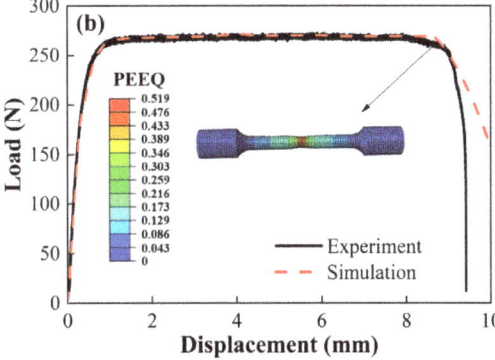

Figure 8. *Cont.*

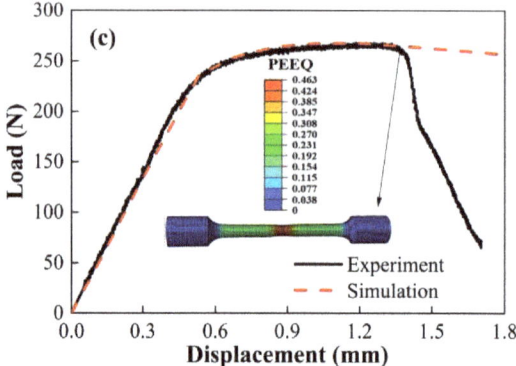

Figure 8. Equivalent plastic strain cloud diagram of smooth round bar sample at fracture time. (a) M20, (b) M50 and (c) M80.

According to the simulations and Equation (7), the relationship between the stress triaxiality, the equivalent plastic strain and the fracture strain was obtained, as shown in Figure 9. The stress triaxiality shows an increasing trend upon continuous loading. The sudden drop in stress triaxiality may be attributed to the transition from elasticity to plasticity. The stress triaxiality gradually increases with the decrease in notch radius, which is consistent with the properties of metal materials. To describe the stress triaxiality and the strain to fracture quantitatively, the average value of stress triaxiality from the initial time to fracture was calculated. Furthermore, the quasi-static compressive experimental data were added and its corresponding stress triaxiality is about −1/3 [25], as shown in Figure 10.

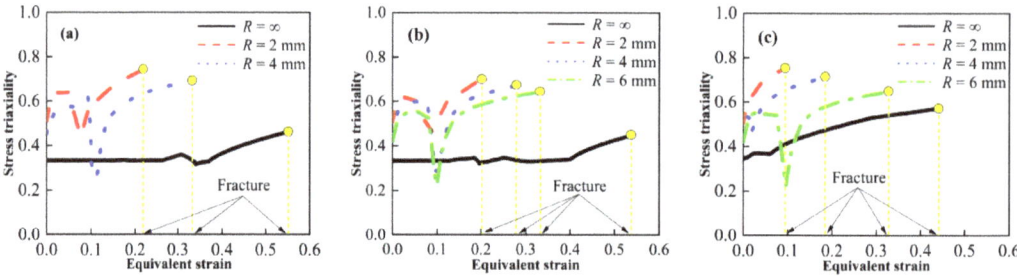

Figure 9. Stress triaxiality varies with equivalent strain for (a) M20, (b) M50 and (c) M80.

3.2. Strain Rate and Thermal Effect under Dynamic Compression

The true stress–strain curves of the Al-PTFE-W materials at elevated strain rates and temperatures are shown in Figures 11 and 12, respectively. Note that the curves in Figure 12 are approximately the same at elevated temperatures, except for M20 at 100 °C. This may be because the specimen does not fit tightly with the incidence bar and the transmission bar, which affects the wave propagation. Similar to quasi-static compression, the materials also exhibit typical elastic–plastic properties under dynamic compression. Compared to quasi-static compression, obvious yield behavior and more significant strain hardening effect can be observed from the dynamic compression curves. In particular, the yield strength under dynamic compression is about 2 to 3 times that under quasi-static compression. All of the specimens' tested fractures and typical recycled residues are presented in Figure 13. Clearly, the materials fracture more seriously with increased strain rate. Relevant material characteristic parameters obtained from Figures 11 and 12 are listed in Tables 3 and 4. Herein, the yield strength is determined by the lower yield point. From Table 3, the yield strength, the ultimate strength and the critical failure strain all show an increasing trend with the elevated strain rate, demonstrating the strain rate strengthening effect. In contrast, the material characteristic parameters in Table 4 all present a decreasing trend with the elevated temperature, indicating the thermal softening effect. On the one hand, the strength of each component in the materials decreases at elevated temperatures. On the other hand, the elevated temperature decreases the viscosity between the matrix and metal particles. As a result, the debonding failure is facilitated by elevated temperatures.

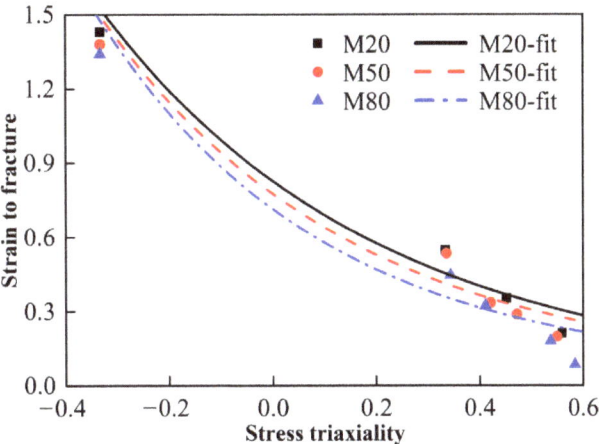

Figure 10. Strain to fracture versus the stress triaxiality.

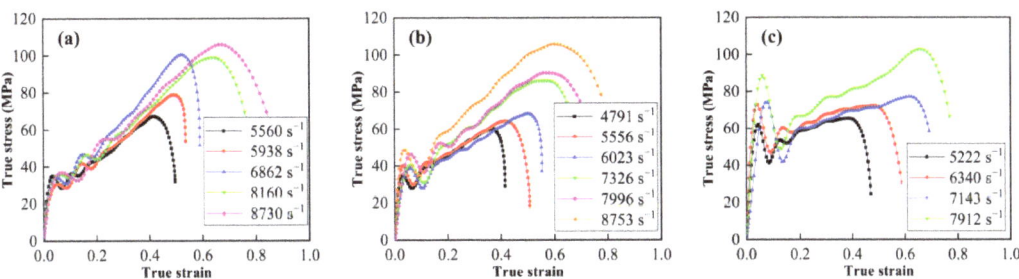

Figure 11. True stress–strain curves of Al-PTFE-W materials at elevated strain rates at 25 °C. (**a**) M20, (**b**) M50 and (**c**) M80.

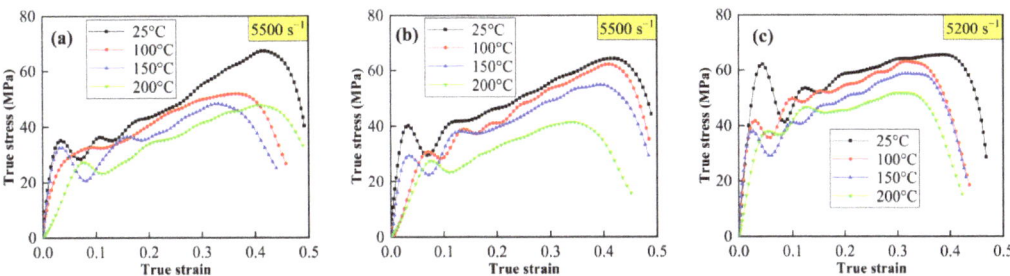

Figure 12. True stress–strain curves of Al-PTFE-W materials at elevated temperatures. (**a**) M20, (**b**) M50 and (**c**) M80.

Materials **2022**, *15*, 5167

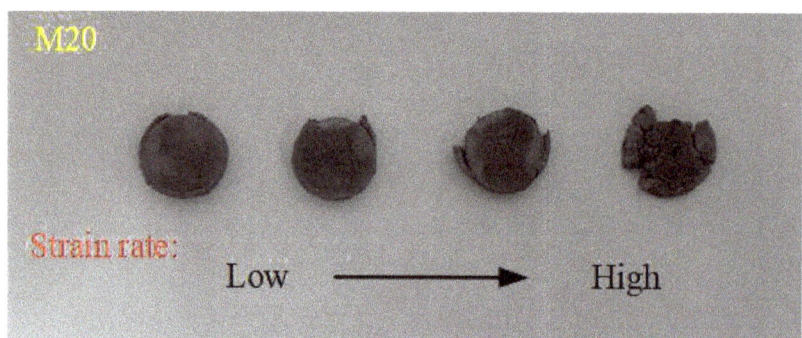

Figure 13. Typical recycled residues after dynamic compression.

Table 3. Characteristic parameters of the Al-PTFE-W materials at elevated strain rates.

Material Type	Strain Rate (s⁻¹)	Yield Strength (MPa)	Ultimate Strength (MPa)	Critical Failure Strain
M20	5560	28.5	67.3	0.42
	5983	28.7	78.9	0.49
	6862	30.8	100.3	0.52
	8160	34.4	98.7	0.63
	8730	32.6	105.9	0.67
M50	4971	27.9	60.2	0.36
	5556	29.6	64.3	0.41
	6023	27.9	68.3	0.5
	7326	30.7	85.9	0.56
	7996	36.5	90.2	0.57
	8753	34.8	105.6	0.6
M80	5222	41.3	65.4	0.38
	6340	46.7	72.3	0.48
	7143	41.8	77.2	0.61
	7912	48.9	102.6	0.66

Table 4. Characteristic parameters of the Al-PTFE-W materials at elevated temperatures.

Material Type	Temperature (°C)	Strain Rate (s⁻¹)	Yield Strength (MPa)	Ultimate Strength (MPa)	Critical Failure Strain
M20	100		27.3	51.9	0.37
	150	5500	20.8	48.3	0.33
	200		23.1	47.6	0.4
M50	100		28.3	62.1	0.41
	150	5500	22.5	54.9	0.39
	200		23.3	41.3	0.34
M80	100		35.7	62.9	0.32
	150	5200	29.2	58.7	0.31
	200		36.6	51.5	0.31

To study the effect of W content on strain rate sensitivity, the data in Table 3 were linearly fitted, as shown in Figure 14. It is observed from Figure 14a that with the increase in W content, the yield strength gradually increases while the ultimate strength decreases first and then increases, which is consistent with the quasi-static compression characteristics. The linear slope of ultimate strength proves to be close to the strain rate for the materials with different W contents. The linear slope of yield strength with respect to strain rate shows a slightly increasing trend with increasing W content, indicating that the sensitivity of the materials to strain rate increases slightly with increasing W content. This should be attributed to the dislocation density difference between the deformed and undeformed zone. However, the critical failure strain shows a complex trend in the tested strain rate range. When the strain rate is less than 6000 s⁻¹, the variation in the critical failure strain with W content is essentially similar to that under quasi-static compression. When the strain rate is higher than 6000 s⁻¹, the critical failure strain follows the following order: M80 > M20 > M50, and the difference between them becomes more significant with the increasing strain rate, as shown in Figure 14b. This may be attributed to the temperature rise during plastic deformation. Most of the plastic work is converted to heat under high strain rate loading [6], which significantly reduces the matrix strength. As the strain rate increases, the matrix softening effect becomes more obvious. As a result, the force chain strengthening effect dominates. Relevant studies have indicated that the bearing capacity of force chains is not only dependent on the number of metal particles, but also on the number and spatial distribution of force chains [26]. Furthermore, the strength of the metal particles also decreases due to the temperature rise. Therefore, the force chain strengthening effect may not be enough to compensate for the softened matrix when W content is less than 50%.

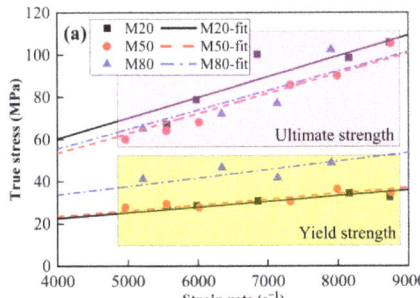 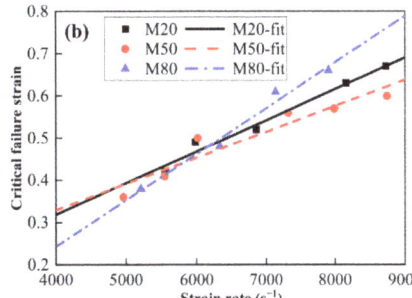

Figure 14. Strain rate effect on yield strength, ultimate strength and critical failure strain of the material with different W contents. (**a**) Strength and (**b**) critical failure strain.

3.3. Constitutive and Failure Modeling

The Johnson–Cook model, which includes a constitutive model and failure model, was applied to describe the constitutive behavior and failure criteria of the materials [27]. The Johnson–Cook constitutive model expresses the equivalent flow stress as a function of the equivalent plastic strain, the strain rate and the temperature, which can be written as

$$\sigma = \left[A + B(\varepsilon_p)^n\right]\left[1 + C\ln\left(\frac{\dot{\varepsilon}}{\dot{\varepsilon}_0}\right)\right]\left[1 - \left(\frac{T - T_r}{T_m - T_r}\right)^m\right] \tag{8}$$

where σ is the equivalent stress, ε_p is the effective plastic strain, $\dot{\varepsilon}$ is the effective strain rate, $\dot{\varepsilon}_0$ is the reference strain rate (0.001 s⁻¹), T is the absolute temperature, T_r is the reference temperature (25 °C), T_m is the melting temperature of the materials (600 °C) and A, B, C, n and m are material constants. Based on the method described in reference [5], the compressive constitutive parameters (A, B, n, C and m) for the materials are obtained, as listed in Table 5.

Table 5. Johnson–Cook constitutive parameters of the materials.

Material Type	A (MPa)	B (MPa)	n	C	m
M20	10.6	45.9	0.81	0.062	1.19
M50	11.4	41.6	0.89	0.074	1.16
M80	16.5	24.2	0.43	0.067	1.32

Based on the damage accumulation during the deformation, the Johnson–Cook failure model presents the strain to fracture as a function of the stress triaxiality, the strain rate and the temperature. The damage to an element is defined as

$$D = \sum \frac{\Delta\varepsilon_p}{\varepsilon_f} \qquad (9)$$

where D is the damage parameter and fracture is then allowed to occur when $D = 1.0$. $\Delta\varepsilon_p$ is the increment of equivalent plastic strain and ε_f is the equivalent strain to fracture, which can be expressed as

$$\varepsilon_f = [D_1 + D_2 \exp(D_3\sigma*)]\left[1 + D_4 \ln\left(\frac{\dot{\varepsilon}}{\dot{\varepsilon}_0}\right)\right]\left[1 + D_5\left(\frac{T - T_r}{T_m - T_r}\right)\right] \qquad (10)$$

where $D_1 \sim D_5$ are five material constants and σ^* is the stress triaxiality. The constants D_1, D_2 and D_3 were determined by the fitted curves in Figure 10. The constants D_4 and D_5 were determined by dynamic compression tests at various strain rates and temperatures, respectively. Note that the critical failure strain under dynamic compression is approximately regarded as the strain to fracture in this paper. The Johnson–Cook failure parameters for the materials are listed in Table 6. Comparison between the experimental data and the JC model is shown in Figure 15. It can be observed that a good agreement is achieved between the predicted and experimental stress–strain curves. The Johnson–Cook model obtained in this paper can provide effective help to the numerical studies of the materials.

Table 6. Johnson–Cook failure parameters of the materials.

Material Type	D_1	D_2	D_3	D_4	D_5
M20	0.02	0.807	−1.873	−0.0455	−0.488
M50	0.043	0.731	−2.061	−0.0461	−0.399
M80	0.049	0.664	−2.3	−0.0459	−0.4

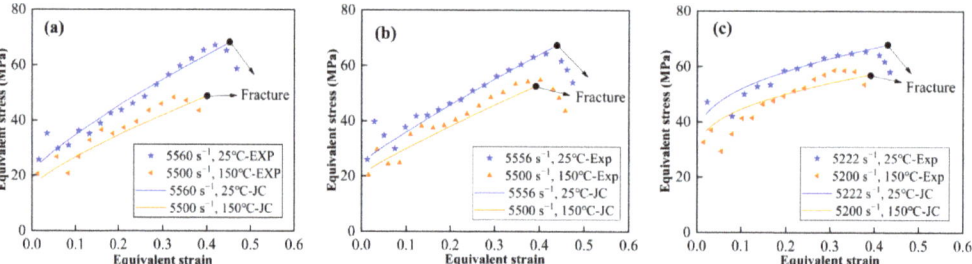

Figure 15. Comparison between the tested and JC model predicted stress–strain curves: (**a**) M20, (**b**) M50 and (**c**) M80.

4. Conclusions

Quasi-static compression tests, quasi-static tension tests and dynamic compression tests were conducted to investigate the mechanical properties, constitutive behaviors and failure criteria of Al-PTFE-W reactive materials with W content from 20% to 80%. The following conclusions were drawn:

(1) Under quasi-static (10^{-3} s^{-1}) condition, the strength of the materials may be independent of stress state and W content. Regardless of compression or tension, the strength of the materials with W content from 20% to 80% ranges from 10 MPa to 20 MPa.

(2) Under quasi-static condition, the compression plasticity of the materials is significantly superior to its tension plasticity. W content has no obvious influence on the

compression plasticity, while tension plasticity is extremely sensitive to W content. As W content increases from 20% to 80%, the compression failure strain decreases from 1.43 to 1.34 with an amplitude of 6.2%, while the tension failure strain decreases from 0.3 to 0.036 with an amplitude of 88%.

(3) The materials show an obvious strain rate strengthening effect and thermal softening effect under dynamic compression. However, the dynamic compression strengths and strain rate sensitivities of the materials with different W contents show no obvious difference. For the materials with a W content of 50%, the dynamic compression strength improves from 60.2 MPa to 105.6 MPa as the strain rate increases from $4971\ \text{s}^{-1}$ to $8753\ \text{s}^{-1}$ at ambient temperature; meanwhile, it decreases from 64.3 MPa to 41.3 MPa as the material temperature increases from 25 °C to 200 °C at the strain rate of $5500\ \text{s}^{-1}$.

(4) The Johnson–Cook constitutive (A, B, n, C and m) and failure parameters ($D_1{\sim}D_5$) were well-determined and predicted stress–strain curves are in good agreement with the experimental results. The results of this research would prove beneficial to the numerical studies, design and application of reactive materials.

Author Contributions: Investigation, T.S., A.L. and Y.Y.; Methodology, A.L., C.G., Y.Y. and H.W.; Software, A.L.; Validation, C.G.; Writing—original draft, T.S.; Writing—review & editing, T.S. and H.W. All authors have read and agreed to the published version of the manuscript.

Funding: This research was funded by Beijing Municipal Natural Science Foundation, grant number 1214022.

Institutional Review Board Statement: Not applicable.

Informed Consent Statement: Not applicable.

Data Availability Statement: Not applicable.

Acknowledgments: The authors are very grateful for the support received from the Natural Science Foundation of Beijing Municipality (No.1214022).

Conflicts of Interest: The authors declare that they have no known competing financial interests or personal relationships that could have appeared to influence the work reported in this paper.

References

1. Wang, L.; Liu, J.; Li, S.; Zhang, X. Investigation on reaction energy, mechanical behavior and impact insensitivity of W-PTFE-Al composites with different W percentage. *Mater. Des.* **2016**, *92*, 397–404. [CrossRef]
2. Wu, Q.; Zhang, Q.; Long, R.; Zhang, K.; Guo, J. Potential space debris shield structure using impact-initiated energetic materials composed of polytetrafluoroethylene and aluminum. *Appl. Phys. Lett.* **2016**, *108*, 135–183. [CrossRef]
3. Liu, S.; Zheng, Y.; Yu, Q.; Ge, C.; Wang, H. Interval rupturing damage to multi-spaced aluminum plates impacted by reactive materials filled projectile. *Int. J. Impact Eng.* **2019**, *130*, 153–162. [CrossRef]
4. Joshi, V. Process for Making Polytetrafluoroethylene-Aluminum Composite and Product Made. U.S. Patent 6,547,993 B1, 15 April 2003.
5. Ge, C.; Yu, Q.; Zhang, H.; Qu, Z.; Wang, H.; Zheng, Y. On dynamic response and fracture-induced initiation characteristics of aluminum particle filled PTFE reactive material using hat-shaped specimens. *Mater. Des.* **2020**, *188*, 108472. [CrossRef]
6. Raftenberg, M.N.; Mock, W., Jr.; Kirby, G.C. Modeling the impact deformation of rods of a pressed PTFE/Al composite mixture. *Int. J. Impact Eng.* **2008**, *35*, 1735–1744. [CrossRef]
7. Wang, H.-F.; Geng, B.-Q.; Guo, H.-G.; Zheng, Y.-F.; Yu, Q.-B.; Ge, C. The effect of sintering and cooling process on geometry distortion and mechanical properties transition of PTFE/Al reactive materials. *Def. Technol.* **2020**, *16*, 720–730. [CrossRef]
8. Tang, L.; Ge, C.; Guo, H.-G.; Yu, Q.-B.; Wang, H.-F. Force chains based mesoscale simulation on the dynamic response of Al-PTFE granular composites. *Def. Technol.* **2021**, *17*, 56–63. [CrossRef]
9. Feng, B.; Fang, X.; Li, Y.-C.; Wang, H.-X.; Mao, Y.-M.; Wu, S.-Z. An initiation of Al-PTFE under quasi-static compression. *Chem. Phys. Lett.* **2015**, *637*, 38–41. [CrossRef]
10. Feng, B.; Li, Y.C.; Wu, S.Z.; Wang, H.X.; Tao, Z.M.; Fang, X. A crack-induced initiation mechanism of Al-PTFE under quasi-static compression and the investigation of influencing factors. *Mater. Des.* **2016**, *108*, 411–417. [CrossRef]
11. Osborne, D.T.; Pantoya, M.L. Effect of Al particle size on the thermal degradation of Al/Teflon mixtures. *Combust. Sci. Technol.* **2007**, *179*, 1467–1480. [CrossRef]

12. Mock, W., Jr.; Holt, W.H. Impact initiation of rods of pressed polytetrafluoroethylene (PTFE) and aluminum powders. In *Shock Compression of Condensed Matter-2005: Proceedings of the Conference of the American Physical Society Topical Group on Shock Compression of Condensed Matter*; AIP Publishing: New York, NY, USA, 2006; pp. 1097–1100.
13. Ames, R.G. Energy release characteristics of impact-initiated energetic materials. *Mater. Res. Soc. Symp. Proc.* **2005**, *896*, 321–333. [CrossRef]
14. Xu, F.Y.; Liu, S.B.; Zheng, Y.F.; Yu, Q.B.; Wang, H.F. Quasi-static compression properties and failure of PTFE/Al/W reactive materials. *Adv. Eng. Mater.* **2017**, *19*, 1600350. [CrossRef]
15. Herbold, E.; Nesterenko, V.F.; Benson, D.J.; Cai, J.; Vecchio, K.S.; Jiang, F.; Addiss, J.W.; Walley, S.M.; Proud, W.G. Particle size effect on strength, failure, and shock behavior in polytetrafluoroethylene-Al-W granular composite materials. *J. Appl. Phys.* **2008**, *104*, 103903. [CrossRef]
16. Cai, J.; Walley, S.; Hunt, R.; Proud, W.; Nesterenko, V.; Meyers, M. High-strain, High-strain-rate flow and failure in PTFE/Al/W granular composites. *Mater. Sci. Eng. A* **2008**, *472*, 308–315. [CrossRef]
17. Zhou, J.; He, Y.; He, Y.; Wang, C.T. investigation on impact initiation characteristics of fluropolymer-matrix reactive materials. *Propellants Explos. Pyrotech.* **2017**, *42*, 603–615. [CrossRef]
18. Ren, H.L.; Li, W.; Liu, X.J.; Chen, Z. Reaction behaviors of Al/PTFE materials enhanced by W particles. *Acta Armamentarii China* **2016**, *37*, 872–878.
19. Ge, C.; Maimaitituersun, W.; Dong, Y.; Tian, C. A study on the mechanical properties and impact-induced initiation characteristics of brittle PTFE/Al/W reactive materials. *Materials* **2017**, *10*, 452. [CrossRef]
20. Xu, F.Y.; Yu, Q.B.; Zheng, Y.F.; Lei, M.A.; Wang, H.F. Damage effects of aluminum plate by reactive material projectile impact. *Int. J. Impact Eng.* **2017**, *104*, 38–44. [CrossRef]
21. Xu, F.Y.; Yu, Q.B.; Zheng, Y.F.; Lei, M.A.; Wang, H.F. Damage effects of double-spaced aluminum plates by reactive material projectile impact. *Int. J. Impact Eng.* **2017**, *104*, 13–20. [CrossRef]
22. Zhang, H.; Wang, H.; Ge, C. Characterization of the dynamic response and constitutive behavior of PTFE/Al/W reactive materials. *Propellants Explos. Pyrotech.* **2020**, *45*, 788–797. [CrossRef]
23. Senthil, K.; Iqbal, M.; Chandel, P.; Gupta, N. Study of the constitutive behavior of 7075-T651 aluminum alloy. *Int. J. Impact Eng.* **2017**, *108*, 171–190. [CrossRef]
24. Joun, M.; Eom, J.G.; Lee, M.C. A new method for acquiring true stress-strain curves over a large range of strains using a tensile test and finite element method. *Mech. Mater.* **2008**, *40*, 586–593. [CrossRef]
25. Bridgeman, P.W. *Studies in Large Plastic Flow and Fracture*; McGraw-Hill: New York, NY, USA, 1952.
26. Johnson, G.R.; Cook, W.H. Fracture characteristics of three metals subjected to various strains, strain rates, temperatures, and pressures. *Eng. Fract. Mech.* **1985**, *21*, 31–48. [CrossRef]
27. Hu, H.; Xu, Z.; Dou, W.; Huang, F. Effects of strain rate and stress state on mechanical properties of Ti-6Al-4V alloy. *Int. J. Impact Eng.* **2020**, *145*, 103689. [CrossRef]

Article

Theoretical Model for the Impact-Initiated Chemical Reaction of Al/PTFE Reactive Material

Guancheng Lu, Peiyu Li, Zhenyang Liu, Jianwen Xie ⓘ, Chao Ge ⓘ and Haifu Wang *ⓘ

State Key Laboratory of Explosion Science and Technology, Beijing Institute of Technology, Beijing 100811, China; 3120195177@bit.edu.cn (G.L.); 1120181461@bit.edu.cn (P.L.); 3120215137@bit.edu.cn (Z.L.); 3120205130@bit.edu.cn (J.X.); gechao@bit.edu.cn (C.G.)
* Correspondence: wanghf@bit.edu.cn

Abstract: Reactive material (RM) is a special kind of energetic material that can react and release chemical energy under highly dynamic loads. However, its energy release behavior is limited by its own strength, showing unique unsustainable characteristics, which lack a theoretical description. In this paper, an impact-initiated chemical reaction model is proposed to describe the ignition and energy release behavior of Al/PTFE RM. The hotspot formation mechanism of pore collapse was first introduced to describe the decomposition process of PTFE. Material fragmentation and PTFE decomposition were used as ignition criteria. Then the reaction rate of the decomposition product with aluminum was calculated according to the gas-solid chemical reaction model. Finally, the reaction states of RM calculated by the model are compared and qualitatively consistent with the experimental results. The model provides insight into the thermal-mechanical-chemical responses and references for the numerical simulation of impact ignition and energy release behavior of RM.

Keywords: reactive material; impact initiated chemical reaction; theoretical model

Citation: Lu, G.; Li, P.; Liu, Z.; Xie, J.; Ge, C.; Wang, H. Theoretical Model for the Impact-Initiated Chemical Reaction of Al/PTFE Reactive Material. *Materials* 2022, *15*, 5356. https://doi.org/10.3390/ma15155356

Academic Editor: Aniello Riccio

Received: 9 June 2022
Accepted: 22 July 2022
Published: 3 August 2022

Publisher's Note: MDPI stays neutral with regard to jurisdictional claims in published maps and institutional affiliations.

1. Introduction

Reactive material, fabricated by pressing/sintering polymer matrix (typically polytetrafluoroethylene) and active metal powders, has metal-like strength and explosive-like reaction capability [1]. Due to its unique impact energy release characteristics, RM has been widely studied and has developed novel military applications over the past decades [2,3].

Significantly different from that of traditional energetic materials such as explosives and propellants, the energy release behavior of RM is closely related to its strength [4]. Under highly dynamic loads, RM is fragmented and scattered, and a local deflagrate reaction occurs. However, the chemical reaction cannot spread in the dense material [5], eventually causing the chemical reaction to be extinguished in the material. Therefore, the existing models for explosives and propellants are not suitable for characterizing the impact ignition and unsustainable chemical reaction process of RMs, and a new theoretical model is urgently needed.

Due to the complex impact ignition and the chemical reaction process of RM, experimental studies are widely used. For the impact ignition mechanism, Ames [6] suggested that material fragmentation is a prerequisite for chemical ignition, and the energy release behavior of RMs is determined by loading conditions, which is proved through vented chamber experiments. Mock [7,8] proposed a stress-delay time mechanism as an ignition criterion based on the Taylor impact test in a vacuum environment according to the first flare time under different loading conditions. On this basis, Ge C [9,10] further found that the impact ignition behavior of RM is related to the stress and strain rate of dynamic loading conditions. Recently, Tang Le [11] proposed that the shock ignition of RM starts from the local hotspot region based on the explosive loading test, and Jiang [12] calculated the relationship between impact ignition and porosity of RM based on the thermal behavior of plastic work generated by pore collapse. To describe the chemical reaction of RM, the Arrhenius equation based on the combustion rate measured by differential scanning calorimetry (DSC) is widely used to phenomenologically characterize the reaction process [13–16]. However, although this increasing understanding can be used for reference, the evolution that considers both the dynamic response and the energy release behavior of RM still lacks description using a theoretical model.

In this paper, the dynamic response and energy release process of RM is divided into two stages: impact ignition and chemical reaction. In the impact ignition stage, the local temperature rise of the material was calculated based on the hotspot theory of pore collapse, and the thermal decomposition rate of PTFE was calculated based on the hotspot temperature. Complete failure and fragmentation of the material were taken as the ignition criteria for reaction calculation. In the chemical reaction stage, the reaction rates of PTFE decomposition products and aluminum were calculated using the gas-solid chemical reaction model. According to the above analysis, a mechanical-thermal-chemical coupling model was developed to describe the impact-initiated chemical reaction of RM. Based on the data of inert numerical simulation, the reactive states of the RM rod were calculated using the model and compared with the experimental results. This research can provide a reference for further studies of numerical simulation on impact ignition and energy release characteristics of fluoropolymer-based RM.

2. Evolution from Impact Ignition to Chemical Reaction

The impact ignition and energy release behavior of RM is a complicated process of dynamic responses and chemical reactions. As experiments have revealed, the stable propagation of chemical reactions can only take place in a powder RM with a low density (less than 0.5 theoretical density) [5]. On the other hand, explosives [11] and lasers [17] cannot cause a sustainable chemical reaction in condensed RM, and only high-speed impact and fragmented materials lead to a wide range of reactions [18–21]. These results indicate that material fragmentation is a prerequisite for large-scale and sustainable chemical reactions. Therefore, in this paper, RM impact fragmentation was used as the criteria for reaction initiation, and the impact-initiated chemical reaction of RM was divided into two stages: impact ignition and chemical reaction, and the concept of the decomposition extent of PTFE was used in both stages.

In the impact ignition stage, the temperature rise induced by uniform plastic deformation of the material, could not heat the entire RM rod to a fire temperature or ignite the chemical reaction. Ames [6] proposed that some additional energy related to the crack propagation properties and the associated void collapse of RM is the key point to the ignition process. Cai [22] analyzed the pore compression and temperature rise of porous aluminum-rich PTFE/Al energetic materials under dynamic loads. They thought that during the compression process, the temperature rise of the RMs near the pores is mainly affected by the hole's inner diameter and loading pressure. Since the RM is a void-rich compressive sintering composite, it is a reasonable ignition mechanism that the chemical reaction is initiated from local hotspots.

In the chemical reaction stage, considering that the chemical reaction cannot directly take place between solid PTFE and Al, it is assumed that the chemical reaction occurs between the gaseous decomposition products of PTFE and the solid Al granules. It should be noted that the thermal decomposition process of PTFE has a variety of channels, among which the main channel is to generate CF_2 gas in the absence of oxygen [23]. Therefore, it is assumed that all the thermal decomposition product of PTFE is CF_2 gas, and the chemical reaction can be simplified to the combustion reaction between gas and solid reactants.

Overall, a schematic of the typical impact-initiated chemical reaction process of RMs is shown in Figure 1. In the impact ignition stage, there are some pores inside the initial RM. Then, under highly dynamic loads, the material around the pores generates a local hot region (Figure 1a) due to the work of plastic deformation and collapse shear. The PTFE around the hotspot decomposes (Figure 1b) to produce oxidizing gas reactant CF_2 after exceeding the threshold decomposition temperature. Subsequently, the material is fractured (Figure 1c) and the CF_2 gas is released while exposing the aluminum particles to the gaseous reactant atmosphere and taking on a chemical reaction. In the chemical reaction stage, there is a reaction boundary layer on the surface of the solid aluminum particle. The decomposition product CF_2 is transferred through the boundary layer to the surface of the aluminum reaction core as a reactant (Figure 1d). The Al reaction core is consumed at a consumption rate ν (Figure 1e), and the reaction product AlF_3 flows out through the boundary layer. Then, the energy released by the chemical reaction will push up the temperature of the surrounding PTFE. If the temperature could exceed the decomposition temperature of PTFE (approximately 750 K from Ref. [15]), more CF_2 gas would be released, thus providing enough gaseous reactants for further chemical reactions in the reaction region, and finally, all materials could react completely. Otherwise, the chemical reaction stops when all the CF2 gas is consumed. Next, we will establish theoretical models for the impact ignition stage and the chemical reaction stage of RM, respectively.

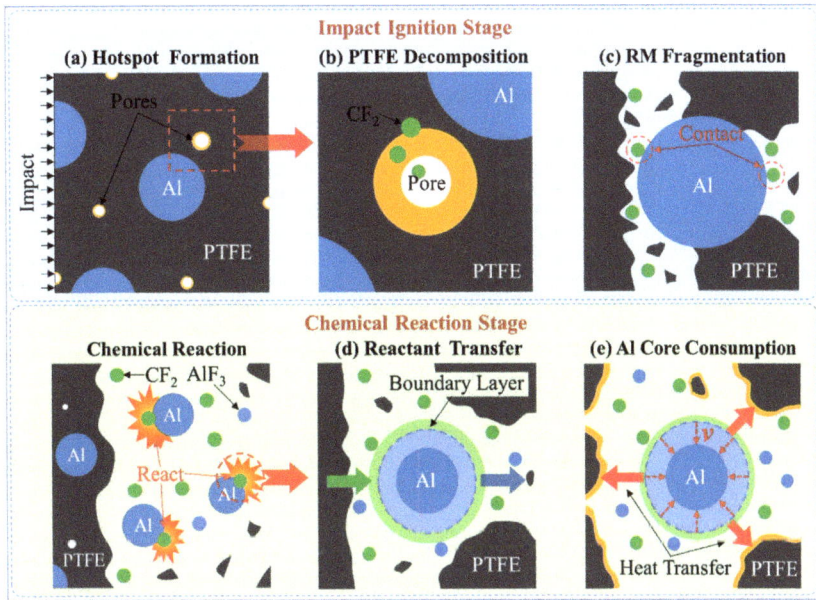

Figure 1. Typical impact ignition and chemical reaction process of RM.

3. Impact Ignition Behavior of RM

3.1. Hotspot Formation Caused Temperature Rise

To characterize the impact ignition behavior described above, the elastic-viscoplastic single spherical shell collapse model was adopted to describe the hotspots formation inside the RM during the dynamic loads. In the study of the explosive impact ignition problem, Kim [24,25] proposed that the temperature rise of materials around the hotspots is caused by mechanical deformation, heat conduction, and chemical reaction energy release. For fluoropolymer-based RMs, according to the SEM of PTFE/Al (73.5/26.5) RM [26] (shown in Figure 2a), the pores which will collapse under impact compression and form hotspots mainly appear inside the PTFE matrix. Meanwhile, in condensed RM without fragmentation, only extremely high temperatures (over 900 K [27]) can lead to direct reactions between PTFE and Al. Therefore, in the impact ignition stage, the calculation of the hotspots effect is assumed to exclude the energy released by chemical reactions and the mechanical deformation and heat transfer are counted for the temperature rise of the RM.

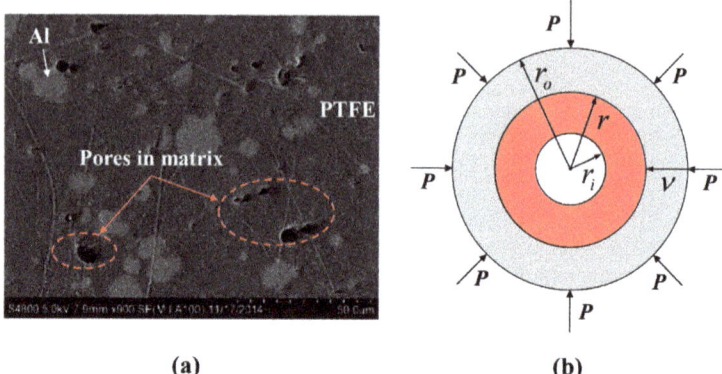

Figure 2. Image and schematic of pores inside the Al/PTFE microstructure: (**a**) SEM of Al/PTFE microstructures from Ref. [26] and (**b**) schematic of one-dimensional elastic-viscoplastic cavity collapse model.

Figure 2b shows the schematic of the one-dimensional elastic-viscoplastic pore compression model where P is the periodic boundary condition, r_i is the inner radius of the pore, r_o is the outer radius of the spherical shell, and r is the current radius of a random position. At the beginning, the initial porosity of a cell is,

$$\alpha = \frac{\rho_t - \rho}{\rho_t} = \frac{r_i^3}{r_o^3} \tag{1}$$

where ρ_t is the theoretical density which is determined by the mass ratios of the components, and ρ is the actual density of the RM. According to Geng [28], the actual density of the RM prepared under the cold press-sintering process is related to the molding pressure and sintering temperature, and under ideal preparation conditions, the porosity of PTFE/AL (73.5/26.5) is about 5%. The collapse velocity of the spherical shell under uniform external pressure loading can be calculated as follows:

$$v = \left[\frac{P - P_g}{4G\left(r_o^{-3} - r_i^{-3}\right)r^2}\right]\delta(t) + \frac{\gamma}{2\left(r_o^{-3} - r_i^{-3}\right)r^2\tau}\left[(P - P_g)H(t) - 2\sqrt{3}\ln\frac{r_o}{r_i}\right] \tag{2}$$

where v is the motion velocity of the localized material, P_g is the initial gas pressure in the pores, t is time, $\delta(t)$ is a delta function at $t = 0$, $H(t)$ is a step function, τ is the shear yield strength and $\tau = \sigma_0/\sqrt{3}$. G is the shear modulus, and $G = 0.0233$ Mbar. γ is the adiabatic exponent, $\gamma = \gamma_1\sigma_0$, σ_0 corresponds to the static yield strength and γ_1 is constant. The motion velocity v can be used to calculate the compression state of pores. When the pores are completely closed, plastic compression can be continued as homogeneous materials, but hotspot-caused temperature rise is no longer calculated. On the other hand, when the material is fractured, the pore structure is destroyed and the calculation of hotspot-caused temperature rise stops. The temperature rise caused by mechanical deformation of spherical shell collapse is:

$$\left(\frac{dT^*}{dt}\right)_{M.D.} = \left(\frac{9}{4\rho_{PTFE}C_p}\right)\frac{\left(P - P_g - 2\sqrt{3}\tau\ln\frac{r_o}{r_i}\right)^2}{\left(r_o^{-3} - r_i^{-3}\right)^2 r^6} \cdot \frac{\gamma}{\tau} \tag{3}$$

where T^* is the temperature of each position inside the spherical shell. The subscript M.D. represents the mechanical deformation, ρ_{PTFE} and C_p is the density and specific heat of

PTFE, respectively. After mechanical deformation, there is a heat conduction process inside the material, so the total temperature change can be obtained as follows:

$$\frac{dT^*}{dt} = \left(\frac{dT^*}{dt}\right)_{M.D.} + \frac{1}{\rho_{PTFE} C_p} \frac{1}{r^2} \frac{\partial}{\partial r}\left(r^2 k^* \frac{\partial T^*}{\partial r}\right) \tag{4}$$

where k^* is the heat transfer efficiency of PTFE. The first term is the temperature rise caused by mechanical deformation and the second term is the temperature change caused by heat conduction. Then, the total temperature change in the spherical shell can be calculated by the following equation:

$$T_{n+1}^*(r,t) = T_n^*(r,t) + \int_{t_n}^{t_{n+1}} \frac{dT^*}{dt}(r,t)dt \tag{5}$$

3.2. PTFE Decomposition Rate

Under highly dynamic loads, the PTFE around the pores is heated to the decomposition temperature and begins to depolymerize. According to the experimental measurement of He [29], the pyrolysis rate of spherical PTFE powders with average radii of 10 μm and 70 μm in the temperature range of 1400~3300 K is:

$$W(T) = 2.23 \times 10^3 \exp(-77.0/RT)s^{-1} \tag{6}$$

where R is the molar gas constant, and its value is 8.314 J/mol·K. Finally, the decomposition extent of PTFE in the hot spot stage can be calculated according to the temperature-time history:

$$\Lambda_{n+1} = \Lambda_n + \frac{1}{M} \int_{r_i}^{r_o} \int_{t_n}^{t_{n+1}} 2\pi r \rho(r) W(T) dt dr \tag{7}$$

where M is the total mass of the material in the spherical shell, r is the spherical coordinate radius, and $\rho(r)$ is the density of each position in the shell.

Figure 3 shows the typical temperature rise of pore collapse and the decomposition extent of PTFE at 20 kbar constant pressure loads. The results show that with the increase of time, the pores are compressed and their radius is continuously reduced. The high-temperature region inside the PTFE keeps expanding while the extent of fully decomposed PTFE keeps increasing. The decomposition extent of PTFE in the spherical shell was obtained by weighing the decomposition extent at all positions along the diameter direction. Assuming that all hotspots formed in RM under highly dynamic loads are similar, the characteristics of all hotspots in RM can be represented by studying the formation process of the single spherical shell collapse model described above. Therefore, the geometric parameters of the single spherical shell collapse model can be determined according to the physical parameters of the RM itself, the particle size, and density/porosity.

3.3. Ignition Criteria

According to the above analytical process, the decomposition extent of PTFE caused by hotspot temperature rise under highly dynamic loads is obtained. However, the decomposition products of PTFE around the hotspots cannot effectively contact the active aluminum particles, thus, material fragmentation is required to provide sufficient contact opportunities for reactants. In addition, some studies [30] have shown that the material crack tip also provides high temperature and further promotes the chemical reaction. Therefore, the ignition criteria of RMs can be summarized as follows:

1. PTFE depolymerizes to release gaseous reactant.
2. The material is fragmented.

The first criterion for the hotspot effect has been fully discussed in the two sections above. The second criterion concerns the dynamic response behavior of the RM, which has been studied extensively [10,31,32]. The Johnson-Cook strength model, which describes the strength behavior of materials subjected to large strains, high strain rates, and high temperatures, is employed for the RM in this paper. The model defines the yield stress as

$$\sigma = \left(A + B\varepsilon_p^N\right)\left(1 + C\ln\dot{\varepsilon}_p^*\right)(1 - T_H^m) \tag{8}$$

where ε_p is the effective plastic strain, $\dot{\varepsilon}_p^*$ is the normalized effective plastic strain rate, T_H is the homologous temperature, $T_H = (T - T_{room})/(T_{melt} - T_{room})$, A, B, C, N, and m are five material constants. Raftenberg [31] determined the parameters for the 74 wt.% PTFE/26 wt.% Al RM experimentally.

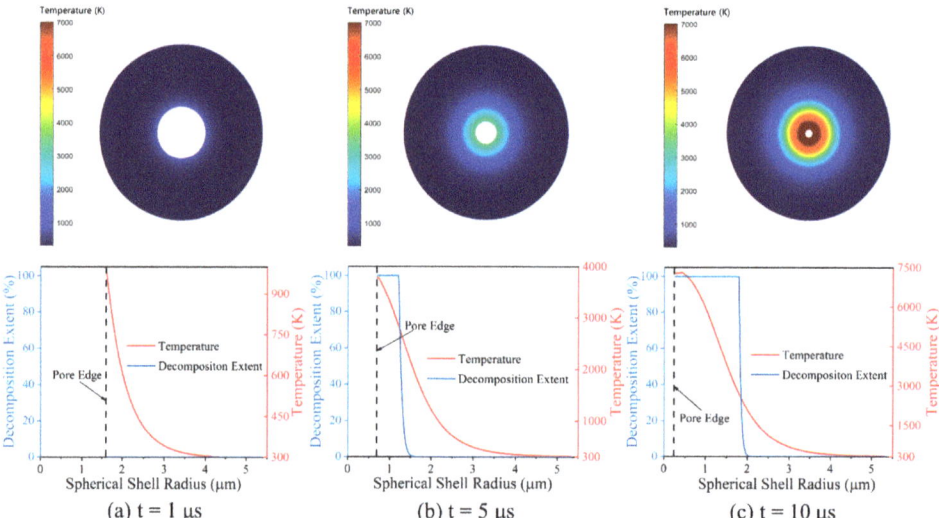

Figure 3. Typical temperature rise of pore collapse and decomposition extent of PTFE at 20 kbar constant pressure loads.

After high-speed impact, the RM may suffer large deformation or even failure. The concept of failure means that the material can no longer withstand tensile loads and is often used to simulate the ejection behavior of fractured debris. The Johnson-Cook failure model is often used to model ductile failure of materials experiencing large pressures, strain rates, and temperatures. It consists of three independent terms that define the dynamic fracture strain as a function of pressure, strain rate, and temperature.

$$\varepsilon^f = \left(D_1 + D_2 e^{D_3\sigma^*}\right)\left(1 + D_4\ln\dot{\varepsilon}^*\right)(1 + D_5 T_H), D = \sum\frac{\Delta\varepsilon}{\varepsilon^f} \tag{9}$$

where $\Delta\varepsilon$ is the increment of effective plastic strain, ε^f is the failure strain, σ^* is the mean stress normalized by the effective stress, and D_1, D_2, D_3, D_4, and D_5 are constants. The failure accumulation factor D is incremented and stored as the ratio of the effective fracture strain. When $D < 1$, the material is assumed to be intact. Once $D = 1$, the failure occurs and the material is assumed to be fractured, then the calculation can turn into the chemical reaction stage.

4. Impact-Initiated Chemical Reaction of RMs

4.1. Transfer Efficiency of Gaseous Reactants

The hotspots formed when the pores inside the material were subjected to impact compression and shearing, which led to a sharp increase of internal energy in the PTFE matrix around the pores, and soon the matrix temperature heats up to the decomposition temperature of PTFE. The PTFE matrix depolymerizes and produces CF_2, CF_3, and other gas products [23]. When the material is fragmented after impact loading, the gaseous decomposition products contact with Al particles and undergo a chemical reaction. Under oxygen-free conditions, the overall reaction process between PTFE and Al is as follows [33]:

$$C_2F_4 \Rightarrow 2CF_2(g)$$

$$2CF_2 \Rightarrow CF(g) + CF_3(g)$$

$$3C_2F_4 + 4Al(s) \Rightarrow 4AlF_3(g) + 6C(s)$$

We assume the solid aluminum particles are surrounded by gaseous decomposition products of PTFE when calculating the combustion reaction process. Therefore, the overall combustion reaction of RM can be regarded as a gas-solid two-phase flow reacting around several spherical Al particles.

As shown in Figure 4, there is a chemical reaction boundary layer between the two reactants, and only when the gaseous decomposition products of PTFE cross the boundary layer and reach the surface of Al particles can they undergo a chemical reaction. Hence the mass transfer process of gaseous reactant through the boundary layer needs to be calculated first. Considering that the gas environment and the gas flow around aluminum particles are relatively limited, the boundary layer mass transfer can be regarded as the diffusion process of gas molecules from high concentration to low concentration. Then the dimensionless relation of gas boundary layer mass transfer theory is used to describe the mass transfer process between a single particle and gas [34]:

$$Sh = 2.0 + 0.6Re^{\frac{1}{2}}Sc^{\frac{1}{3}}, Re = 0 \sim 200 \tag{10}$$

$$Sh \equiv \frac{2h_D r_{particle}}{D_T}$$

$$Re \equiv \frac{2\rho_g u_g}{\mu} \tag{11}$$

$$Sc \equiv \frac{\mu}{\rho_g D_T}$$

where Sh is the Sherwood constant, Re is the Reynold constant, Sc is the Schmidt constant, h_D is the mass transfer efficiency, ρ_g is the gas density, and u_g is the airflow velocity. Both Sh and Sc need to be calculated using diffusion coefficient D_T, which can be estimated using Chapman-Enskog empirical formula:

$$D_T = 0.001858T^{\frac{3}{2}}\frac{(\frac{1}{M_{CF_2}} + \frac{1}{M_{AlF_3}})^{\frac{1}{2}}}{P_{out}\sigma_{AB}^2\Omega_{AB}} \tag{12}$$

where T is the gas Kelvin temperature surrounding the Al particle, M_{CF_2} and M_{AlF_3} are the relative molar masses of gaseous reactants and gaseous products, respectively. P_{out} is the pressure of the principal part of the gas phase (Bar), σ_{AB} is the average collision radius, and Ω_{AB} is the collision integral (0.417 from ref. [34]). The mass transfer efficiency can be expressed by combining the Equations (10)–(12),

$$h_D = \frac{D_T}{r_{particle}} \cdot \frac{1 + a(\frac{r_{core}}{r_{particle}})^{0.5}}{\frac{r_{core}}{r_{particle}}} \tag{13}$$

where r_{core} is the current Al core radius, $r_{particle}$ is the original Al core radius, and $a = 0.3S_c^{1/3}R_e^{1/2}$. In this model, the decomposed gas environment is relatively closed, and the airflow velocity can be approximated to $u_g = 0$, then $R_e = 0$ and $a = 0$. Thus, the mass transfer efficiency h_D can be converted into:

$$h_D = 0.001858T^{\frac{3}{2}} \frac{\left(\frac{1}{M_{CF_2}} + \frac{1}{M_{AlF_3}}\right)^{\frac{1}{2}}}{P_{out}\sigma_{AB}^2 \Omega_{AB} r_{core}} \tag{14}$$

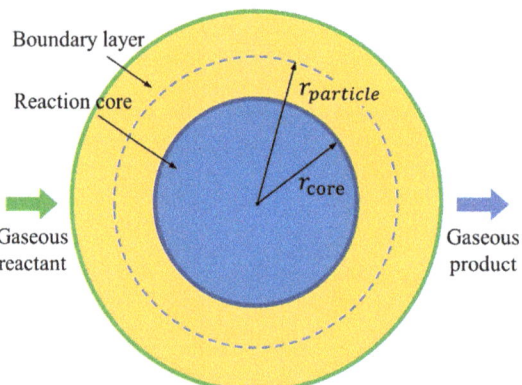

Figure 4. Gas-solid chemical reaction process in RM.

4.2. Aluminum Core Consumption Rate

When the PTFE decomposition products were transferred to the surface of aluminum particles, the gas-solid chemical reaction began to take place. According to the quasi-steady state hypothesis, the reaction rate in this chemical reaction depends only on the substance concentration on the particle surface. The consumption rate of the incoming material is much faster than the gaseous reactants transfer efficiency, which means that the gaseous reactants transferred into the reaction zone will be consumed in time, so as to achieve a steady-state:

$$kC_{CF_2core} = h_D(C_{CF_2out} - C_{CF_2core}) \tag{15}$$

where k is the reaction rate constant, and according to the Arrhenius equation, $k = Ae^{-\frac{Ea}{RT}}$. C_{CF_2} represents the concentration of gaseous reactants, and the subscripts out and core represent the principal part of the gas phase outside the boundary layer and the surface of the Al reaction core, respectively. The CF_2 concentration outside the boundary layer can be calculated as follows:

$$C_{CF_2out} = \frac{2\Lambda_{PTFE}m\omega_{PTFE}M_{CF_2}}{VF + m\omega_{PTFE}(F - \Lambda_{PTFE})/\rho_{PTFE}} \tag{16}$$

Here m is the unit mass, ω_{PTFE} is the mass fraction of PTFE, V is the unit volume and F is the reaction content. During the reaction process, the size of the Al core continuously shrinks, so the reaction rate can be characterized by the linear velocity of the Al core interface (the consumption velocity of the Al core along the diameter direction):

$$-\frac{\rho_{particle}}{4} \cdot \frac{dr_{core}}{dt} = kC_{CF_2core} \tag{17}$$

Here, $\rho_{particle}$ is the aluminum particle density. Combining the Equations (15) and (16),

$$-\frac{dr_{core}}{dt} = \frac{4C_{CF_2out}/\rho_{particle}}{1/h_D + 1/k} \tag{18}$$

The total combustion reaction rate is defined as follows,

$$R = \frac{dF}{dt} = \frac{dF}{dr_{core}} \frac{dr_{core}}{dt} \tag{19}$$

where R is the reaction rate and there is a relationship between F and r_{core}:

$$F = 1 - \left(\frac{r_{core}}{r_{particle}}\right)^3 \tag{20}$$

$$\frac{dF}{dr_{core}} = -\frac{3}{r_{particle}}\left(\frac{r_{core}}{r_{particle}}\right)^2 = -\frac{3}{r_{particle}}(1-F)^{\frac{2}{3}}$$

Substituting Equations (18) and (20) into Equation (19), it can be obtained:

$$R = \frac{12C_{CF_2out}}{r_{particle}\rho_{particle}} \cdot (1-F)^{\frac{2}{3}} \cdot \left(\frac{r_{particle}P_{out}\sigma_{AB}^2\Omega_{AB}}{0.001858T^{1.5}\left(\frac{1}{M_{CF_2}} + \frac{1}{M_{AlF_3}}\right)^{\frac{1}{2}}} \cdot \frac{\sqrt[3]{1-F}}{1+a(1-F)^{\frac{1}{6}}} + \frac{1}{A}e^{\frac{E_a}{RT}}\right)^{-1} \tag{21}$$

where E_a is the activation energy and its value is 50.836 kJ·mol^{-1} [11]. Finally, Equation (21) is the reaction rate of the combustion reaction between PTFE decomposition products and Al under highly dynamic loads.

5. Validation of Impact-Initiated Chemical Reaction Model for RMs

5.1. Calculation Process

The impact-initiated chemical reaction model for RMs was calculated based on the simulation results of the inert collision behavior of the RM rod. Mock [7,8] performed experiments to investigate the impact ignition of the RM rods impacted by steel anvils in a vacuum. To simulate the inert dynamic response of RM to mechanical shock, the Smooth Particle Hydrodynamics (SPH) method was adopted to develop the finite element models. As shown in Figure 5, the finite element model consists of a RM rod (φ 7.6 mm × 50.8 mm) and a steel anvil (φ 50 mm × 25.4 mm), and a quarter symmetric model was used to shorten the computation duration. The RM rod was constructed using 0.38 mm diameter SPH particles, and the 1 mm × 1 mm Lagrange cell was adopted for the steel anvil.

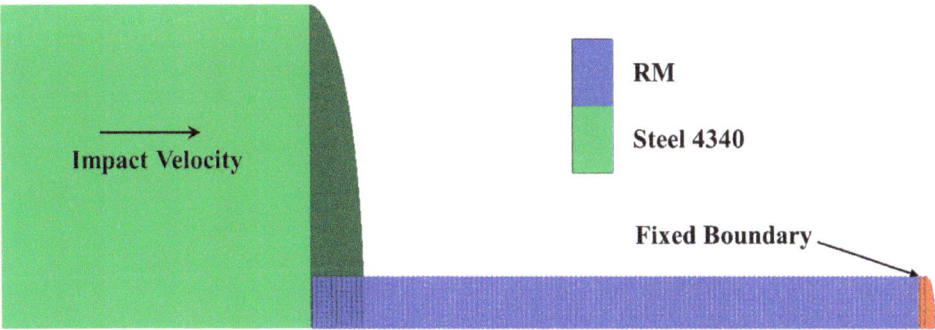

Figure 5. Scheme of simulation model.

In this paper, the shock equation of state (EOS) is used to describe the behavior of RM and steel. In the Autodyn program, the shock EOS is established from the Mie–Gruneisen form of EOS based on shock Hugoniot,

$$P = P_H + \Gamma\rho(E - E_H) \tag{22}$$

where it is assumed that $\Gamma\rho = \Gamma_0\rho_0 = $ constant and

$$P_H = \frac{\rho_0 c_0 u (1 + u)}{[1 - (s - 1)u]^2} \tag{23}$$

$$E_H = \frac{1}{2}\frac{P_H}{\rho_0}\left(\frac{u}{1 + u}\right) \tag{24}$$

here, Γ_0 is the Gruneisen coefficient, $u = (\rho/\rho_0) - 1$, ρ is the current density, ρ_0 is the initial density, s is a linear Hugoniot slope coefficient, and c_0 is the bulk sound speed. The Johnson–Cook strength and failure model, which is the form in Equations (8) and (9), is used to represent the strength and failure behavior of RM. The main material model parameters with the basic units of cm, g, and μs for RM used in the simulation are listed in Tables 1 and 2. The parameters of RM are from reference [10,11,31] and the parameters of steel 4340 are from the Autodyn material libraries.

Table 1. Material model parameters for RMs.

Hotspot Stage		Chemical Reaction Stage	
σ_0(Mbar)	1.95×10^{-6}	R_g(J/mol·K)	8.314
γ_1	13	σ_{AB}(Å)	4.35
ρ_{PTFE}(g/cm^3)	2.23	Ω_{AB}	0.417
k^*(cm/μs·g)	2.40×10^{-14}	C_p(cm^2/μs^2·K)	1.20×10^{-5}

Table 2. Strength and failure model parameters for RMs.

A (Mbar)	B (Mbar)	N	C	m
8.044×10^{-5}	2.506×10^{-3}	1.8	0.4	1
D1	D2	D3	D4	D5
0.02	0.807	−1.873	−0.0392	−0.488

After calculation based on the simulation model, the pressure-time history data of RM for each selected time were obtained from the Autodyn program using the print function [35]. These data were input into Equations (3)–(5) as loading conditions to calculate the hotspot formation induced temperature rise. Then the PTFE decomposition extent was calculated through Equation (7). It should be noted that at the same time as calculating the decomposition extent, Equation (2) is used to calculate the pore compression velocity. When the pore is closed, the calculation of the hotspot stage stops even if the material does not reach a failure state. Temperature, pressure, volume, and mass of RM particles at the failure time (failure factor D reaches 1) are obtained through numerical simulation as input variables in the calculation of the chemical reaction process.

For the chemical reaction calculation, the decomposition extent Λ_{PTFE} can be used to obtain the concentration of the decomposition product C_{CF_2out} according to Equation (16). By substituting these variables into Equation (21), the reaction content of RM with time can be calculated iteratively. It should be noted that the temperature used in calculating the chemical reaction rate R is the average temperature of the material particle. This is because the chemical reaction occurs on the surface of Al particles, while the heat transfer efficiency of Al particles is much higher than that of PTFE, leading to the result that the temperature can be evenly distributed to the whole Al particle instantaneously. Therefore, the average temperature of the whole particle is used as the reference temperature in the chemical reaction stage, and it is assumed that the energy released by the chemical reaction is converted to temperature only based on the initial specific heat of the RM.

5.2. Calculation Results for Impact Ignition Behavior

A typical simulation result of a RM rod at 30 μs and impact velocity of 775 m/s is shown in Figure 6. After being impacted by the target plate, the top area of the RM rod is deformed and thickened to a "mushroom" shape. Radial cracking occurs at the edge of the mushroom part of the RM rod. The outer materials completely fail (D = 1), and are extruded from the mushroom and dispersed outward, forming debris clouds. At the same time, the radially expanding mushroom involves circumferential shear bands inside the RM rod, which will result in the subsequent failure of nearby materials.

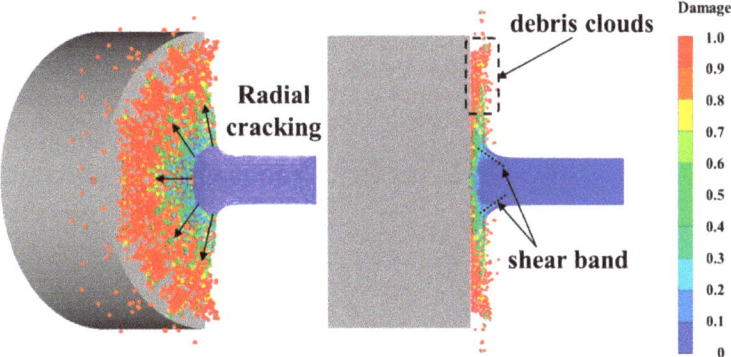

Figure 6. Simulation result of RM rod with impact velocity of 775 m/s at 30 μs.

Based on the results of the numerical simulation, the pressure-time history, failure time, and other parameters of each particle can be obtained. These parameters were substituted into the impact-initiated chemical reaction model, and the chemical reaction process of each particle was calculated. The profiles of typical completely reacted and partially reacted particles are shown in Figure 7. It can be seen from the results that the two particles accumulated a certain PTFE decomposition extent due to the hotspot effect before particle failure. After the complete failure of the material, the chemical reaction of the two particles occurred and pushed up the average temperature of the particles.

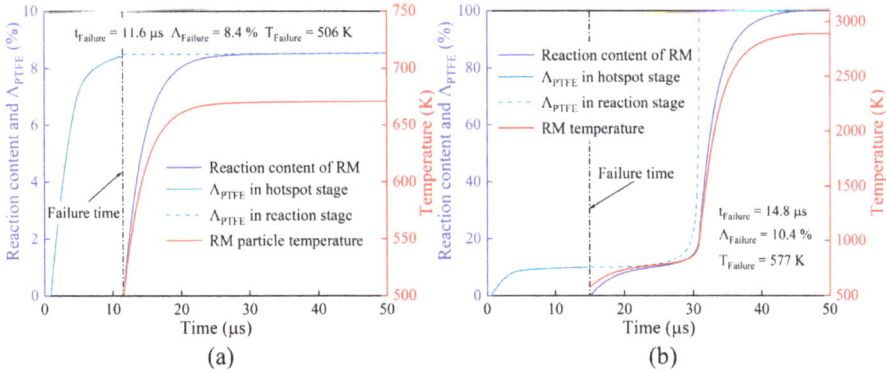

Figure 7. Reaction content profiles of two typical kinds of reacting particles: (**a**) partially reacted particle, and (**b**) completely reacted particle. Note: $t_{Failure}$, $\Lambda_{Failure}$, $T_{Failure}$ represent failure time, PTFE decomposition extent, and particle average temperature at the failure time, respectively.

However, the partially reacted particle (Figure 7a) accumulated a lower PTFE decomposition extent as well as the average temperature (temperature rise induced by plastic work from the compression of the uniform particles from the simulation) at the failure time.

After the failure of the material, the average temperature of the particle (a) can only increase to approximately 660 K due to the energy released from the chemical reaction. Because the temperature cannot maintain the further decomposition of PTFE, the chemical reaction stops. The decomposition extent of PTFE and the average temperature accumulated by a particle (Figure 7b) at the hotspot stage were higher because the particle failed later and the energy released by the initial chemical reaction pushed up the particle average temperature to above the PTFE decomposition temperature. Thus, the chemical reaction was sustained, and finally, all the material reacted completely.

The chemical reaction content of all particles in the RM rod was calculated, and the images of the RM rod from different perspectives at 30 μs and impact velocity of 775 m/s are shown in Figure 8. In the figure, the particles are painted gray, blue, and red, which represent the materials that were unreacted, partially reacted, and completely reacted, respectively. To show the reaction states more clearly, different rotation angles are used to present the images.

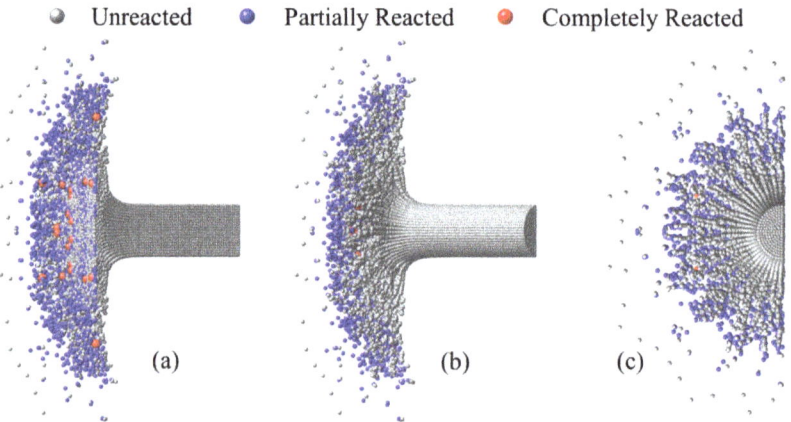

Figure 8. Images of RM rod from different perspectives at 775 m/s impact velocity and 30 μs: (**a**) cross-section view with a rotation angle of 30°; (**b**) surface view with a rotation angle of −30° and (**c**) top view.

As can be seen from Figure 8a, the hotspot reaction mainly occurred on the contact surface between the RM rod and the steel anvil. With the radial diffusion of the RM, a large amount of partially and completely reacted particles appeared at the contact surface. However, from Figure 8b,c, radial cracking occurs at the surface of the RM rod, resulting in petal-shaped cracks on the surface material. Meanwhile, compared with the materials on the contact surface of the steel anvil, partially/completely reacted particles on the surface of the RM rod are greatly reduced. This is because the surface material of the RM rod suffered from low loading intensity and only a few particles failed, so almost no reaction occurs.

The reaction morphologies of the RM rods at different times are shown in Figure 9. As can be seen from the figures, with the continuous compression of the steel anvil, the top area of the RM rod gradually thickened to a mushroom shape. Then the radial cracking spread in the mushroom-shaped rod, and internal material fragmented and extruded, forming the radial debris clouds. The chemical reaction mainly takes place where the material has broken up and is flying outward. At 20 μs, completely reacted particles occur, and as time goes on, the number of partially reacted materials keeps increasing as well as the completely reacted ones. After 40 μs, although the material rod is further fragmented, the number of reactive particles changes little and becomes more dispersed. This is because some materials cannot completely fail and react, so the number of particles that can ignite a chemical reaction is limited.

Materials **2022**, *15*, 5356

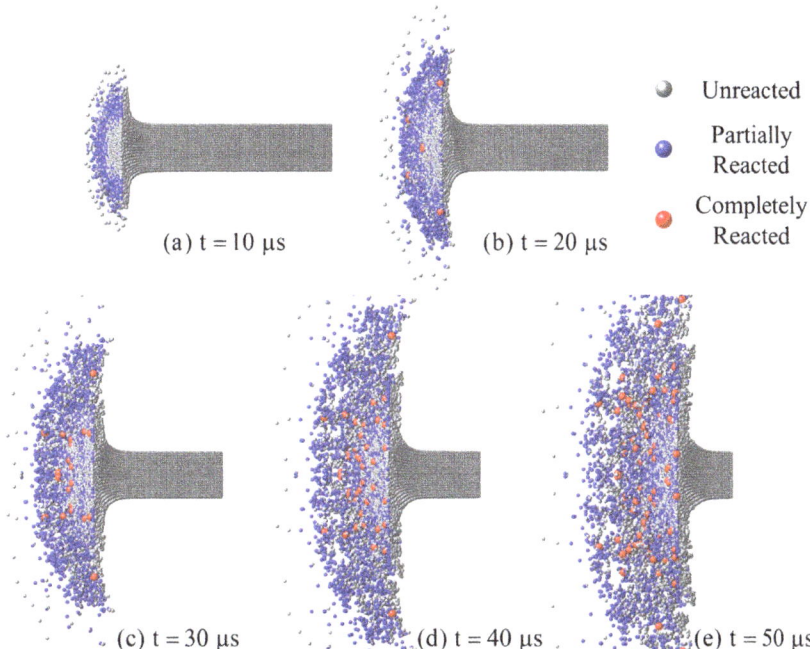

Figure 9. Reaction morphologies (cross-section view with a rotation angle of 30°) of RM rods with the impact velocity of 775 m/s at different times.

In general, the chemical reaction began at the contact surface between the RM rod and the steel anvil. The material from the outer ring of the mushroom-shaped part of the RM rod extruded and a hotspot reaction first appeared. Then, fully reacted particles began to appear in the radial expansion part. As time goes on, chemical reactions took place at various locations on the contact surface between the RM rod and the anvil.

To further analyze the failure and chemical reaction process of the RM rod under highly dynamic loads, the particles with different failure times and corresponding reaction states at an impact velocity of 775 m/s were plotted. As shown in Figure 10, the particles with different failure times are all compared with the corresponding reaction states of 60 µs. This is because the chemical reaction falls behind the failure of the material in time, and the duration of the complete chemical reaction is approximately 30 µs according to the result of Figure 7b. At the same time, only the front third part of the RM rod was cut for morphology to highlight the failure and reaction characteristics since the materials of other parts of the RM rod have not failed at 30 µs.

As shown in Figure 10a, the material that failed was first located in the shear band of the RM rod. During the impact loading process, these particles (wathet blue) first reach their tensile rupture strain. Then the green particles on both sides of the shear band were further compressed, reaching a failure state. At the same time, the RM rod was compressed into a mushroom shape, and the material on the outer surface also showed radial cracking, but failed particles only appeared at the cracks of the outer surface. Subsequently, the yellow particles continued to be compressed until they were extruded by the subsequent material and reached the failure state.

In terms of the chemical reaction, the material in the shear band failed the earliest, but the loading duration was short, so only partially reacted particles occurred, but no completely reacted particles were observed. For the subsequently failed material, only part of the hotspot reaction occurred near the outer surface of the rod, and completely failed particles appeared near the core of the rod. Finally, the failed core material accumulated the most PTFE decomposition extent at the hotspot stage, and produced the most completely reacted particles after failure because of the long duration and highly dynamic loads. This indicates that the particles which can partially and completely react need to meet sufficient loading intensity and duration to achieve a higher reaction content.

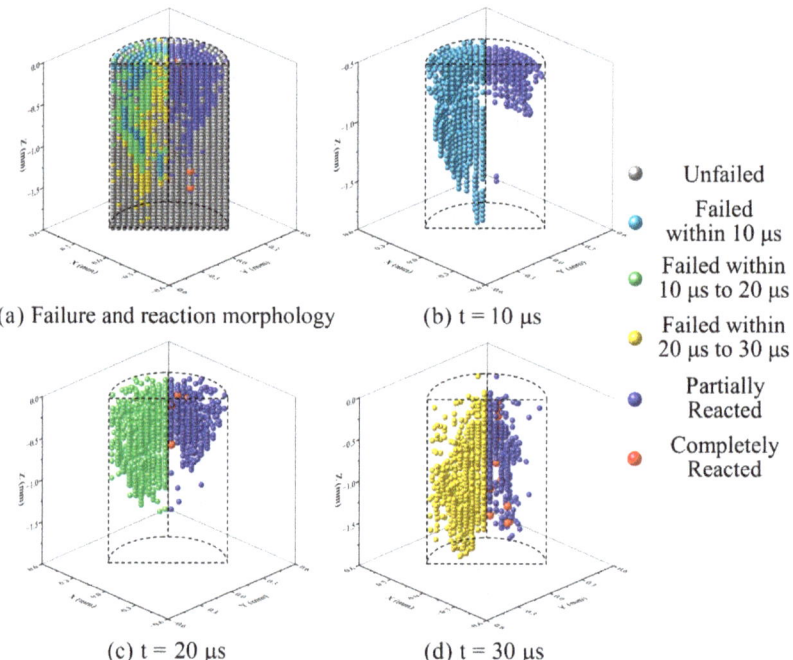

(a) Failure and reaction morphology   (b) t = 10 μs

(c) t = 20 μs   (d) t = 30 μs

Figure 10. Failure and reaction morphology of the front third part of RM rod of 775 m/s impact velocity at different times: (**a**) overall failure image at 30 μs and corresponding reaction morphology at 60 μs; (**b**) particles failed within 10 μs; (**c**) particles failed within 10 μs to 20 μs; (**d**) particles failed within 20 μs to 30 μs.

5.3. Comparison with Experiment

To validate the impact-initiated chemical reaction model, the calculation results are compared with the vacuum collision test of Mock [8]. As shown in Figure 11, at the impact velocity of 775 m/s and 30 μs, the RM rod fractured under the impact loads, creating a scattering cloud of debris. In the experimental result of the same loading conditions in Figure 4a of Ref. [8], the part near the RM rod in the debris cloud showed obvious reaction light, and the impact light gradually weakened from the rod to the periphery. In the calculation results, the hotspot reactions occurred in a large number of the extruded material in the mushroom-shaped region of the RM rod. Many completely reacted particles appeared in the annular region close to the rod (yellow area in Figure 11), while only a few particles completely reacted in the outer ring.

This is because the material in the top area of the RM rod was the first to be crushed under dynamic loads. Although the load strength was high, the loading duration was short, and material failure occurred before the hotspot reaction temperature accumulation, so the complete reaction could not be achieved. For the materials subsequently loaded, the loading strength and duration are enough to ignite hotspot reactions and achieve a more adequate chemical reaction. Overall, for the macroscopic phenomenon, the chemical reaction fire gradually diminishes from the rod to the periphery, which is very consistent with the experimental results.

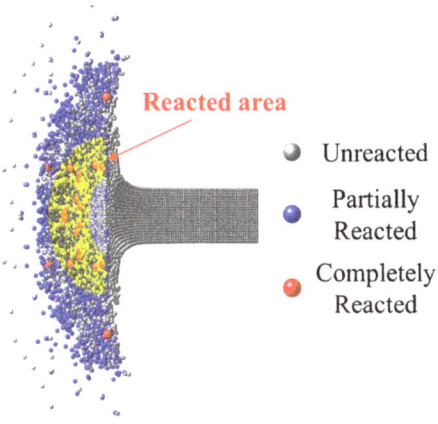

Figure 11. Calculated result (cross-section view with a rotation angle of 30°) at 775 m/s impact velocity and 30 µs (corresponding to the experimental result of the same loading conditions in Figure 4a of Ref. [8]).

Figure 12 shows the reaction morphologies of the RM rod at the time of the first light of different impact velocities in the experiments. In the case of higher velocity, the first light appeared earlier, and the mushroom-shaped part of the RM rod was smaller in size. The partially reacted particles were mainly concentrated in the outer region of the debris cloud, and the completely reacted particles appeared at the contact surface of the steel anvil and near the core of the rod. With the decrease in impact velocity (Figure 12a–c), the length of the compression part of the RM rod increased when the first light was observed, and the distribution of partially reacted materials in the debris cloud became more diffuse. This is because lower impact velocity corresponds to lower loading strength, while fewer materials will be ignited at the same time. When the fire light is observed, the material has accumulated a certain reaction content. Therefore, the lower the impact velocity is, the longer the loading time (until the time for the first light) will be, and the more completely reacted particles appear in the reaction morphology. However, when the impact velocity is further reduced (Figure 12d), the loading intensity will be insufficient, leading to no completely reacted particles in the material.

The relationship between the reaction content and time of all particles in the RM rod at different impact velocities was statistically weighted according to mass, and the results are shown in Figure 13. The results suggest that with the increase of impact velocity, the time of the beginning of reaction content accumulation in materials is slightly earlier, and the reaction content increases successively. This is because, under different dynamic loads, RM particles will completely fail and ignite hotspot reaction, so the reaction content of the material accumulates to a certain degree. However, these reactions are too weak to be observed through macroscopic phenomena. When the total reaction content of the RM rod continues to accumulate to a certain extent, fully reacted particles begin to appear, leading to a higher probability of macroscopically visible firelight.

Assume that the overall reaction content of the RM rod with macroscopically visible flame is 0.4%, as shown in Figure 13. The red marks represent the time for the first light from the experiments. The result indicates that the higher the impact velocity is, the earlier the material accumulates to the threshold chemical reaction content, and the higher the probability of observing firelight in the macroscopic phenomena will be. This is qualitatively consistent with the experimental results. However, it should be pointed out that the calculation in this paper is based on inert collision simulation and does not consider that debris clouds will be more dispersed after the chemical reaction occurred.

At the same time, the chemical reaction transfer between adjacent material particles is not considered, so the overall reaction content calculated is low.

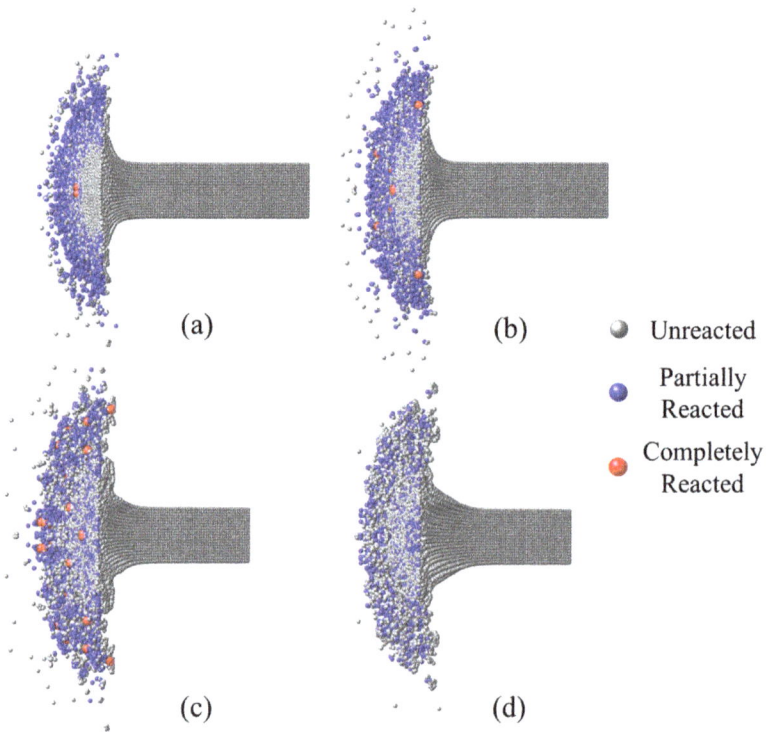

Figure 12. Reaction morphology of RM rod at the time for the first light of different impact velocities: (a) 969 m/s at 14 μs; (b) 775 m/s at 22 μs; (c) 617 m/s at 38 μs; (d) 468 m/s at 50 μs (no light).

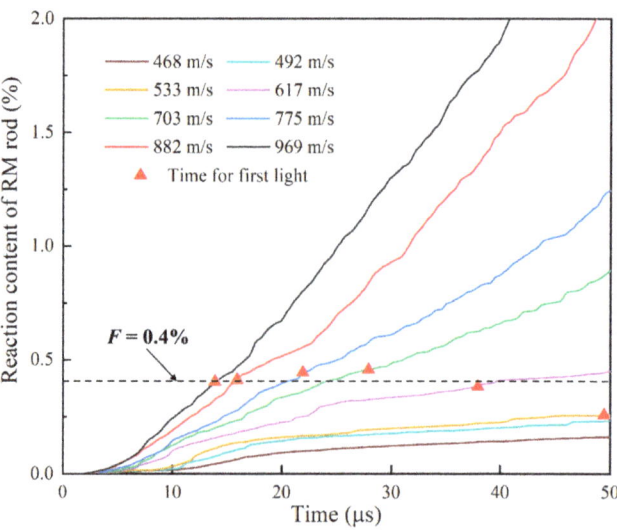

Figure 13. Reaction content of RM rods at different impact velocities.

6. Conclusions

In this paper, an impact-initiated chemical reaction model for Al/PTFE reactive material is proposed. Different from the phenomenological numerical model, the model can well characterize the impact of the unsustainable reaction behavior of RMs and can provide a reference for a numerical simulation of the impact ignition and energy release behavior of fluoropolymer-based RM. The main conclusions are as follows:

(a) Based on the evolution from impact ignition to chemical reaction, the PTFE decomposition and material fragmentation were chosen as the impact ignition criteria. The hotspot formation mechanism of pore collapse was introduced to describe the temperature rise as well as the decomposition process of PTFE. The reaction rate equation was established based on the gas-solid chemical reaction model.

(b) The decomposition products accumulated before the material fragmentation contact with Al particles and ignite the chemical reaction. The energy released by the initial chemical reaction pushes up the material temperature. When the material temperature exceeds the PTFE decomposition temperature, PTFE continues to decompose and react until the material is completely consumed. Otherwise, the chemical reaction stops, causing the RM to show unsustainable chemical reaction characteristics.

(c) The material which can completely react needs to meet sufficient loading intensity and duration. The material in the shear band of the RM rod failed earliest, but the loading duration was short, hence only partially reacted particles occurred. The failed core material accumulated the most PTFE decomposition extent at the hotspot stage and produced the most completely reacted particles after the material fragmentation because of the long loading duration.

(d) Based on the numerical simulation of the inert dynamic response of RM, the chemical reaction process of the Taylor rod is calculated using the model in this paper. The results are compared and qualitatively consistent with the experimental ones.

Author Contributions: Conceptualization, G.L., P.L., Z.L. and H.W.; Investigation, G.L. and J.X.; Resources, Z.L. and C.G.; Writing—original draft, G.L. and P.L.; Writing—review and editing, G.L., C.G. and H.W. All authors have read and agreed to the published version of the manuscript.

Funding: This research was funded by the State Key Program of National Natural Science Foundation of China (Grant No. 12132003).

Institutional Review Board Statement: Not applicable.

Informed Consent Statement: Not applicable.

Data Availability Statement: Not applicable.

Conflicts of Interest: The authors declare no conflict of interest.

References

1. Xu, F.Y.; Yu, Q.B.; Zheng, Y.F.; Lei, M.A.; Wang, H.F. Damage effects of double-spaced aluminum plates by reactive material projectile impact. *Int. J. Impact Eng.* **2017**, *104*, 13–20. [CrossRef]
2. William, J.F. *Reactive Fragrant Warhead for Enhanced Neutralization of Mortar, Rocket, and Missile Threats*; ONR: Arlington, VA, USA, 2006.
3. Baker, E.L.; Daniels, A.S.; Ng, K.W.; Martin, V.O.; Orosz, J.P. Barnie: A unitary demolition warhead. In Proceedings of the 19th International Symposium on Ballistics, Interlaken, Switzerland, 7–11 May 2001; pp. 7–11.
4. Wang, H.; Zheng, Y.; Yu, Q.; Liu, Z.; Yu, W. Impact-induced initiation and energy release behavior of reactive materials. *J. Appl. Phys.* **2011**, *110*, 074904. [CrossRef]
5. Dolgoborodov, A.; Makhov, M.N.; Kolbanev, I.V.; Streletskii, A.N.; Fortov, V.E. Detonation in an aluminum-Teflon mixture. *J. Exp. Theor. Phys. Lett.* **2005**, *81*, 311–314. [CrossRef]
6. Ames, R. Energy Release Characteristics of Impact-Initiated Energetic Materials. *MRS Online Proc. Libr.* **2005**, *896*, 0896-H03-08. [CrossRef]
7. Mock, W., Jr.; Holt, W.H. Impact Initiation of Rods of Pressed Polytetrafluoroethylene (PTFE) and Aluminum Powders. In *AIP Conference Proceedings*; American Institute of Physics: College Park, MD, USA, 2006; Volume 845, pp. 1097–1100. [CrossRef]

8. Mock, W.; Drotar, J.T.; Elert, M.; Furnish, M.D.; Chau, R.; Holmes, N.; Nguyen, J. Effect of Aluminum Particle Size on the Impact Initiation of Pressed Ptfe/Al Composite Rods. In *AIP Conference Proceedings*; American Institute of Physics: College Park, MD, USA, 2008; pp. 971–974. [CrossRef]
9. Ge, C.; Dong, Y.; Maimaitituersun, W.; Ren, Y.; Feng, S. Experimental Study on Impact-induced Initiation Thresholds of Polytetrafluoroethylene/Aluminum Composite. *Propellants Explos. Pyrotech.* **2017**, *42*, 514–522. [CrossRef]
10. Ge, C.; Yu, Q.; Zhang, H.; Qu, Z.; Wang, H.; Zheng, Y. On dynamic response and fracture-induced initiation characteristics of aluminum particle filled PTFE reactive material using hat-shaped specimens. *Mater. Des.* **2020**, *188*, 108472. [CrossRef]
11. Tang, L.; Wang, H.; Lu, G.; Zhang, H.; Ge, C. Mesoscale study on the shock response and initiation behavior of Al-PTFE granular composites. *Mater. Des.* **2021**, *200*, 109446. [CrossRef]
12. Jiang, C.; Cai, S.; Mao, L.; Wang, Z. Effect of Porosity on Dynamic Mechanical Properties and Impact Response Characteristics of High Aluminum Content PTFE/Al Energetic Materials. *Materials* **2019**, *13*, 140. [CrossRef] [PubMed]
13. Qiao, L.; Zhang, X.F.; He, Y.; Zhao, X.N.; Guan, Z.W. Multiscale modelling on the shock-induced chemical reactions of multifunctional energetic structural materials. *J. Appl. Phys.* **2013**, *113*, 173513. [CrossRef]
14. Zhou, J.; He, Y.; He, Y.; Wang, C.T. Investigation on Impact Initiation Characteristics of Fluoropolymer-matrix Reactive Materials. *Propellants Explos. Pyrotech.* **2017**, *42*, 603–615. [CrossRef]
15. Wang, J.; Zhang, L.; Mao, Y.; Gong, F. An effective way to enhance energy output and combustion characteristics of Al/PTFE. *Combust. Flame* **2020**, *214*, 419–425. [CrossRef]
16. Guo, B.; Ren, K.; Li, Z.; Chen, R. Modelling on Shock-Induced Energy Release Behavior of Reactive Materials considering Mechanical-Thermal-Chemical Coupled Effect. *Shock Vib.* **2021**, *2021*, 1–12. [CrossRef]
17. Enling, T.; Hongwei, L.; Yafei, H.; Chuang, C.; Mengzhou, C.; Kai, G. Temperature evolution of Al/PTFE reactive materials irradiated by femtosecond pulse laser. *Mater. Chem. Phys.* **2020**, *254*, 123443. [CrossRef]
18. McGregor, N.M.; Sutherland, G.T. Plate Impact Experiments on a Porous Teflon-Aluminum Mixture. In *AIP Conference Proceedings*; American Institute of Physics: College Park, MD, USA, 2004; pp. 1001–1004. [CrossRef]
19. Ames, R. Vented Chamber Calorimetry for Impact-Initiated Energetic Materials. In Proceedings of the 43th AIAA Aerospace Sciences Meeting and Exhibit, Reno, NV, USA, 10–13 January 2005; American Institute of Aeronautics and Astronautics: Las Vegas, NV, USA, 2005; pp. 10–13. [CrossRef]
20. Zhang, X.F.; Shi, A.S.; Qiao, L.; Zhang, J.; Zhang, Y.G.; Guan, Z.W. Experimental study on impact-initiated characters of multifunctional energetic structural materials. *J. Appl. Phys.* **2013**, *113*, 083508. [CrossRef]
21. Xu, F.; Zheng, Y.; Yu, Q.; Wang, Y.; Wang, H. Experimental study on penetration behavior of reactive material projectile impacting aluminum plate. *Int. J. Impact Eng.* **2016**, *95*, 125–132. [CrossRef]
22. Shangye, C.; Chunlan, J.; Liang, M.; Zaicheng, W.; Rong, H.; Sheng, Y. Impact Temperature Rise Law of Porous Aluminum-rich PTFE/Al Energetic Material. *Acta Armamentarii* **2021**, *42*, 225–233.
23. Losada, M.; Chaudhuri, S. Theoretical Study of Elementary Steps in the Reactions between Aluminum and Teflon Fragments under Combustive Environments. *J. Phys. Chem. A* **2009**, *113*, 5933–5941. [CrossRef]
24. Kim, K.; Sohn, C.H. Modeling of reaction buildup processes in shocked porous explosives. In Proceedings of the 8th Symposium (International) on Detonation, Albuquerque, NM, USA, 15–19 July 1985; Volume 926.
25. Kim, K. Development of a model of reaction rates in shocked multicomponent explosives. In Proceedings of the 9th Symposium (International) on Detonation, Portland, OR, USA, 28 August–1 September 1989; pp. 593–603.
26. Ge, C.; Dong, Y.; Maimaitituersun, W. Microscale Simulation on Mechanical Properties of Al/PTFE Composite Based on Real Microstructures. *Materials* **2016**, *9*, 590. [CrossRef]
27. Liang, M.; Sheng, Y.; Wanxiang, H.; Chunlan, J.; Zaicheng, W. Thermochemical Reaction Characteristics of PTFE/Al Reactive Material. *Acta Armamentarii* **2020**, *41*, 1962–1969.
28. Geng, B.; Wang, H.; Yu, Q.; Zheng, Y.; Ge, C. Bulk Density Homogenization and Impact Initiation Characteristics of Porous PTFE/Al/W Reactive Materials. *Materials* **2020**, *13*, 2271. [CrossRef]
29. Yuzhong, H.; Bingcheng, F.; Jiping, C. Single Pulse Shock Tube Studies on the Decomposition of Fluoropolymers. *Chin. J. Chem. Phys.* **1993**, *6*, 199–205.
30. Feng, B.; Li, Y.-C.; Hao, H.; Wang, H.-X.; Hao, Y.-F.; Fang, X. A Mechanism of Hot-spots Formation at the Crack Tip of Al-PTFE under Quasi-static Compression. *Propellants Explos. Pyrotech.* **2017**, *42*, 1366–1372. [CrossRef]
31. Raftenberg, M.N.; Mock, W., Jr.; Kirby, G.C. Modeling the impact deformation of rods of a pressed PTFE/Al composite mixture. *Int. J. Impact Eng.* **2008**, *35*, 1735–1744. [CrossRef]
32. Zhang, H.; Wang, H.; Ge, C. Characterization of the Dynamic Response and Constitutive Behavior of PTFE/Al/W Reactive Materials. *Propellants Explos. Pyrotech.* **2020**, *45*, 788–797. [CrossRef]
33. Ames, R.G.; Waggener, S.S. Reaction efficiencies for impact-initiated energetic materials. In Proceedings of the 32nd International Pyrotechnics Seminar, Karlsruhe, Germany, 28 June–1 July 2005; Volume 28.
34. Szekely, J.; Evans, J.W.; Sohn, H.Y. *Gas-Solid Reactions*; Elsevier: Amsterdam, The Netherlands, 1976.
35. Lu, G.; Ge, C.; Liu, Z.; Tang, L.; Wang, H. Study on the Formation of Reactive Material Shaped Charge Jet by Trans-Scale Discretization Method. *Crystals* **2022**, *12*, 107. [CrossRef]

Article

Failure Mechanism of the Fire Control Computer CPU Board inside the Tank under Transient Shock: Finite Element Simulations and Experimental Studies

Xiangrong Li [1,2], Guohui Wang [2], Yongkang Chen [2,*], Bo Zhao [3] and Jianguang Xiao [1]

[1]   College of Mechatronic Engineering, North University of China, Taiyuan 030051, China; lxr118@163.com (X.L.); xiaojg@nuc.edu.cn (J.X.)
[2]   Department of Arms and Control, Academy of Army Armored Forces, Beijing 100072, China; guohui305@126.com
[3]   Unit 32612 of the PLA, Guangyuan 628000, China; zhaobo8855@126.com
*   Correspondence: chenyk0305@163.com; Tel.: +86-131-2160-6365

Abstract: The electronic components inside a main battle tank (MBT) are the key components for the tank to exert its combat effectiveness. However, breakdown of the inner electronic components can easily occur inside the MBT due to the strong transient shock and large vibration during artillery fire. As a typical key electronic component inside an MBT, the fault mechanism and fault patterns of the CPU board of the fire control computer (FCC) are discussed through numerical simulation and experimental research. An explicit nonlinear dynamic analysis is performed to study the vibration features and fault mechanism under instantaneous shock load. By using finite element modal analysis, the first six nature frequencies of the CPU board are calculated. Meanwhile, curves of stress–frequency and strain–frequency of the CPU board under different harmonic loads are obtained, which are applied to further identify the peak response of the structure. Validation of the finite element model and simulation results are performed by comparing those obtained from the modal with experiments. Based on the dynamic simulation and experimental analysis, fault patterns of CPU board are discussed, and some optimization suggestions were proposed. The results shown in this work can provide a potential technical basis and reference for the optimization design of the electronic components that are commonly used in the modern weapon equipment and wartime support.

Keywords: dynamic analysis; modal analysis; harmonic response analysis; electronic component; fault mechanism

Citation: Li, X.; Wang, G.; Chen, Y.; Zhao, B.; Xiao, J. Failure Mechanism of the Fire Control Computer CPU Board inside the Tank under Transient Shock: Finite Element Simulations and Experimental Studies. *Materials* **2022**, *15*, 5070. https://doi.org/10.3390/ma15145070

Academic Editor: Giovanni Garcea

Received: 30 May 2022
Accepted: 18 July 2022
Published: 21 July 2022

Publisher's Note: MDPI stays neutral with regard to jurisdictional claims in published maps and institutional affiliations.

1. Introduction

With the extensive application of the high-performance electronic devices and control technologies, the automation and power of modern tanks have been greatly improved. However, the strong instantaneous shock and vibration during artillery firing can easily cause faults in electronic devices. The fault patterns are usually complicated, which involves mechanical deformation, contacts dislocation, sudden circuit interruption, devices failures, and so on. This makes it very difficult to maintain and support the main battle tank (MBT). Thus, understanding the fault mechanisms and fault patterns of these key components during tank artillery firing are essential to resolve these problems.

Dynamics studies on electronic devices under shock and vibration were started in the 1950s. Dave [1] systematically presents the vibration characteristics of all kinds of electronic components, and is still an important reference in the field related to shock and vibration of electronic devices [2–4]. During the 1980s, the finite element method (FEM) was introduced, and became the primary method to carry out vibration analysis of electronic devices. Pitarresi et al. [5,6] proposed and further improved the one region equivalent method, which effectively increased the computation accuracy and speed of calculating the natural frequencies and vibration modes of printed circuit boards (PCBs), and they also proposed five ways to set up the PCB finite element model. Yang [7] carried out research on the adaptability design in mechanical environment and dynamic reliability of space computer, and studied the structural dynamic reliability of device pins with random vibration excitation. Liu [8] proposed the methods of parameterized substructure modeling and PCB equivalent density modeling, which can greatly improve the simulation efficiency. Jeon et al. [9] built an isotropic elastoplastic FEM model for cell phone circuit boards and carried out both simulations and experiments with drop impact. Shtennikov and Budai [10] analyzed the circuit board welding points failure under vibration and proposed an effective improvement method. To enhance the reliability of electronic devices, Chen [11] investigated model transformation, boundary conditions and meshing on PCB modeling techniques in detail. Li [12] established a self-propelled artillery dynamics model to study the vibration characteristics and failure mechanism of the FCC under the launch shock and proposed the anti-vibration and cushion measures. Ding [13] carried out statistical analysis and failure cause analysis of impact failure cases of electronic equipment in aerospace equipment, and focused on the impact of important parameters such as peak value and frequency of impact response spectrum on electronic equipment failure. Xu [14] used impact dynamic response analysis to construct the impact damage boundary of plug-in components. The results showed that when the dominant frequency of the shock environment is higher than the first-order natural frequency of the SMA connector, its impact damage boundary is the relative displacement response asymptote of the shock environment. Xiang [15] explored a design method combining theoretical analysis and software simulation based on the mechanical environment stress design simulation, and performed a comprehensive simulation analysis and evaluation of the equipment's structural strength.

In addition to the finite element method, the statistical energy method and transient statistical energy method are also widely used in impact response analysis [3,4]. These two methods do not require fine mesh division to be performed for the system, so there is no problem whereby the finite element method needs to subdivide the mesh when solving high-frequency problems, and can better deal with the high-frequency impact response of complex systems. However, these two methods are based on the statistical average of energy, so it is difficult to obtain the impact response at a specific location, and these two methods may fail when the modal density of the structure is small.

The shock from tank artillery firing is extremely strong and instantaneous, which can result in serious damage to internal electronic components inside the MBT. However, research on this topic has been rarely reported. Based on dynamics theory and FEM simulations, the computational analysis of faults mechanism of the CPU board in FCC (which is one of the important electronic components in the MBT) were performed. The results in this paper can be used as a helpful guide for the follow-up research on other electronic devices under strong and instantaneous shock during MBT artillery firing.

2. Basic Theory of Dynamics Analysis

Dynamics analysis is one of the important methods for structural analysis under instantaneous/dynamic load, and includes modal analysis, harmonic response analysis and transient dynamics analysis. Modal analysis is usually used to identify frequencies and vibration types of a structure. Harmonic response analysis is used to identify the response of the structure to steady-state harmonic load. Meanwhile, the transient dynamics analysis is used to identify the response of structure to load changing with the time [16].

The dynamics equilibrium equation is:

$$M\ddot{u} + I - F = 0 \tag{1}$$

where M represents mass, $\ddot{u}$ represents acceleration, I represents internal forces which is determined by structure deformation and damping, and F represents external forces.

2.1. Modal Analysis

The mode is the inherent vibration characteristic of a mechanical structure. Each mode has a corresponding natural frequency, damping ratio and modal vibration type.

The mode can be calculated by

$$\left([K] - \omega_i^2[M]\{\phi_i\}\right) = 0 \tag{2}$$

where K represents stiffness, M represents mass, ϕ_i represents the mode at stage i, ω represents vibration frequency of the mode at stage i. Please note that the modal analysis is the basis of harmonic response analysis and transient dynamics analysis.

2.2. Harmonic Response Analysis

Harmonic response analysis is used to identify the steady-state response of a linear object under simple harmonic loads. By calculating modal values for different frequencies and performing modal superposition method, the peak value can be obtained, and then the stress corresponding to the frequency can be analyzed.

The motion equation of harmonic response analysis is:

$$\left(-\omega^2[M] + i\omega[C] + [K]\right)(\{\phi_1\} + i\{\phi_2\}) = (\{F_1\} + i\{F_2\}) \tag{3}$$

where C represents damping, F represents loads, and ω represents vibration frequency.

The harmonic response analysis can be used to ensure that the research objects can withstand a variety of sinusoidal load with different frequencies. In addition, it can be used to obtain the resonant response which is either intended to avoid or make it happen, depending on each specific case.

2.3. Transient Dynamics Analysis

Transient dynamics analysis is a method used to identify the dynamic response of the structure subject to arbitrary loads changing over time. The stress, strain, and displacement of the structure can be obtained, which are usually changing with the time under transient loads.

The explicit nonlinear dynamic calculation in this work was performed in the FEM framework using the ABAQUS code. The central differential method is used to implement explicit time integration of the equation of motion (EOM), and the dynamic condition of the next step is calculated using that on the previous step. It also shows a high speed and good convergence of computation [17,18]. The process of transient dynamic calculation using the explicit dynamic analysis method is shown in the following.

A. Node calculation

The dynamic equilibrium equation is:

$$\ddot{u}\big|_{(t)} = (M)^{-1}(P - I)\big|_{(t)} \tag{4}$$

where t represents time.

The explicit integration to time is:

$$\dot{u}\big|_{(t+\frac{\Delta t}{2})} = \dot{u}\big|_{(t-\frac{\Delta t}{2})} + \frac{\Delta t\big|_{(t+\Delta t)} + \Delta t\big|_{(t)}}{2}\ddot{u}\big|_{(t)} \tag{5}$$

B. Cell calculation

According to the strain rate, Δ calculates the strain increment $d\varepsilon$ of one cell. Then, according to the constitutive relation, the stress can be calculated as

$$\sigma\big|_{(t+\Delta t)} = f(\sigma_{(t)}, d\varepsilon) \tag{6}$$

where σ represents stress.

The internal forces of cell nodes are integrated as $I(t + \Delta t)$.

C. Set time t as $t + \Delta t$, then go back to step A.

D. When t is equal to or larger than the preinstalled time, the calculation will stop.

3. Fault Mechanism Simulation and Analysis of CPU Board

The host module of FCC is a very important electronic device for the modern MBT, and includes the CPU board, I/O board, A/D board, power board, control board, display board and master unit. In this paper, a CPU board with a high fault rate is chosen as a typical research object, which will be simulated using FEM. The CPU board mainly consists of one CPU, one memory, data bus, control bus, interruption system circuit and reset system circuit, which is considered to be the "traffic light" in the FCC. It determines what functions should be stopped or started in the next step. It outputs signals to complete the determination by receiving the interruption or starting signal from other components such as the laser power counter, the I/O board, the A/D board or the surface board buttons. In addition, the operation and self-test programs of fire control and reset circuit of computer are stored in the CPU.

3.1. Modeling of CPU Board

Generally, there are five kinds of methods for PCB finite element modeling, including the simple method, the overall mass equivalent method, the overall mass and stiffness equivalent method, the overall and important parts equivalent method, and the complete modeling method [19]. The first three methods ignore the influence of small components on the board. Although the presence or absence of these small components has tiny influence on the vibration types of PCB, it can actually affect the value of the natural frequencies. The complete modeling method can yield simulation results with greater accuracy, but some shortcomings such as increased complexity of modeling and much more time-consuming simulation usually accompany this method. Therefore, the overall and important parts equivalent method is used in the paper. To make the simulation results more reliable, all components that are bigger than 10 mm are modeled using FEM, while other smaller components are ignored, such as the electric resistance and welding joints.

The established three-dimensional (3D) finite element model of the CPU board is shown in Figure 1. The size of the CPU board is $176 \times 139 \times 2$ mm. There are nineteen chips, each of which includes one DSP chip, two CPU chips, and some small electric resistance and welding joints. Since the pins are not considered, connectors are modeled using an approximated cuboid geometrical model, so the circuit board and connectors are connected by using the conditions of multi-points constraint (MPC). Each element node represents a pin. The material properties for different parts are listed in Table 1.

Table 1. Material parameters of the CPU board model.

Parts	Density (kg/m^3)	Elasticity Modulus (GPa)	Poisson's Ratio
PCB	1800	11	0.28
The chip substrate	1350	18.62	0.18
Welding joints	7500	15	0.32

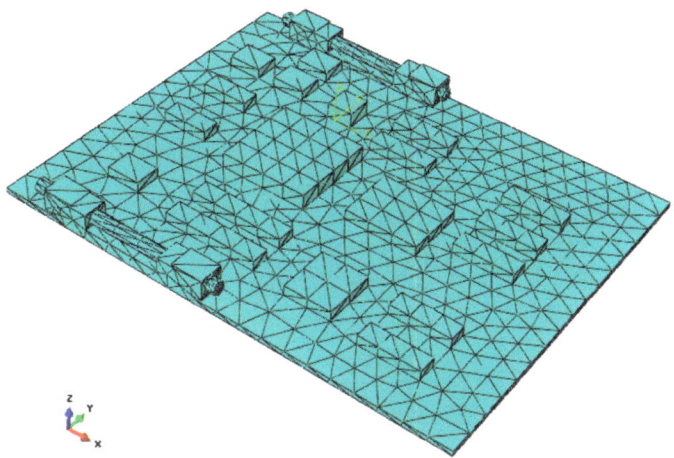

Figure 1. Finite element model of the CPU board.

3.2. Results of Modal Analysis

According to the actual boundary conditions of the circuit board, the needle-type plug base and the locking devices on the left and right sides are completely fixed. The first six natural frequencies and vibrational modes obtained by the modal analysis are shown in Table 2 and Figure 2, respectively.

Table 2. First six natural frequencies of CPU board.

Vibration type	I	II	III
Frequency/Hz	415.41	998.65	1325.7
Vibration type	IV	V	VI
Frequency/Hz	1935.7	2501.3	2814.1

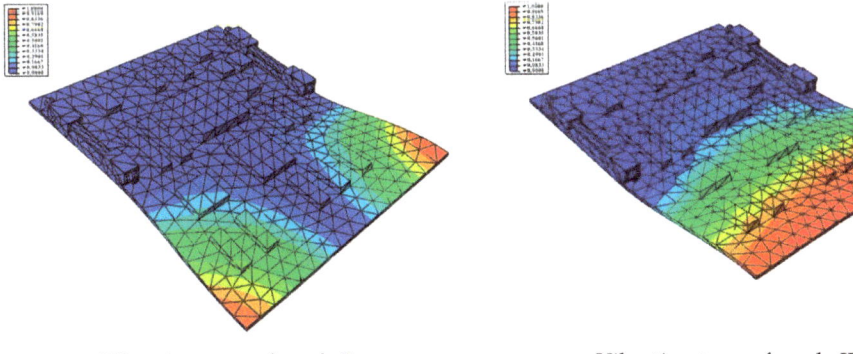

Vibration type of mode I   Vibration type of mode II

Figure 2. *Cont.*

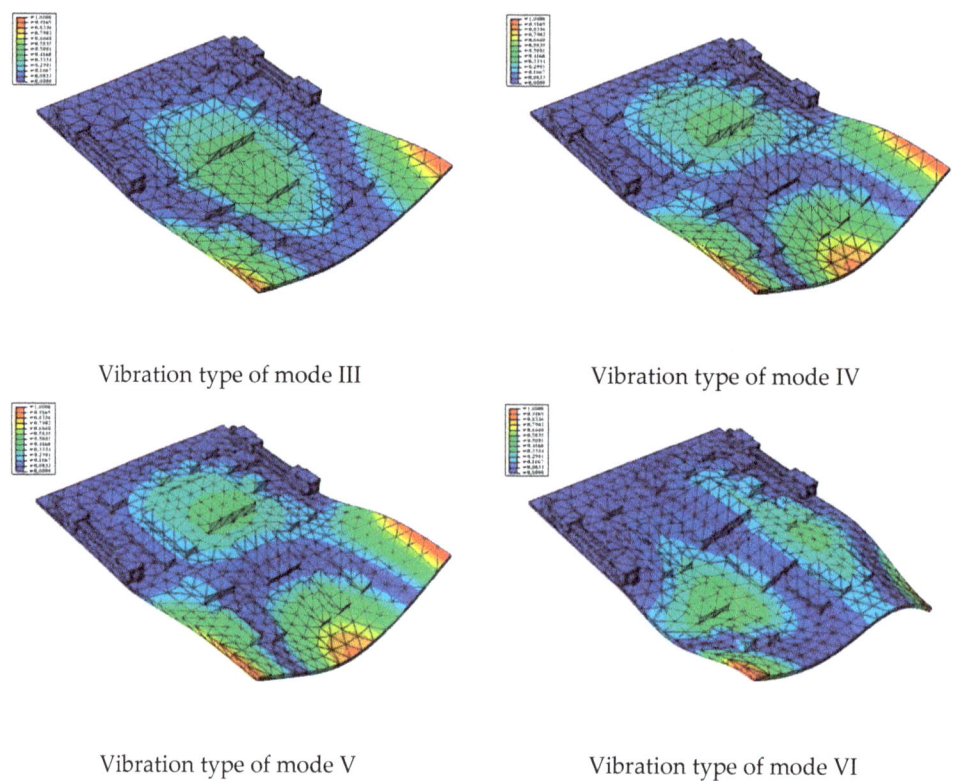

Vibration type of mode III   Vibration type of mode IV

Vibration type of mode V   Vibration type of mode VI

Figure 2. The first 6 modes vibration type based on stress of CPU finite element model.

A discussion regarding these simulation results is presented below.

(1) The first-order natural frequency of the PCB calculated in this work is 463.41 Hz, while the first-order resonance oscillation frequency of the computer case is 956.44 Hz, as shown in the previous computational study [20]. Thus, the design meets the multiplier rule. The previous modal analysis showed that large deformation occurred in some locations on the computer case that are used to fix the circuit board; thus, it is not easy for the CPU board and the computer case to resonate. However, a gap between the PCB and the slot can easily appear, leading to vibration.

(2) The second-order natural frequency of the PCB reaches up to 998.65 Hz, which is slightly higher than the first order of the computer case. This indicates that the second mode or higher-order modes are not easily activated. Therefore, the effect of the first mode vibration should be mainly focused on for the design of the structure.

(3) As shown in Figure 2, it can be found that the largest strain is located in the fore-end area of the board at low mode, and the strain gradually decreases toward the rear end. The minimum value of strain is located in the area near the rear end and the locking devices. Therefore, when designing boards in the future, important and relatively large parts should be placed as far away as possible from the board front, and should be settled close to the fixed areas.

3.3. Results of the Experimental Modal Analysis

To verify the simulation models and the corresponding results, experimental modal analysis was performed to obtain the natural frequencies and modal types of the CPU board. The principle of the modal experiment is to collect the excitation input and its corresponding output data through experimental measurement, and thus to obtain the modal parameters of the structure by fitting the experiment data points into the theoretical model. The flowchart of the modal experiment system is shown in Figure 3.

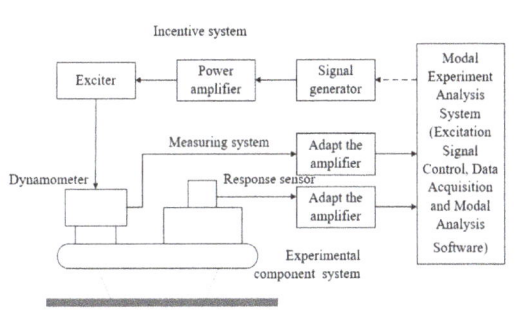

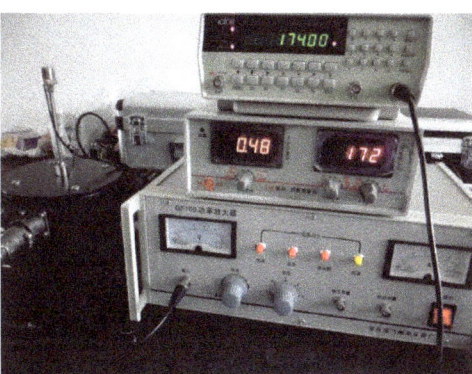

Schematic diagram of the experiment setup   The Spots photo of the experiment setup

Figure 3. Flowchart of the mode experiment system.

To obtain more accurate data, the original circuit board is fixed on the experiment platform according to the practical situation. Meshing of the board is shown in Figure 4.

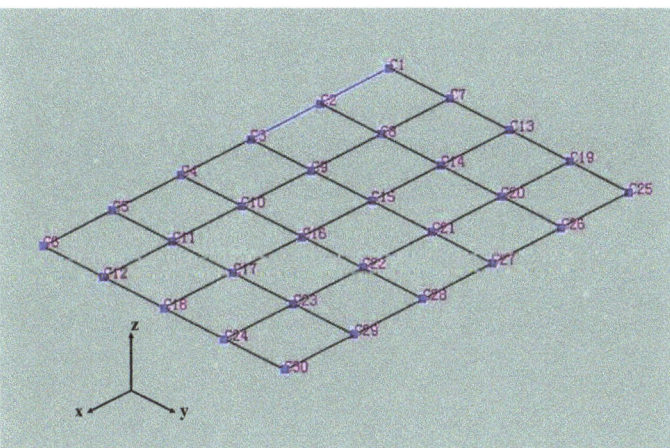

Figure 4. Meshing during the mode experiment.

A transient excitation method with a single point pulse hammer was used to motivate each grid node and obtain the output signal of the vibration response. For example, the output signal of node C21 is shown in Figure 5. Then, the modal types and natural frequencies were calculated by parameter identification for the single-reference-point frequency domain [21]. The comparison between the simulation and experiment results is shown in Table 3.

Table 3. Comparison between experiment and simulation of PCB mode analysis.

Frequency (Hz)	Simulation Mode	Frequency (Hz)	Experimental Mode	Relative Error
415.41		401.70		3.3%
998.65		969.29		2.94%
1325.7		1296.93		2.17%
1935.7		1808.33		6.58%

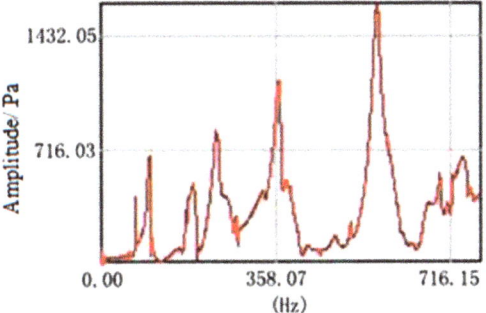

Figure 5. Response output of node C21.

The coherence value of excitation and response signal is taken in the interval of (0.1). It is one of the important indicators of test quality and is a comprehensive evaluation parameter for the nonlinear impact of the structure, excitation force, noise pollution and frequency resolution. Generally, the coherence function should be greater than 0.8. By performing the orthogonality check for the coherence and vibration types of the response signal, it was found that the coherence value was 0.92 and the orthogonality value was 0.11. Both of these values are in accordance with the standard (GJB 2706A-2008, Modal test method), thus verifying that the experiment was effective and reliable.

3.4. Harmonic Response Analysis

Based on the results of the modal analysis, harmonic response analysis was carried out by modal superposition to obtain the curves of stress vs. frequency and strain vs. frequency of the CPU board. Different harmonic loads with different frequencies were considered, and the peak value of the dynamic response was found.

To obtain the response peak of the locations at which faults occur more easily, such as the chip pins and welding points, ten reference points (RPs) were defined on the board, as shown in Figure 6. RP 1, 4 and 7 are welding points, and the others are chip pins. The ten nodes corresponding to the ten RPs are listed in Table 4.

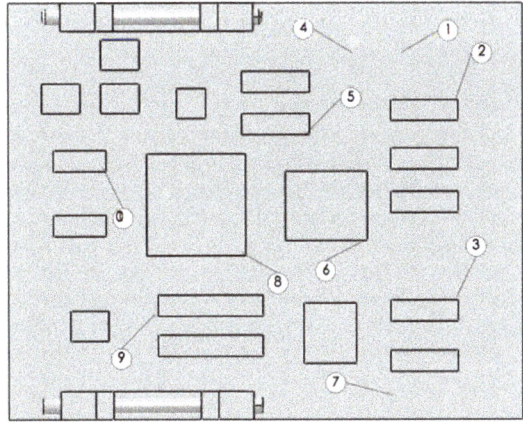

Figure 6. Ten reference points defined in the CPU board.

Table 4. Ten reference points corresponding to the nodes.

Reference Point	Node	Reference Point	Node
1	354	6	1403
2	733	7	2544
3	1014	8	4368
4	437	9	5629
5	1355	10	7323

Three vibration curves of RP2 (UX_2, UY_2, and UZ_2) are shown in Figure 7, and the harmonic responses in the vertical direction (the Y-direction) of all RPs are obtained. The range of the frequency is 0~2000 Hz. The results show that the smaller thickness and lower strength of the circuit board can lead to higher amplitude in the vertical direction, and all peak values appear at the location of natural frequencies. The maximum amplitude is found at the first mode. Thus, it is clear that the most serious deformation is perpendicular to the circuit board surface.

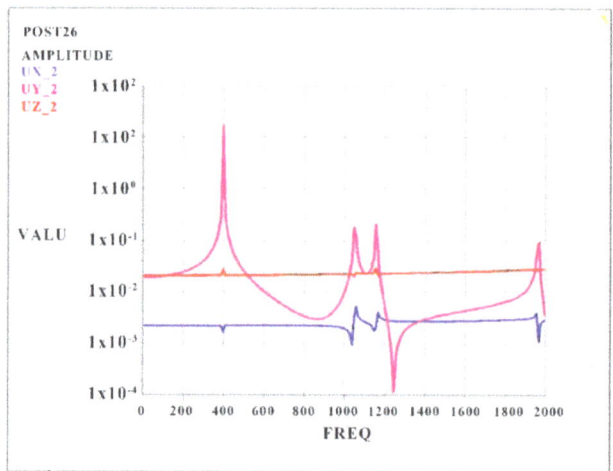

Figure 7. Harmonic response curve of RP2.

3.5. Transient Dynamics Analysis

The time history of the acceleration in the Y direction of the FCC experiencing the shock and vibration caused by MBT artillery firing is shown in Figure 8. The acceleration was converted into a force that was changing over time, and then applied to the PCB. The integral time step was set as 0.001 s, and transient dynamics analysis was carried out using the full simulation method [22,23]. The stress response curve of RP 2 is shown in Figure 9. By comparing the curves between Figures 8 and 9, it can be found that the peak values of both curves appear almost simultaneously at the time of 0.48, 0.8 and 1.3 s. The stress of RP 2 reaches its highest value at 1.3 s, which is close to 25 MPa.

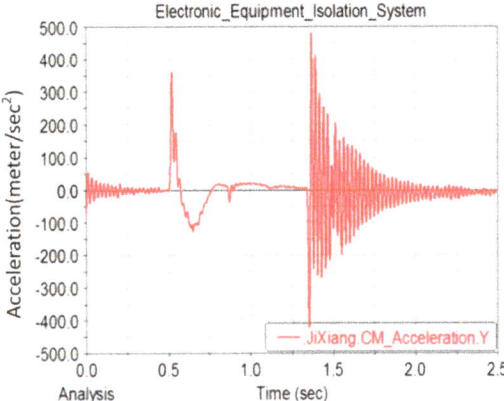

Figure 8. Acceleration in Y direction of FCC under shock and vibration caused by MBT artillery firing.

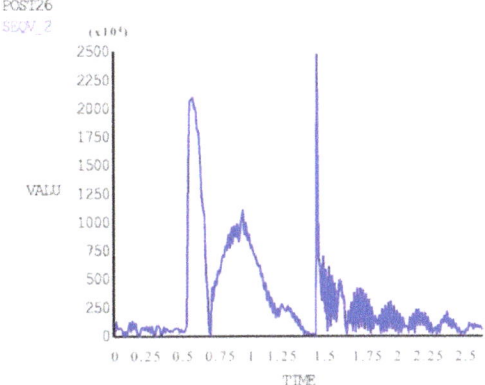

Figure 9. Stress response curve of RP 2.

3.6. Suggestions for Design Optimization

According to the results of modal analysis, harmonic response analysis and transient dynamics analysis of the PCB, we were able to gain some insight into the dynamic properties and fault features of PCB of FCC under the assumed strong and transient shock during artillery firing.

The PCB and the computer case meet the multiplier rule at the first natural frequency, and they do not resonate easily. However, some locations on the case used to fix the circuit board deform seriously, which can easily result in a gap between the PCB and the slot, leading to vibration.

The locking devices are located at the positions on the rear end on both transverse sides of the board. In addition, there is only a connector in the fore-end as a stiffened support. The faults statistics show that the circuit board connector would become loose or pop under the shock [24]. Therefore, the fore-end of the board, especially both transverse sides, can be considered to be in a suspended state. On the basis of the dynamics simulation results shown above, it can be found that the stress and strain responses of the fore-end area are much stronger, and the deformation is much more severe. This is in good agreement with the actual measured fault statistics, which show that faults like cracks and punctures evidently appear more easily at the fore-end area.

In regard to the above fault features of PCB, some optimization suggestions are proposed. In the future design of this structure, it would be better to:

(1) Increase the fixing points on the board and set locking devices on both sides of the fore-end area to make it no longer suspended.

(2) Add high-strength tendons on the fore-end area to enhance the toughness of the PCB, and increase the thickness of the PCB or use a kind of higher elasticity and intensity material to enhance the bending resistance performance.

(3) Place the components on the board closer to the locking devices or the tendons. The dynamic analysis shows that the closer the component is to the fix point, the weaker the vibration response of board becomes, and the less damage occurs to the chips and resistors.

Taking the first suggestion as a research object, verified simulation analysis was further implemented. As shown in Figure 10, each locking device on both sides was divided into three spatially separated parts placed evenly, with average intervals along the edge of the board. The results of the modal analysis are shown in Figure 11, and the first six natural frequencies are listed in Table 5.

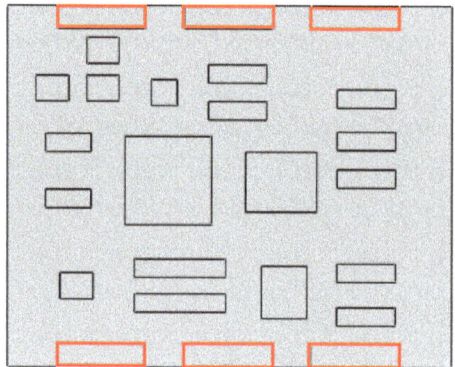

Figure 10. Circuit board locking device optimization.

Table 5. The first six frequencies of the CPU board.

Vibration type	I	II	III
Frequency/Hz	306.40	768.26	1532.48
Vibration type	IV	V	VI
Frequency/Hz	1842.25	2611.13	3436.28

From the vibration types of mode shown in Figure 11, it can be seen that for the vibration types of I, IV and V, the structure deforms longitudinally, while for vibration types of II, III and VI, it deforms transversally. Considering the first natural frequency and its corresponding mode as an example, the deformation trend after optimization decreased, while still meeting the multiplier rule. The deformation trend of the board displacement after optimization changes from transverse to longitudinal. The stress contours in Figure 11 show that the region with the most severe deformation is significantly reduced, and is mainly located at the center of the fore-end area. There is still some deformation at the center of the board. However, this can be ignored, since the value of displacement deformation is tiny. The analysis results indicate that the deformation of the board can be reduced effectively, and the reliability can be improved through designing a reasonable distribution of locking devices.

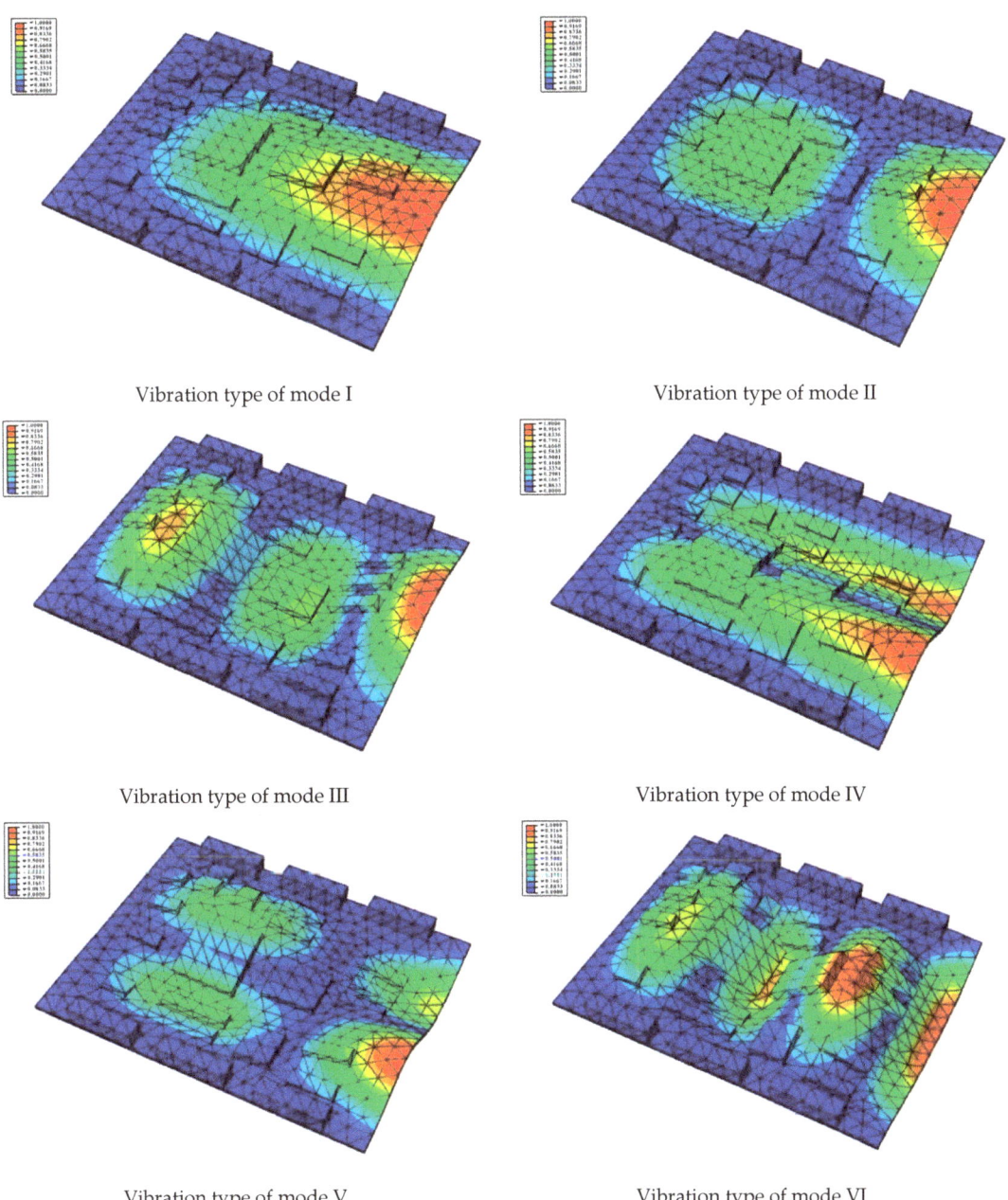

Figure 11. The first six vibration types of CPU board after optimization.

4. Conclusions

In summary, the dynamic characteristics and fault mechanism of the CPU board under strong and transient shock during MBT artillery firing were studied by utilizing FEM simulations and dynamic analysis. The conclusion remarks are summarized as follows,

(1) The first six natural frequencies and vibration modes were calculated in the FEM simulations and verified by experiments. The results of harmonic response analysis show that smaller thickness and lower strength of the PCB can result in larger response amplitude in the vertical direction. All peak values appear at the natural frequencies, and reach their maximum at the first frequency (463.41 Hz), and the most serious deformation appears in the direction perpendicular to the board surface.

(2) The results of transient dynamics analysis show that peak stresses appear at the time of 0.48, 0.8 and 1.3 s and the maximum value is close to 25 MPa at 1.3 s. The first peak is directly resulted from firing shock, and the second and the third peak appear during aftereffect. Therefore, more attention should be paid to the aftereffect so as to avoid faults in electrical components.

(3) The dynamic simulation shows that a gap between the CPU board and the slot can easily occur, leading to large vibrations throughout the entire structure. The fore-end area exhibits much stronger stress and strain responses with a larger deformation, which is consistent with the actual fault statistics [14].

(4) On the basis of this research, some optimization suggestions are proposed. Increase the fixing points and set locking devices on both sides of the fore-end area so that it is no longer suspended. Add some high-strength tendons to enhance the PCB's toughness. Increase the thickness or use a kind of higher elasticity and intensity material to enhance the bending resistance performance. Settle the components on the board closer to the locking devices or the tendons.

(5) Due to technical and time constraints, this paper only analyzes the mechanical failure mechanism of the CPU board. However, the number of tank electronic control components is large, and the types are miscellaneous. More electronic control components, such as the program control box of the automatic loader and the night vision device of the gunner, should be discussed. At the same time, secondary fault modes such as short circuit, ablation and breakdown should be considered, and the transmission fault mechanism of the internal circuit of the electronic control components under the impact should be studied by electromechanical joint simulation.

Author Contributions: Conceptualization, X.L. and G.W.; formal analysis, J.X.; investigation, B.Z.; writing—original draft preparation, X.L. and Y.C.; writing—review and editing, Y.C.; supervision, G.W.; project administration, X.L. and Y.C.; funding acquisition, X.L. All authors have read and agreed to the published version of the manuscript.

Funding: This research was funded by National Nature Science Foundation of China, and the project No. is 11502302.

Institutional Review Board Statement: Not applicable.

Informed Consent Statement: Not applicable.

Data Availability Statement: The data presented in this study are available on request from the corresponding author. The data are not publicly available due to the privacy of the research object.

Conflicts of Interest: The authors declare no conflict of interest.

References

1. Steinberg, S. *Vibration Analysis for Electronic Equipment*, 3rd ed.; John Wiley & Sons: New York, NY, USA, 2000.
2. Lall, P.; Pandurangan, A.R.R.; Dornala, K.; Suhling, J.; Deep, J.; Lowe, R. Effect of shock pulse variation on surface mount electronics under high-g shock. In Proceedings of the 18th IEEE Intersociety Conference on Thermal and Thermomechanical Phenomena in Electronic Systems (ITherm), Las Vegas, NV, USA, 28–31 May 2019; pp. 586–594.
3. Xu-Dong, S.; Zi-Jun, N.; Bing-Wei, L.I. A Review of Shock Response Analysis and Damage/Failure Evaluation of Aerospace Products. *Struct. Environ. Eng.* **2021**, *48*, 36–45.
4. Runming, H.; Xiang, L.; Jifeng, D. Damage boundary evaluation for honeycomb sandwich panel using pseudo-velocity shock response spectrum. *J. Astronaut.* **2020**, *41*, 1151–1157.
5. Pitarresi, J.M.; Caletka, D.V.; Caldwell, R. The Smeared Property Technique for the FE Vibration Analysis of Printed Circuit Cards. *J. Electron. Packag.* **1991**, *113*, 250–257. [CrossRef]

6. Pitarresi, J.M.; Primavera, A.A. Comparison of Modeling Techniques for the Vibration Analysis of Printed Circuit Cards. *J. Electron. Pack.* **1992**, *114*, 378–383. [CrossRef]
7. Yang, Y. Research on Dynamics Simulation and Adaptive Design of Space Computer Structure. Ph.D. Thesis, Xi'an University of Electronic Science and Technology, Xi'an, China, 2009.
8. Liu, X. *Research on Dynamic Performance Finite Element Modeling and Optimization Method of Electronic Equipment*; University of Electronic Science and Technology: Chengdu, China, 2011.
9. Jeon, H.; Park, M.; Hyongwon, S.; Kim, M.; Lee, Y. Finite Element Analysis of Printed Circuit Boards Using Isotropic Elasto-plastic Model and Application to Drop Simulation for Mobile Phone. In Proceedings of the ASME 2012 International Design Engineering Technical Conferences & Computers and Information in Engineering Conference, Chicago, IL, USA, 12–15 August 2012; p. DETC2012-70781.
10. Shtennikov, V.N.; Budai, B.T. The development of the physical fundamentals of contact soldering as a factor for reducing the number of defects in electronic devices. *Russ. J. Nondestruct. Test.* **2013**, *49*, 178–183. [CrossRef]
11. Rui, C. Finite Element Modeling Technology and Dynamics Analysis of Airborne Electronic Equipment. Ph.D. Thesis, Xi'an University of Electronic Science and Technology, Xi'an, China, 2013.
12. Li, B. *Influence Research about Shooting Vibration and Shock on the Gun Mount Electronic Equipment Performance*; Academy of Ordnance Engineering: Shijiazhuang, China, 2013.
13. Pei, D.; Wenhui, L.; Jianan, W. Failure study of aerospace electric equipment under pyroshock. *Shanxi Arch.* **2019**, *45*, 40–42.
14. Xu, Z.; Nangong, Z.; Li, B.; Zhang, Z.; Yu, M. Impact damage boundary of spacecraft plug-in components. *J. Vib. Shock.* **2020**, *39*, 71–75.
15. Yixin, X.; Xuexin, Z. Structural Design and Analysis of an Airborne Communication Electronic Equipment. *Mech. Eng.* **2022**, *6*, 29–33.
16. Chang, T. *Research on Accurate Finite Element Modeling Method of Electronic Equipment Structure*; Nanjing University of Aeronautics and Astronautics: Nanjing, China, 2013.
17. Yong-Zheng, Q.I. Realization of numerical simulation of vacuum preloading with ABAQUS. *J. Wenzhou Univ.-Nat. Sci.* **2010**, *31* (Suppl. 1), 111–117.
18. Wang, C.; Dong, X.; Ding, J.; Nie, B. Numerical investigation on the spraying and explosibility characteristics of coal dust. *Int. J. Min. Reclam. Environ.* **2014**, *28*, 287–296. [CrossRef]
19. Wang, G.; Zhao, B. Dynamic Analysis of Electrical Control Components under Tank Gun Firing Impact. *Mech. Mater.* **2014**, *628*, 235–239. [CrossRef]
20. Bo, Z. *Study on the Failure Mechanism of the Control Components of Main Battle Tank's Weapon Systems under Strong Shock*; Academy of Armored Forces Engineering: Beijing, China, 2014.
21. Fan, S.; Min, Z.; Huai-Hai, C.; Ming, B. Modal parameter extraction using frequency domain poly-reference method under operational conditions. *Acta Aeronaut. Astronaut. Sin.* **2002**, *23*, 294–299.
22. Fengren, J. Simulation and experiment research of dynamic characteristics of PCB components. *Electron. Mech. Eng.* **2010**, *26*, 13–17.
23. National Defense Committee of Science and Technology. *GJB 2706A-2008*; The Spacecraft Modes Test Method. Army Standard Publication Distribution Department: Beijing, China, 2008.
24. Defu, S. *Armored Vehicle Fault Diagnosis of Complex System*; National Defense Industrial Press: Beijing, China, 2013.

Article

A Simulation and an Experimental Study of Space Harpoon Low-Velocity Impact, Anchored Debris

Wei Zhao [1], Zhaojun Pang [1],*, Zhen Zhao [2], Zhonghua Du [1] and Weiliang Zhu [1]

[1] School of Mechanical Engineering, Nanjing University of Science and Technology, Nanjing 210094, China; zhao0219@njust.edu.cn (W.Z.); duzhonghua1971@163.com (Z.D.); bywzwl@163.com (W.Z.)

[2] Aerospace System Engineering Shanghai, Shanghai 201109, China; zhaozhen101@163.com

* Correspondence: pangzj@njust.edu.cn

Abstract: The space harpoon is a rigid-flexible, coupled debris capture method with a simple, reliable structure and a high adaptability to the target. For the process of impacting and embedding the harpoon into the target plate, the effect of friction at a low-velocity impact is studied, and the criteria for effective embedding of the harpoon and the corresponding launch velocity are determined. A simulation model of the dynamics of the harpoon and the target plate considering tangential friction is established, and the reliability of the numerical simulation model is verified by comparing the impact test, focusing on the kinetic energy change and embedding length during the impact of the harpoon. The results show that the frictional effect in the low-velocity impact is more obvious for the kinetic energy consumption of the harpoon itself, and the effective embedding of the harpoon into the anchored target ranges from 50~90 mm, corresponding to a theoretical launch initial velocity between 88.4~92.5 m/s.

Keywords: space harpoon; impact; friction; embedding length; launch initial velocity

Citation: Zhao, W.; Pang, Z.; Zhao, Z.; Du, Z.; Zhu, W. A Simulation and an Experimental Study of Space Harpoon Low-Velocity Impact, Anchored Debris. *Materials* 2022, 15, 5041. https://doi.org/10.3390/ma15145041

Academic Editors: Aniello Riccio and Guozheng Quan

Received: 26 May 2022
Accepted: 16 July 2022
Published: 20 July 2022

Publisher's Note: MDPI stays neutral with regard to jurisdictional claims in published maps and institutional affiliations.

1. Introduction

With the increasing frequency of space launch activities, the amount of space debris has dramatically increased [1]. According to data from the U.S. Space Surveillance Network (SSN) long-term monitoring of targets above 10 cm in low Earth orbit and targets above 1 m in geostationary orbit, it is known that the number of objects in Earth orbit has surged from approximately 11,000 in 2000 to nearly 20,000 in 2020 over the last two decades [2,3]. Currently, most policies, standards, and other regulations are reflected in curbing the increase, but the existing space debris in orbit still poses a threat to spacecraft operating in orbit. Therefore, active debris removal (ADR) has received more attention in recent years. Studies have shown that when five pieces of space debris are cleaned up each year, the probability of collision of debris and satellites can be effectively mitigated and reduced [4,5]. Space harpoon capture is a rigid-flexible, coupled debris capture method with a simple, reliable structure and a high adaptability to the target [6,7].

The satellite platform carried with the harpoon payload launches the harpoon by electromagnetic spring or pyrotechnics, which impacts the debris and anchors it to the target, where it is dragged through the atmosphere and burned up by a tow rope between the platform and the harpoon. Space debris is mostly non-cooperative targets. In order to solve the problem of difficult identification and capture caused by the complex attitude of non-cooperative targets and to achieve effective anchoring of the harpoon to the target, a higher launch velocity is usually required to increase the kinetic energy of the harpoon when it hits the target. Excessive launch velocity on the one hand will produce a large recoil force on the platform that will cause the platform to become unstable, on the other hand, the harpoon may penetrate the debris and anchor poorly. In order to reduce the impact on the platform during launch and to ensure that the harpoon impacts and anchors the debris effectively, the launch velocity must be controlled as much as possible. The problem of harpoon impact and anchoring of space debris can be considered as kinetic energy projectile penetration and embedding in thin plates.

The current research on the impact penetration problem still mainly concerns target plate failure and ballistic limiting velocity involved in different head shape projectiles hitting different types and thicknesses of targets vertically or inclined, and the velocity problem involved in low-velocity impact and anchored and embedded targets is less studied. Campbell et al. [8] analyzed the effect of mesh sensitivity on ballistic terminal velocity by comparing the ballistic terminal velocity problem of flat-headed and ogive harpoons impacting aluminum honeycomb panels through simulation and experiment; Dudziak et al. [9] studied the change in ballistic limiting velocity of blunt projectiles and conical projectiles impacting 3 mm steel plates, and they analyzed the cracking pattern of steel plates at low temperatures; Aglietti et al. [10] analyzed the velocity and the impact of harpoons during harpoon impact experiments on aluminum honeycomb panels in the RemoveDEBRIS project; Fras et al. [11], Wang et al. [12], Kpenyigba et al. [13], and Deng et al. [14] studied the ballistic terminal velocities and the target failure forms of three structures with blunt projectile, hemispherical projectile and ogival projectile impacting high-strength armor steel, 2024-T351 aluminium alloy plate, mild steel sheets, and 6061-T651 aluminium alloy thin plates by conducting simulations and experiments; Rusinek [15] analyzed the effect of bullet diameter on the ballistic terminal velocity under the same initial kinetic energy and the failure form of the target plate.

This paper compares the current situation at home and abroad, combines the designed harpoon structure, uses simulation and experimental verification methods to discuss the influence caused by the friction coefficient between the harpoon and the target plate at low-velocity impact, studies the relationship between the velocity and the embedding length when the harpoon impacts the 2A12 aluminum alloy plate under different conditions of initial velocity, and further obtains the launch velocity problem corresponding to the harpoon within the effective anchoring interval length.

2. Impact Experiment

The working scene of harpoon capture is shown in Figure 1. A satellite platform detects the orbital position of the debris and approaches the target debris after maneuvering to change the orbit, and the platform launches the harpoon, which is connected to the platform by a high-strength tether. The harpoon impacts the debris and effectively embeds into the debris by the kinetic energy provided by the launch, after which the platform drags the debris down to burn. This paper focuses on the stage of harpoon impacting and embedding debris.

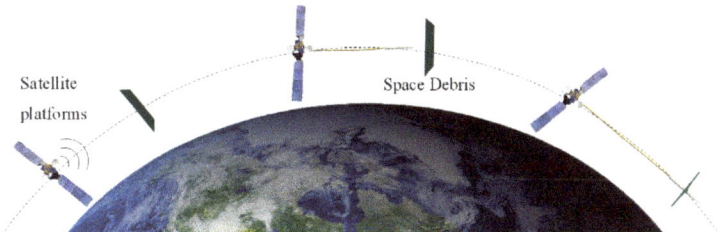

Figure 1. Work scene.

2.1. Harpoon and Target Plate Model

The structure of the harpoon and the target plate used in the test is shown in Figure 2. The harpoon used in the figure is 100 mm in length and 10 mm in diameter—the head is the harpoon penetration part with an ovoid structure design—the length is 45 mm, and there is a columnar structure at the bottom. The material of the harpoon is S45C steel, with a HRC hardness between 55 and 58 after heat treatment. The total mass of the harpoon is 48.4 g. The target plate material in the simulation process is 2A12 aluminum alloy plate commonly used in spacecraft skin and other components, and the target plate is a homogeneous plate with a size of 200 mm × 200 mm and a thickness 2 mm, with 16 M5 bolt holes evenly distributed around.

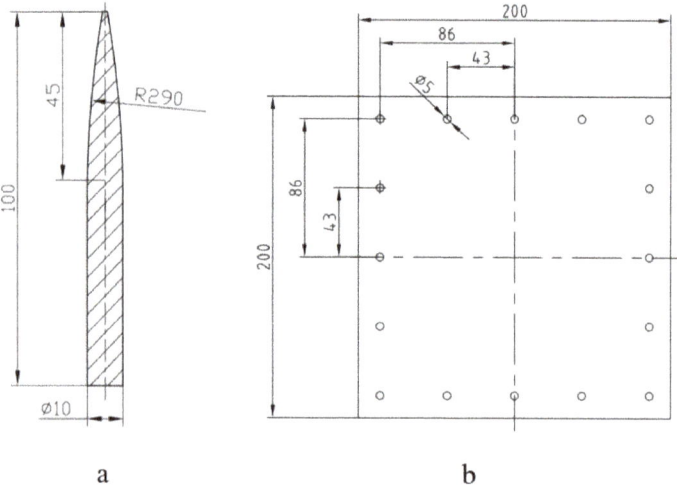

a   b

Figure 2. Structural model. (**a**) Harpoon, (**b**) Target plate.

The physical drawing of the harpoon is shown in Figure 3a, the white part is the nylon sabot, which is used as the support part to ensure the stability of the harpoon during launch, and the harpoon is embedded in the sabot when launch, and put into the air gun together with the sabot. The target plate is fixed to the target frame by 16 evenly distributed M5 bolts, which is used to ensure that the target plate will not be displaced axially and radially after being hit by the harpoon during the test, and its configuration is shown in Figure 3b.

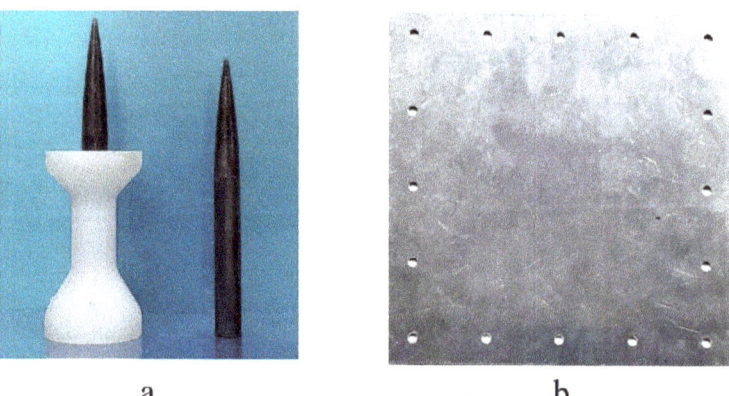

a   b

Figure 3. Harpoon and target plate physical drawing. (**a**) Harpoon, (**b**) Target plate.

2.2. Experimental Work

Design the test plan shown in Figure 4, the test equipment mainly includes a gas chamber, 37 mm caliber air gun, velocimeter, target frame, etc. In the test, the velocity of the harpoon launch is changed by controlling the nitrogen pressure provided by the gas chamber, and the distance from the mouth of the air gun to the target plate is 1.2 m. There is a recoil sleeper fixed after the target frame to prevent the target plate from being impacted, which will drive the target frame to move back and affect the test results.

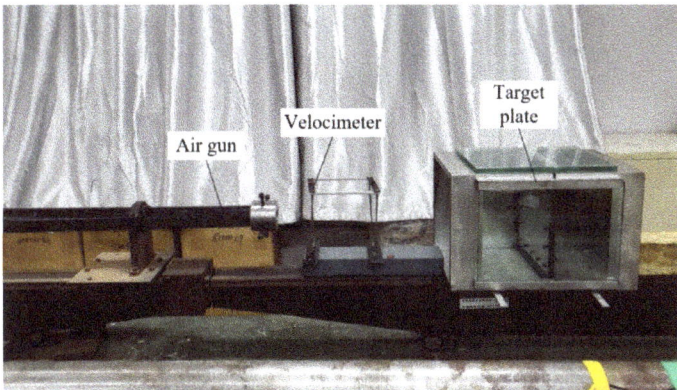

Figure 4. Impact experiment equipment.

2.3. Experimental Results and Analysis

The process of the harpoon impacting the target plate is mainly the harpoon pushing the target material through the "ductile hole expansion" mechanism, which makes the harpoon pass through, causing the thin plate near the impact point to be stretched and bent, and the phenomenon of petal cracking will occur under certain conditions. The distance from the tip of the harpoon embedded in the target plate to the maximum height of the projection is defined as the maximum embedding length, which is shown as the length of *l* in Figure 5. The test was conducted using the above test scheme, and the test results are shown in Figure 6. After the test, there was no obvious structural damage to the harpoon itself, but there were obvious scratches on the harpoon, which was caused by strong friction during the process of harpoon impact, and finally the harpoon was embedded in the target plate, and the harpoon was closely fitted to the target plate without obvious movement in the radial direction. The length of harpoon embedment was 22 mm, 37 mm, and 51.5 mm at an impact velocity of 50.7 m/s, 65 m/s, and 76.4 m/s, respectively.

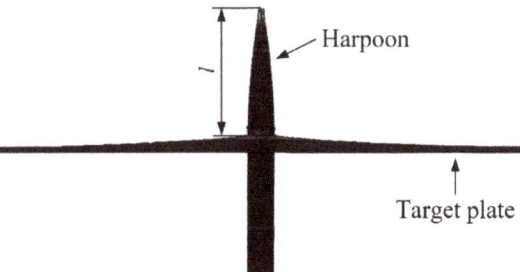

Figure 5. Maximum embedding length of harpoon.

50.7m/s   65m/s   76.4m/s

Figure 6. Experimental results.

3. Numerical Simulation

Based on the above test conditions and results, this paper uses ABAQUS/Explicit module to establish the dynamics simulation model of harpoon and target plate, and further uses the simulation test to study the relationship between velocity and friction coefficient on the length of the harpoon embedded in the target plate, and analyzes the results.

3.1. Harpoon–Target Plate Material Model and Equation of State

During the high-velocity impact, the harpoon and the target plate materials are subjected to large deformations and high strain rates, while the temperature of the material also dramatically increases. The Johnson–Cook constitutive model is well suited to characterize the dynamic mechanical behavior of materials under impact, in particular the characteristic behavior of the mechanical response of the material during intrusion, such as the adiabatic shear phenomena [16]. The model uses the von Mises yield surface and its flow law to take into account the strain, strain rate hardening, and temperature rise softening of the material, assuming that the isotropic strain, strain rate strengthening, and temperature rise softening factors of the material are decoupled. The Johnson–Cook material model was used for both the harpoon and the target plate in the simulation [17], and the expression of the Johnson–Cook material model is as follows.

$$\sigma_e = (A + B\varepsilon_e^n)\left(1 + C\ln\dot{\varepsilon}_e^*\right)(1 - T^{*m}) \tag{1}$$

where σ_e is the equivalent effect force; A is the yield strength of the material at the reference strain rate and reference temperature; B and n are strain strengthening coefficients; C is the strain rate sensitivity coefficient; ε_e is the equivalent plastic strain; $\dot{\varepsilon}_e^*$ is the dimensionless equivalent plastic strain rate; $\dot{\varepsilon}_e^* = \dot{\varepsilon}_e/\dot{\varepsilon}_0$, $\dot{\varepsilon}_0$ is the reference strain rate, usually takes the value of $1.0\,\text{s}^{-1}$; T^* is the dimensionless temperature, $T^* = (T - T_r)/(T_m - T_r)$, T_r, T_m are room temperature and material melting points, respectively.

The strain at fracture is [12]:

$$\varepsilon_f = [D_1 + D_2\exp(D_3\sigma^*)]\left(1 + D_4\ln\dot{\varepsilon}_e^*\right)(1 + D_5T^*) \tag{2}$$

where $D_1 \sim D_5$ are material parameters, σ^* is the stress triaxiality, defined as $\sigma^* = \sigma_m/\sigma_{eq}$, where σ_m is the hydrostatic pressure and σ_{eq} is the von Mises equivalent force.

The Johnson–Cook failure model uses the theory of cumulative damage to consider the effects of stress states, strain rates, and temperature changes on material damage. It is considered that the damage does not affect the strength of the material and that the damage variable has an initial value of 0, when it reaches 1, the material fails. The damage evolution of a unit is defined as [18]:

$$D = \sum(\Delta\varepsilon_{eq}/\varepsilon_f) \tag{3}$$

where D is the damage parameter, ε_{eq} is the equivalent plastic strain increment during the integration cycle.

The parameters of the harpoon and the target plate materials are shown in Table 1, Reprinted with permission from Refs. [19,20].

Table 1. Harpoon and target plate material parameters.

Model	Density/g·cm⁻³	Elastic Modulus/Gpa	Poisson Ratio	Johnson-Cook Model Parameters				
				A/MPa	B/Mpa	n	C	m
Harpoon	7.85	210	0.33	714	563	0.518	0.064	0.698
Target plate	2.77	71.7	0.33	375	592	0.42	0.001	1.426

3.2. Harpoon–Target Plate Simulation Model

ABAQUS/Explicit module is used to establish the finite element simulation model of harpoon impact, as shown in Figure 7. The model consists of two solid parts, harpoon and target plate, and the landing angle of the harpoon is 0° during the simulation. The contact between the harpoon and the target plate is modeled using the penalty method with a finite sliding formulation. The four boundaries of the target plate are set as fixed constraints, and the harpoon is simulated by setting different initial velocities in predefined fields to simulate the impact results.

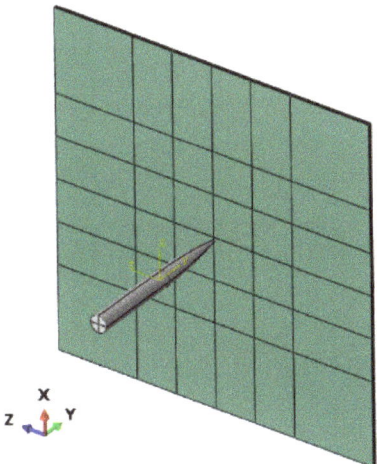

Figure 7. Finite element model.

Both the harpoon and the target plate use the structural mesh shown in Figure 8, and the mesh at the center of the target plate is encrypted to simulate the impact effect and to improve the computational efficiency. The harpoon is divided into 49,800 cells and the target plate cell number is 51,200. Meanwhile, in order to avoid distortion of the mesh during the simulation, which may affect the accuracy of the analysis, both mesh cell types are C3D8R (eight-node linear hexahedral cells) [21].

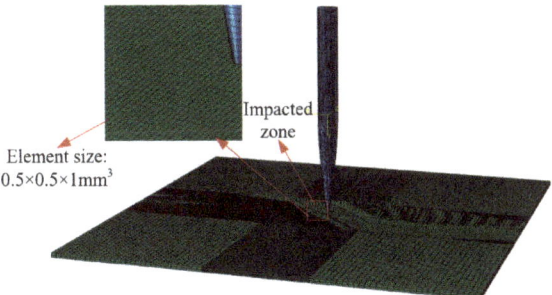

Element size: $0.5 \times 0.5 \times 1 \text{mm}^3$

Impacted zone

Figure 8. Mesh model.

3.3. Comparison of Experimental and Simulation Conclusions

The accuracy of the simulation model and the material constitutive parameters is verified by comparing the above air gun experimental data through simulation, and the dynamic friction coefficient between the harpoon and the target plate is set in the simulation, and the simulation data are compared with the relevant data derived from the experiments in Section 2.3 of this paper, and the conclusions are shown in Table 2.

Comparing the simulated and the experimental data, the errors of the harpoon embedded length data obtained by the three groups of simulation and the experimental data are 5%, 4.3%, and 0.91% respectively, which are small and within the allowed range. The simulation results of the harpoon embedded length are in good agreement with the actual experimental results. It can be seen that the material model and the material-related parameters used in the simulation are reasonable.

Table 2. Harpoon embedding experiment.

No.	Velocity/m·s^{-1}	Experimental Embedding Length/mm	Simulation Embedding Length/mm
1	50.7	22	23.1
2	65	37	38.58
3	76.4	51.5	51.03

3.4. The Effect of Friction on Harpoon Head Embedding

In order to study the influence of whether there is friction between the penetrating part of the harpoon head and the target plate on the embedded length, the simulation analysis of whether there is friction between the harpoon and the target plate is carried out, respectively. The results of the simulation with and without friction on the embedding length of the harpoon at different velocity are shown in Table 3.

Table 3. Simulation experimental design and results.

No.	Friction Coefficient	Velocity/m·s^{-1}	Embedding Length—With Friction/mm	Embedding Length—Without Friction/mm
1	0.17	37.5	15.48	25.71
2	0.17	40	16.74	29.41
3	0.17	42.5	19.37	33.99
4	0.17	45	21.18	36.63

According to the simulation test results, under the simulation velocity conditions set above, the first 1 ms of the simulation is mainly for the impact of the harpoon head on the target plate. Using the kinetic energy data of the harpoon in the first 1 ms of time, the relationship change is shown in Figure 9.

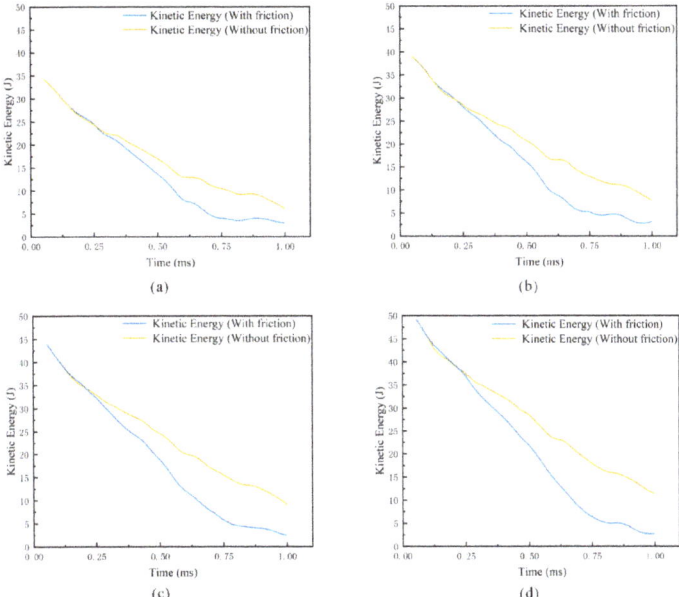

Figure 9. Kinetic energy evolution curve. (**a**) Evolution of kinetic energy at a velocity of 37.5 m/s; (**b**) Evolution of kinetic energy at a velocity of 40 m/s; (**c**) Evolution of kinetic energy at a velocity of 42.5 m/s; (**d**) Evolution of kinetic energy at a velocity of 45 m/s.

As can be seen from Figure 9, the kinetic energy of the harpoon starts to decrease when the head of the harpoon touches and impacts the target plate in the initial 1 ms. With the increasing impact depth of the harpoon, under the simulation condition of adding the friction coefficient, the impact caused by the tangential contact friction between the harpoon and the target plate gradually becomes larger, the harpoon significantly loses kinetic energy, and the impact ability rapidly decreases. When the kinetic energy of the harpoon dropped to 4.5 J, the harpoon basically lost forward ability; the kinetic energy curve was affected by the fluctuation of the target plate, and it finally gradually decreased to 0.

When the initial velocity of the harpoon was 50 m/s, the relationship between the length of the harpoon and the embedded length of the target plate with and without friction was obtained, as shown in Figure 10. At the initial stage of impact, the difference between the length of the embedded target plate with and without the friction coefficient is not obvious; with the depth of the harpoon embedded, when the impact time reaches 0.75 ms, the length of the harpoon embedded length is more and more affected by the tangential friction. At this point in the frictionless conditions, when the impact time reaches 1 ms, the remaining kinetic energy of the harpoon is 17.2 J, and the head of the harpoon can successfully complete the penetration of the target plate. Under the condition of friction coefficient, the kinetic energy of the harpoon gradually decreases; and, at 1 ms, the kinetic energy of the harpoon decreases to 4 J, which is only approximately 1/4 of the frictionless condition; at this time, the kinetic energy of the harpoon itself can no longer continue to penetrate the target plate, and under the effect of friction and fluctuation of the target plate, the kinetic energy of the impact of the harpoon is gradually eliminated until it decreases to 0. It can be seen that for the head penetration part of the harpoon, the tangential friction between the harpoon and the target plate has a greater impact on the impact efficiency; so, for this low-velocity impact problem, the impact of the tangential friction between the harpoon and the target plate needs to be considered in the simulation.

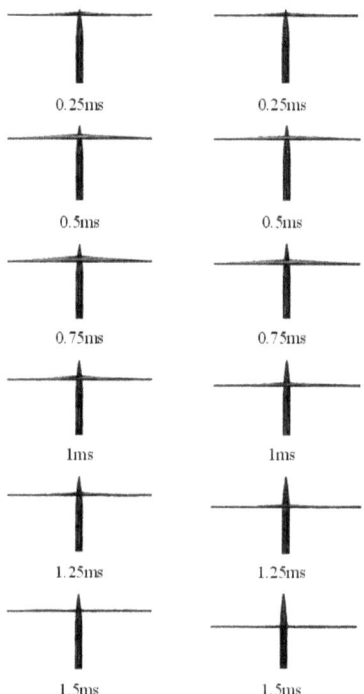

Figure 10. Variation of harpoon impact embedded length (left side is with friction coefficient, right side is without friction coefficient).

3.5. The Effect of Friction on Harpoon Column Surface Embedding

When the head of the harpoon finishes impacting the target plate, the kinetic energy of the harpoon decreases and gradually loses forward ability, and it is embedded in the target plate due to the mutual friction coefficient and the fluctuation of the target plate deformation. Analysis of the above embedding data when the initial velocity of the harpoon is 50 m/s under frictionless conditions reveals that when the time is 1.5 ms, the embedding length of the harpoon is more than 45 mm; at this time the head of the harpoon has penetrated the target plate for the contact stage between the column surface part of the harpoon and the target plate. Additionally, when there is tangential friction between the harpoon and the target plate, the velocity needs to reach more than 70 m/s; after 1 ms, the harpoon penetration part can penetrate the target plate, and the displacement change curve of the harpoon is shown in Figure 11. A comparison of simulation results at a velocity of 75 m/s with friction conditions with the results at 50 m/s without friction conditions was made, the kinetic energy curve changes are shown in Figure 12.

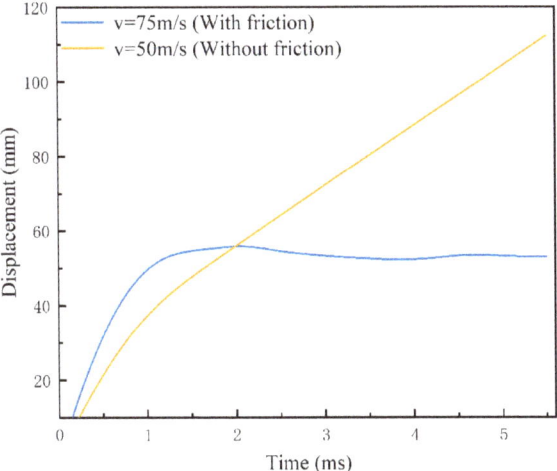

Figure 11. Harpoon displacement curve.

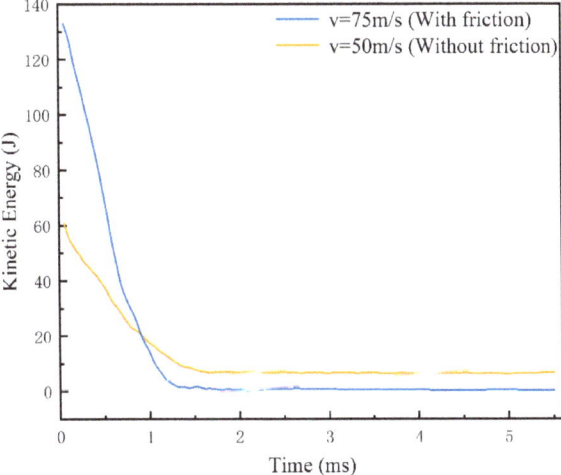

Figure 12. Kinetic energy curve.

From the data analysis of Figures 11 and 12, the kinetic energy of the harpoon dropped to approximately 7.75 J after 1.5 ms under the frictionless condition; there was no obvious change in kinetic energy after 1.5 ms, and the kinetic energy of the harpoon was kept at approximately 6.5 J at all times until the harpoon flew completely through the target plate after 5 ms, and the harpoon was not effectively embedded in the target plate. In the case of friction, the impact velocity of 75 m/s after 1.1 ms of kinetic energy dropped to 7.74 J, after that the displacement of the harpoon column surface in the target plate change rate is relatively flat; the friction between the harpoon and the target plate consumes the kinetic energy of the harpoon itself, the kinetic energy of the harpoon in 1.1~1.25 ms rapidly declines, and it is finally embedded in the target plate. Due to the impact of the target plate fluctuations, the kinetic energy of the harpoon changes slightly and finally decreases to 0, as the fluctuation of the target plate stops.

It can be seen that when the harpoon deeply impacts the stage of contact with the target plate, the kinetic energy of the harpoon decreases obviously under the condition of friction, and the harpoon can be successfully embedded in the target plate under the action of friction. Under the condition of not considering the friction coefficient, the harpoon can always leave the target plate as long as the head penetration part flies away from the target plate and a certain amount of kinetic energy remains. The kinetic energy remains almost the same during the stage of contact with the column surface, and the harpoon can always penetrate through the target plate, but it cannot be embedded in the target.

3.6. Velocity and Embedding Length

According to the above simulation analysis and combined with the structural characteristics of the harpoon itself, after the head of the harpoon is embedded in the target plate, the rebound phenomenon may occur due to the fluctuation of the target plate, and the anchoring is not reliable after embedding; and, considering the thickness of the target plate and its failure form, the embedding and the anchoring effect of 10 mm at the end of the harpoon is not good, and there may be a risk of it falling off in the process of towing. Therefore, the effective embedding part of the harpoon designed in this paper is the bottom column surface position, and the effective embedding length of the harpoon is 50~90 mm.

The impact embedding length at different velocities when tangential contact friction exists between the harpoon and the target plate was studied, and the simulation test design and results are shown in Table 4. The variation curves are shown in Figure 13.

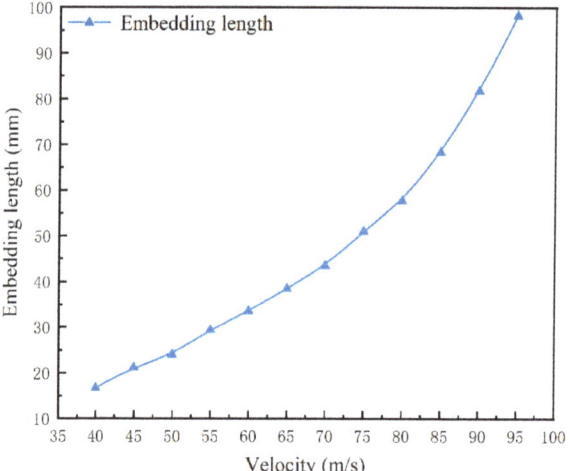

Figure 13. Velocity and embedding length curves.

Table 4. Simulation tests and results in the presence of tangential contact friction.

No.	Friction Coefficient	Velocity/m·s^{-1}	Embedding Length/mm
1	0.17	40	16.74
2	0.17	45	21.18
3	0.17	50	24.05
4	0.17	55	29.49
5	0.17	60	33.67
6	0.17	65	38.58
7	0.17	70	43.57
8	0.17	75	51.03
9	0.17	80	57.80
10	0.17	85	68.42
11	0.17	90	81.78
12	0.17	95	98.32

Combined with the relevant data in Table 4 and Figure 13, it can be concluded that when the harpoon velocity is below 70 m/s, it is mainly for the harpoon head to hit the target plate and embed, and the change of the harpoon embedding length at this time is basically linear; when the velocity is greater than 70 m/s, the final form of the embedding form is the contact stage between the harpoon column surface and the target plate, and the offsetting effect of the fluctuation of the target plate by the impact deformation on the velocity of the forward direction of the harpoon is reduced at this stage. The loss of kinetic energy of the harpoon is mainly due to the frictional effect. The embedding length of the harpoon gradually increases with the increase of the velocity, and the increase rate of the embedding length is obvious.

The friction coefficient involved in the above simulations are all fixed at 0.17; however, in the actual problem of harpoon impact embedding, the area of harpoon and target contact contains complex physical and chemical changes, such as hardening, melting, and the phase change of materials. Therefore, the friction coefficient at the time of impact will not be a constant, and it is related to many coefficients, such as the relative velocity of sliding, the pressure on the surface of the harpoon, and the characteristics of the target material [22]. In order to further obtain the suitable interval velocity of the harpoon launch, a simulation of harpoon impact embedding under different friction coefficient conditions was carried out. As the friction coefficient increases, the kinetic energy consumed by the harpoon during the impact embedding process gradually increases, and the required initial launch velocity becomes higher. The simulation test involves the friction coefficient as well as the harpoon embedding length as shown in Table 5.

Table 5. Harpoon embedding length.

No.	Friction Coefficient	Velocity/m·s^{-1}	Embedding Length/mm
1	0.17	70	43.57
		75	51.03
		80	57.80
		85	68.42
		90	81.78
		95	98.32
2	0.27	80	47.96
		85	55.48
		90	62.96
		95	71.52
		100	78.27
		105	90.37
3	0.37	85	49.42
		90	53.72
		95	61.51
		100	69.88
		105	77.65
		110	91.94
4	0.47	85	46.31
		90	51.37
		95	59.19
		100	64.78
		105	73.61
		110	83.54
		115	93.68

Analyzing the above simulation data, the relationship between harpoon embedment length and velocity when the friction coefficient changes is shown in Figure 14. The analysis shows that, on the one hand, as the friction coefficient increases, the velocity required for the harpoon to impact the target plate is higher for the same embedment length; on the other hand, the increase in the friction coefficient leads to a gradual increase in the length range of the harpoon embedded in the target plate, and a larger friction coefficient is beneficial for the harpoon to be embedded in the target plate. In order to accommodate the effective embedding length range of the harpoon within 50~90 mm for all friction coefficients, the initial velocity of the harpoon at launch needs to be controlled in the range of 88.4~92.5 m/s.

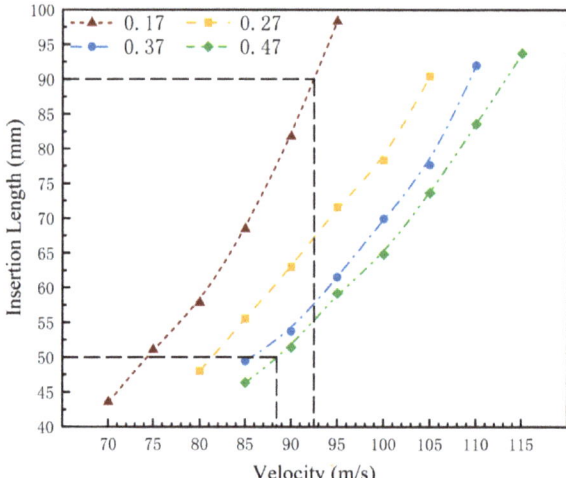

Figure 14. Embedding length and velocity relationship under variable friction coefficient.

4. Conclusions

In this paper, the influence of the law of friction coefficient in the low-velocity impact process is studied from the kinetic energy change and embedding length of the harpoon in the impact process through the impact test, combined with the simulation numerical simulation. The main conclusions are obtained as follows:

1. In the problem of the harpoon impacting and embedding in the target plate at a velocity lower than the ballistic limit, the tangential friction effect is more obvious for the consumption of the kinetic energy of the harpoon itself, and it cannot be ignored;

2. When the head of the harpoon is embedded in the target plate, due to which the fluctuation of the target plate may rebound after the impact of the head of the harpoon, the embedding is not reliable. The effective embedding part of the harpoon is the bottom column surface position. Considering the influence of the target plate deformation on embedding, the effective embedding length range of the harpoon is 50~90 mm;

3. Considering the effect of friction coefficient change, in order to adapt to the effective embedding length range of the harpoon under each friction coefficient condition to within 50~90 mm, the theoretical launch initial velocity of the harpoon should be between 88.4~92.5 m/s.

Author Contributions: Conceptualization, W.Z. (Wei Zhao) and Z.P.; methodology, Z.D.; validation, W.Z. (Wei Zhao) and W.Z. (Weiliang Zhu); formal analysis, W.Z. (Wei Zhao) and Z.P.; investigation, Z.P. and Z.Z.; resources, Z.D. and Z.P.; data curation, W.Z. (Wei Zhao) and W.Z. (Weiliang Zhu); writing—original draft preparation, W.Z. (Wei Zhao); writing—review and editing, Z.P.; visualization, W.Z. (Wei Zhao); supervision, Z.P.; project administration, Z.P. and Z.D.; funding acquisition, Z.D. and Z.Z. All authors have read and agreed to the published version of the manuscript.

Funding: This work was supported by the Funded by Science and Technology on Space Intelligent Control Laboratory (HTKJ2021KL502010), Research Project of Space Debris and Near-Earth Asteroid Defense (KJSP2020010303), National Natural Science Foundation of China (11802130).

Institutional Review Board Statement: Not applicable.

Informed Consent Statement: Not applicable.

Data Availability Statement: Not applicable.

Acknowledgments: We wish to express our gratitude to the members of our research team, Jia Guo, Lizhi Xu, Jiangbo Wang, and Xiaodong Wang.

Conflicts of Interest: The authors declare no conflict of interest.

References

1. Wei, Z.; Zhang, H.; Zhao, B.; Liu, X.; Ma, R. Impact Force Identification of the Variable Pressure Flexible Impact End-Effector in Space Debris Active Detumbling. *Appl. Sci.* **2020**, *10*, 3011. [CrossRef]
2. Adushkin, V.V.; Aksenov OYu Veniaminov, S.S.; Kozlov, S.I.; Tyurenkova, V.V. The small orbital debris population and its impact on space activities and ecological safety. *Acta Astronaut.* **2020**, *176*, 591–597. [CrossRef]
3. Ru, M.; Zhan, Y.; Cheng, B.; Zhang, Y. Capture Dynamics and Control of a Flexible Net for Space Debris Removal. *Aerospace* **2022**, *9*, 299. [CrossRef]
4. Virgili, B.B.; Krag, H. Analyzing the criteria for a stable environment. In Proceedings of the AAS/AIAA Astrodynamics Specialist Conference, Girdwood, AK, USA, 31 July–4 August 2011; Volume 411.
5. Liou, J.-C.; Johnson, N.; Hill, N. Controlling the growth of future LEO debris populations with active debris removal. *Acta Astronaut.* **2010**, *66*, 648–653. [CrossRef]
6. Forshaw, J.L.; Aglietti, G.S.; Fellowes, S.; Salmon, T.; Retat, I.; Hall, A.; Chabot, T.; Pisseloup, A.; Tye, D.; Bernal, C.; et al. The active space debris removal mission RemoveDebris. Part 1: From concept to launch. *Acta Astronaut.* **2019**, *168*, 293–309. [CrossRef]
7. Reed, J.; Barraclough, S. Development of Harpoon System for Capturing Space Debris. *Eur. Conf. Space Debris* **2013**, *723*, 174.
8. Campbell, J.; Hughes, K.; Vignjevic, R.; Djordjevic, N.; Taylor, N.; Jardine, A. Development of modelling design tool for harpoon for active space debris removal. *Int. J. Impact Eng.* **2022**, *166*, 104236. [CrossRef]
9. Dudziak, R.; Tuttle, S.; Barraclough, S. Harpoon technology development for the active removal of space debris. *Adv. Space Res.* **2015**, *56*, 509–527. [CrossRef]
10. Aglietti, G.S.; Taylor, B.; Fellowes, S.; Salmon, T.; Retat, I.; Hall, A.; Chabot, T.; Pisseloup, A.; Cox, C.; Zarkesh, A.; et al. The active space debris removal mission RemoveDebris. Part 2: In orbit operations. *Acta Astronaut.* **2019**, *168*, 310–322. [CrossRef]
11. Fras, T.; Roth, C.C.; Mohr, D. Dynamic perforation of ultra-hard high-strength armor steel: Impact experiments and modeling. *Int. J. Impact Eng.* **2019**, *131*, 256–271. [CrossRef]
12. Wang, Y.; Chen, X.; Xiao, X.; Vershinin, V.V.; Ge, R.; Li, D.-S. Effect of Lode angle incorporation into a fracture criterion in predicting the ballistic resistance of 2024-T351 aluminum alloy plates struck by cylindrical projectiles with different nose shapes. *Int. J. Impact Eng.* **2020**, *139*, 103498. [CrossRef]
13. Kpenyigba, K.; Jankowiak, T.; Rusinek, A.; Pesci, R.; Wang, B. Effect of projectile nose shape on ballistic resistance of interstitial-free steel sheets. *Int. J. Impact Eng.* **2015**, *79*, 83–94. [CrossRef]
14. Deng, Y.; Wu, H.; Zhang, Y.; Huang, X.; Xiao, X.; Lv, Y. Experimental and numerical study on the ballistic resistance of 6061-T651 aluminum alloy thin plates struck by different nose shapes of projectiles. *Int. J. Impact Eng.* **2022**, *160*, 104083. [CrossRef]
15. Rusinek, A.; Rodríguez-Martínez, J.; Arias, A.; Klepaczko, J.; López-Puente, J. Influence of conical projectile diameter on perpendicular impact of thin steel plate. *Eng. Fract. Mech.* **2008**, *75*, 2946–2967. [CrossRef]
16. Johnson, G.R.; Cook, W.H. A constitutive model and data for metals subjected to large strains, high strain rates and high tem-peratures. In Proceedings of the 7th International Symposium on Ballistics, The Hague, The Netherlands, 19–21 April 1983; Volume 21, pp. 541–548.
17. Elek, P.M.; Jaramaz, S.S.; Micković, D.M.; Miloradović, N.M. Experimental and numerical investigation of perforation of thin steel plates by de-formable steel penetrators. *Thin-Walled Struct.* **2016**, *102*, 58–67. [CrossRef]
18. Li, X.; Li, J.; Zhao, Z.; Liu, D.; Ou-Yang, X. Numerical study on penetration of a high-speed-rotating bullet into the moving sheet-metal plate. *Impact Dyn.* **2008**, *28*, 57–62. (In Chinese)
19. Gang, W.; Wei, Z.; Yunfei, D. Identification and validation of constitutive parameters of 45 Steel based on J-C model. *J. Vib. Shock* **2019**, *38*, 173–178. (In Chinese)

20. Chen, J.; Xu, W.; Xie, R.; Zhang, F.; Hu, W.; Huang, X.; Chen, G. Sample size effect on the dynamic torsional behaviour of the 2A12 aluminium alloy. *Theor. Appl. Mech. Lett.* **2017**, *7*, 317–324. [CrossRef]
21. Rusinek, A.; Rodríguez-Martínez, J.; Zaera, R.; Klepaczko, J.; Arias, A.; Sauvelet, C. Experimental and numerical study on the perforation process of mild steel sheets subjected to perpendicular impact by hemispherical projectiles. *Int. J. Impact Eng.* **2009**, *36*, 565–587. [CrossRef]
22. Li, J.; Zhang, X.; Liu, C.; Chen, H.; Wang, J.; Xiong, W. Study on mass erosion model of projectile penetrating concrete at high speed considering variation of friction coefficient. *Explos. Shock. Waves* **2021**, *41*, 114–124. (In Chinese)

MDPI

St. Alban-Anlage 66

4052 Basel

Switzerland

www.mdpi.com

Materials Editorial Office

E-mail: materials@mdpi.com

www.mdpi.com/journal/materials

www.ingramcontent.com/pod-product-compliance
Lightning Source LLC
LaVergne TN
LVHW070424100526
838202LV00014B/1520